Autoimmunity: Experimental Aspects

NATO ASI Series

Advanced Science Institutes Series

A series presenting the results of activities sponsored by the NATO Science Committee, which aims at the dissemination of advanced scientific and technological knowledge, with a view to strengthening links between scientific communities.

The Series is published by an international board of publishers in conjunction with the NATO Scientific Affairs Division

A Life Sciences	Plenum Publishing Corporation
B Physics	London and New York
C Mathematical and Physical Sciences	Kluwer Academic Publishers
D Behavioural and Social Sciences	Dordrecht, Boston and London
E Applied Sciences	
F Computer and Systems Sciences	Springer-Verlag
G Ecological Sciences	Berlin Heidelberg New York
H Cell Biology	London Paris Tokyo Hong Kong
I Global Environmental Change	Barcelona Budapest

NATO-PCO DATABASE

The electronic index to the NATO ASI Series provides full bibliographical references (with keywords and/or abstracts) to more than 30000 contributions from international scientists published in all sections of the NATO ASI Series. Access to the NATO-PCO DATABASE compiled by the NATO Publication Coordination Office is possible in two ways:

- via online FILE 128 (NATO-PCO DATABASE) hosted by ESRIN, Via Galileo Galilei, I-00044 Frascati, Italy.
- via CD-ROM "NATO Science & Technology Disk" with user-friendly retrieval software in English, French and German (© WTV GmbH and DATAWARE Technologies Inc. 1992).

The CD-ROM can be ordered through any member of the Board of Publishers or through NATO-PCO, Overijse, Belgium.

Series H: Cell Biology, Vol. 80

Autoimmunity: Experimental Aspects

Edited by

Moncef Zouali

Institut Pasteur
28, rue du Dr. Roux
75015 Paris, France

Springer-Verlag
Berlin Heidelberg New York London Paris Tokyo
Hong Kong Barcelona Budapest
Published in cooperation with NATO Scientific Affairs Division

Proceedings of the NATO Advanced Research Workshop on Fundamental and Experimental Aspects of Autoimmunity, held at Acquafredda di Maratea, Italy, May 22-26, 1993

ISBN-13:987-3-642-78781-2 e-ISBN-13:987-3-642-78779-9
DOI: 10.1007/987-3-642-78779-9

CIP data applied for.

Softcover reprint of the hardcover 1st edition 1994

Typesetting: Camera ready by authors
31/3145 - 5 4 3 2 1 0 - Printed on acid-free paper

Contents

4. Trigger Mechanisms of autoantibody production

5. T Cell Receptor Genes

6. Animal Models and Immunointervention

Introduction

The origins of autoimmunity have been debated since the beginning of this century and its history, like that of immunology, has been generated by a tradition which is a result of the synthesis of different styles of thinking, ranging from embryology to Darwinian thoughts. It is a history of convergence and cooperation of scientific fields, which Georges Canguilhem has described as a regular factor in scientific progress. Central to this debate is the potential of the immune system to recognize a huge diversity of external determinants, yet avoid high-affinity reactions with self-components. How an autoimmune response is generated in patients with autoimmune diseases remains intriguing with several schools of thought and ideological controversies in a number of critical questions.

The germ of this book on *"Experimental Aspects of Autoimmunity"* was a NATO Advanced Research Workshop held in May 1993 in Acquafredda di Maratea, in the volcanic hills of Southern Italy. The aim of the editor in calling for contributions to this volume was to bring under one cover, data on some major advances in the field of autoimmunity and autoimmune diseases. But more than this, the aim was to go beyond putting together a collection of papers. A sharp focus was placed on innovative approaches and on novel experimental and molecular strategies used to unravel the origin of organ-specific and systemic autoimmunity. In addition, to convey the creative sense that permeates research on the development of autoimmunity, some non-conservative ideas are presented. It was not, after all, until the last two decades that the revived dogma of the *"Horror Autotoxicus"* of the previous century was successfully overturned!

When mature T cells encounter tissue-specific antigens not present in the thymus, tolerance mechanisms operating in the periphery are triggered. Using transgenic mice, Arnold *et al.* show here that, depending on the tolerogenic signals, T cells reach various levels of tolerance either in one or in several consecutive steps. The question of how T cells escape tolerance induction in autoimmune diseases is investigated in three model systems. First, the autoimmune disease of NOD mice parallels that of progressive type-1 diabetes with destruction of insulin-producing beta cells in the Langerhans islets of the pancreas. To investigate the T cells involved in type-1 diabetes, Katz *et al.* generated a strain of mice carrying T cell receptor (TCR) α and -β transgenes from a diabetogenic T cell clone. They provide evidence that diabetogenic T cells can evade several forms of tolerance induction. Second, autoimmune gastritis is another example of organ-specific autoimmune disease where autoantibodies to the proton pump of parietal cells correlate with the severity to the disease. It can be induced by neonatal thymectomy of BALB/c mice. Toh *et al.* show that transgenic expression of a subunit of the pump in the thymus prevents induction of autoimmune gastritis, implying that a T cell response to this autoantigen is required for disease development and that tolerance to this autoantigen occurs at an extra-thymic location. It is now critical to design strategies to induce peripheral tolerance to various antigens in a predictible manner. Third, an autoimmune response to the self-antigen myelin basic protein is responsible for a T cell-mediated demyelating disease of the central nervous system, experimental allergic encephalomyelitis, an animal model of multiple sclerosis. Kumar and Sergarz asked whether the TCR itself can be processed and presented in the context of either class I or class II molecules *in vivo* and whether the activated T cells become involved in regulation of the autoimmune response. They conclude that a subset of T cells is essential in the regulatory network that maintains tolerance to the autoantigen tested.

It is generally accepted that clonal selection of the B cell repertoire occurs at the level of mature B cells. Here, Schwartz offers the interesting view that, in addition to mature B cells which proliferate when they are positively selected, pre-B cells are also the target of selection that leads to the phenomenon of repertoire restriction. In this proposal, which is reminescent of what is seen with T cells, pre-B cells would complete their differentiation if their surface Ig heavy chains are positively selected after binding to self-antigens. Since normal B cells can process and present idiopeptides to T cells, the work of Bartnes and Hannestad opens the possibility that self-peptide-specific T cells may suppress normal B cells.

Identification of the targets of autoimmune attack is the focus of considerable attention. One possibility is to screen expression libraries using antibodies of the patients, as illustrated by the work of Lafyatis and Capron who used antibodies from the synovial fluid of rheumatoid arthritis patients to clone a novel autoantigen. It has been proposed that molecular mimicry between host and exogenous antigens may play an essential role in the specific stimulation of autoreactive B-cell clones. An interesting example is an Epstein-Barr virus protein which mimicks the SmD ribonucleoprotein, a target of lupus antibodies. As Migliorini *et al.* show, anti-EBV antibodies are present in connective tissue disorders, anti-SmD antibodies bind the EBNA antigen, and conversely, immunization with EBNA induces anti-SmD antibodies. To understand the etiological and pathogenic mechanisms that lead to autoimmunity, Muller describes an epitope analysis based on the use of synthetic peptides from antigens recognized by antibodies of autoimmune patients. Autoantibodies frequently bind to sites in proteins that are post-transcriptionally modified. One possibility is that, in predisposed individuals, abnormal levels of such modifications will occur and lead to a change in the antigenicity of altered proteins. In this connection, heat shock proteins represent overlapping sets of self and exogenous molecules that can be elicited by a variety of insults, and as discussed by Dhillon *et al.*, they are at the interface of tolerance and autoimmunity.

To understand the driving forces that underly the production of pathogenic autoantibodies in systemic autoimmunity, Demaison *et al.* and Kalden *et al.* provide substantial information on their variable region genes and argue that somatic selection pressures are responsible for their production. Using a similar approach, Stevenson *et al.* provide useful informations on the tantalizing link between autoimmunity and B-cell tumors. On the other hand, as discussed in the report by Staines *et al.* on mapping of autoantibody-related idiotopes, the study of idiotypes is of potential value in the clinical assessment of the disease, in the analysis of the immunopathological mechanisms and in identifying the targets for immune-based anti-idiotype therapy.

A thread common to several chapters is the probing of the TCR repertoire in relation to autoimmunity. For example, in rheumatoid arthritis, the destruction of the joint tissues is thought to be mediated by T cells. Analyzing the Vβ repertoire in the peripheral blood and in the synovium, Natvig *et al.* show that the T cell response to BCG antigens is rather heterogenous. In another chapter by Sarukhan *et al.* , analysis of the Vβ repertoire in the spleen of NOD mice reveals a deletion of $V\beta^{+}$ T cells. The results rule out the possibility that a Vβ protein polymorphism is the source of autoreactivity in the NOD strain and suggest that the repertoire is the target of local selection processes. In a different context, Shenoy *et al.* use an experimental model to study the relationship between TCR Vβ genes and disease susceptibility to Myasthenia Gravis where a subset of T cells is thought to be involved in the effector phase of the autoimmune response to the acetyl choline receptor. Finally, the genetic background plays an important role in disease susceptibility and clinical expression of autoimmune diseases and, as Youinou *et al.* argue in the case of human lupus, several gene combinations and environmental factors may account for this heterogeneity.

As Hattori *et al.* show in the case of diabetes, the availability of animal models provides an opportunity to unravel the origin of autoimmune diseases. Tight skin mice, discussed by Bona *et al.*, represent an experimental model to study the pathogenesis of the autoimmune disease scleroderma. M. de Alboran *et al.* describe the effect of prophylactic treatment of "undiseased" MRL-*lpr* mice. The beneficial IL-2 effect is only obtained when it is applied before autoimmune manifestations begin to develop. Two molecules targeted to TCR Vβ products that are overexpressed in MRL-*lpr* mice are also considered. Monoclonal antibody to Vβ8 attenuates the autoimmune manifestations of the mice and treatment with a single dose of a murine superantigen inhibits lymphoproliferation and ameliorates the autoimmune parameters. Finally, the association between MHC alleles and autoimmune diseases has prompted research to design new means of immuno-intervention. Passive therapies using MHC class II peptides are successful in the prevention of experimental models of autoimmune diseases and the advantage of actively induced specific effector mechanisms is discussed by Wauben and van Eden.

Developments in technology are, to a large degree, responsible for advances in our understanding of biological processes. This principle applies to autoimmunity which, like other scientific disciplines, has benefited tremendously from the driving force of new techniques. As can be seen in this book, advances in the techniques and analytical concepts of immunology, molecular genetics, cell biology and transgenesis, made in the study of a particular autoimmune disease are usually applicable to others. The present volume, as a whole, concentrates on critical and current topics of contemporary autoimmunity. It raises pertinent questions that will hopefully set the stage for further studies and help in the spread of useful information accross the barriers which sometimes result in compartimentalization of the rapidly expanding world of autoimmunity.

List of Contributors

* Alderuccio, F. - Monash Univ. Medical School, Melbourne, Australia
* Alferink, J. - Deutsches Krebsforshungszentrum, Heidelberg, Germany
* Arnold, B. - Deutsches Krebsforshungszentrum, Heidelberg, Germany
* Bach, J.F. - Hôpital Necker-Enfants Malades, Paris, France
* Bansal, S. - King's College London, London, UK
* Bartnes, K. - University of Tromsø, Tromsø, Norway
* Benoist, C. - INSERM-CNRS, Faculté de Médecine, Strasbourg, France
* Bensimon, C., Institut Pasteur, Paris, France
* Bona, C. - Mount Sinai School of Medicine, New York, USA
* Capra, J.D. - University of Texas, Dallas, USA
* Capron, A. - Institut Pasteur de Lille, Lille, France
* Carnaud, C. - INSERM U25, Hôpital Necker, Paris, France
* Chapman, C.J. - Southampton University Hospitals, Southampton, UK
* Chapman, V.M. - Roswell Park Cancer Institute, New York, USA
* Chastagner, P. - Institut Pasteur, Paris, France
* Christadoss, P. - University of Texas Medical Branch, Galveston, USA
* Daian C. - Mount Sinai School of Medicine, New York, USA
* David, C. - Mayo Clinic, Rochester, USA
* de Alboran, I.M. - Universidad Autonoma de Madrid, Madrid, Spain
* Demaison, C. - Institut Pasteur, Paris, France
* Denbury, A.N. - King's College London, London, UK
* Dhillon, V.B. - Middlesex School of Medicine, London, UK
* Dolcher, M.P. - University of Pisa, Pisa, Italy

* Ercarz E. - University of California, Los Angeles, USA

* Fehr, H.K. - Univ. of Erlangen-Nuremberg, Erlangen, Germany

* Ferber, I. - Deutsches Krebsforshungszentrum, Heidelberg, Germany

* Førre, Ø. - University of Oslo, Oslo, Norway

* Fukuda, M. - Joslin Diabetes Center, Boston, USA

* Garchon, H.-J. - INSERM U25, Hôpital Necker, Paris, France

* Gleeson, P.A. - Monash Univ. Medical School, Melbourne, Australia

* Goluszko, E. - University of Texas Medical Branch, Galveston, USA

* Gonzalo, J.A. - Universidad Autonoma de Madrid, Madrid, Spain

* Guillevin, L. - Hôpital Avicenne, Paris, France

* Hamblin, T.J. - Southampton University Hospitals, Southampton, UK

* Hannestad, K. - University of Tromsø, Tromsø, Norway

* Hämmerling, J. - Deutsches Krebsforshungszentrum, Heidelberg, Germany

* Hattori, M. - Joslin Diabetes Center, Boston, USA

* Hobby, P. - Guy's Hospital, London, UK

* Isenberg, D.A. - Bloomsbury Rheumatology Unit, London, UK

* Jones, D.B. - Southampton University Hospitals, Southampton, UK

* Kalden, J.R. - Univ. of Erlangen-Nuremberg, Erlangen, Germany

* Kalsi, J. - Bloomsbury Rheumatology Unit, London, UK

* Kasturi K. - Mount Sinai School of Medicine, New York, USA

* Katz, J.D. - INSERM-CNRS, Faculté de Médecine, Strasbourg, France

* Kaul, R. - University of Texas Medical Branch, Rochester, USA

* Kjeldsen-Kragh, J. - University of Oslo, Oslo, Norway

* Kroemer, G. - Universidad Autonoma de Madrid, Madrid, Spain

* Kumar, V. - University of California, Los Angeles, USA

* Lafyatis, R. - Institut Pasteur de Lille, Lille, France

* Latchman, D.S. - Middlesex School of Medicine, London, UK

* MacMurray, A. - Whitehead Institute for Biomedical Research, Cambridge, USA

* Marchini, B., University of Pisa, Pisa, Italy

* Martinez-A., C. - Universidad Autonoma de Madrid, Madrid, Spain

* Mathis, D. - INSERM-CNRS, Faculté de Médecine, Strasbourg, France

* Migliorini, P. - University of Pisa, Italy

* Muller, S. - CNRS, Strasbourg, France

* Natvig, J.B. - University of Oslo, Oslo, Norway

* Oleinick, S.R. - VA Medical Center, Oklahoma, USA

* Petruzzelli, M. - Joslin Diabetes Center, Boston, USA

* Piette, J.C. - Hôpital La Pitié-Salpêtrière, Paris, France

* Puccetti, A. - University of Pisa, Pisa, Italy

* Quayle, A. - University of Oslo, Oslo, Norway

* Sabbatini, A. - University of Pisa, Pisa, Italy

* Salmon, D. - CNRS, Paris, France

* Sarukhan, A. - INSERM U25, Hôpital Necker, Paris, France

* Schönrich, G. - Deutsches Krebsforshungszentrum, Heidelberg, Germany

* Schwartz, R.S. - New England Medical Center, Boston, USA

* Semana, G. - Université de Rennes, Rennes, France

* Sergarz, E. - University of California, Los Angeles, USA

* Shenoy, M. - University of Texas Medical Branch, Galveston, USA

* Silberstein, L.E. - Hospital of the Univ. of Pennsylvania, Philadelphia, USA

* Sioud, M. - University of Oslo, Oslo, Norway

* Spellerberg, M.B. - Southampton University Hospitals, Southampton, UK

* Staines, N.A. - King's College London, London, UK

* Stevenson, F.K. - Southampton University Hospitals, Southampton, UK

* Suleyman, S. - University of Oslo, Oslo, Norway

* Sutton, B.J. - King's College London, London, UK

* Tan, S.-S. - Monash Univ. Medical School, Melbourne, Australia

* Toh, B.-H. - Monash Univ. Medical School, Melbourne, Australia

* Treasure, J. - Southampton University Hospitals, Southampton, UK

* Van Driel, I.R. - Monash Univ. Medical School, Melbourne, Australia

* van Eden, W. - Faculty of Veterinary Science, Utrecht, The Netherlands

* Viard, J.P. - Hôpital Necker-Enfants Malades, Paris, France

* Wang, B. - INSERM-CNRS, Faculté de Médecine, Strasbourg, France

* Ward, F.J. - King's College London, London, UK

* Wauben, M.H.M. - Faculty of Veterinary Science, Utrecht, The Netherlands

* Weaver, A. - Whitehead Institute for Biomedical Research, Cambridge, USA

* Williams, D.G. - Guy's Hospital, London, UK

* Wilson, K.B. - University of Texas, Dallas, USA

* Winkler, T.H. - Univ. of Erlangen-Nuremberg, Erlangen, Germany

* Wu, B. - University of Texas Medical Branch, Rochester, USA

* Yamato, E. - Joslin Diabetes Center, Boston, USA

* Youinou, P. - Brest University Medical School Hospital, Brest, France

* Zouali, M. - Institut Pasteur, Paris, France

Tolerance induction in mature peripheral T cells

Bernd Arnold, Günther Schönrich, Iris Ferber, Judith Alferink and Günter J. Hämmerling
Department of Somatic Genetics
Tumor Immunology Program
Deutsches Krebsforschungszentrum
Im Neuenheimer Feld 280
6900 Heidelberg, Germany

I. Introduction

The accurate discrimination of the immune system between pathogens and self-determinants is based on specific recognition events by receptors on B and T lymphocytes. The enormous diversity of the receptor repertoire is achieved by random rearrangement of DNA segments. In this process specificities for non-self as well as for self-determinants are created. T lymphocytes with high affinity for self-antigens are physically eliminated when they develop in the thymus provided the respective self-antigen is present on bone marrow derived cells (von Boehmer, 1990). Clonal deletion appears to be highly sensitive. Even thymocytes with an avidity below the required threshold for activation in the periphery are deleted (Pircher et al., 1991; Knobloch et al., 1992). On the other hand, thymocytes are not deleted but rendered functionally incompetent (clonal anergy) when the self-antigens are exclusively expressed on thymic epithelial cells (Ramsdell et al., 1989; Salaun et al., 1990; Hoffmann et al., 1992; Schönrich et al., 1992a).

Besides thymic tolerance induction one has to expect also tolerance mechanisms operating in the periphery when mature T cells encounter tissue-specific antigens not present in the thymus. Regardless whether extrathymic tolerance mechanisms have a major or only a minor contribution to the acceptance of self determinants in a normal individual it is of crucial importance for therapeutic intervention in allograft transplantations and autoimmune diseases to establish under which circumstances mature peripheral T cells are susceptible for tolerance induction. Furthermore a detailled knowledge of the fate of anergic T cells and their susceptibility for reactivation in vivo is essential to establish strategies for breaking nonresponsiveness to tumor antigens.

NATO ASI Series, Vol. H 80
Autoimmunity: Experimental Aspects
Edited by Moncef Zouali

To prove whether peripheral tolerance induction can occur in an untreated animal many investigators used transgenic mice expressing a "foreign protein" under the control of tissue-specific promoter elements in non-thymic cells (reviewed in Hämmerling et al., 1991; Miller and Morahan, 1992). Most of these animals were tolerant to the transgene product, although antigen expression could not be detected in the thymus and although the respective cells were not deleted. It was therefore concluded that peripheral tolerance induction exists and is based on clonal anergy. However, recent results showed that these criteria were not sufficient to reach these conclusions. For example, in case of the K^b expression under the rat insulin promoter (RIP) more sensitive techniques demonstrated low but significant thymic K^b expression (Morahan et al., 1989; Heath et al., 1992). Transfer of the transgenic RIP-K^b thymus into thymectomized non-transgenic animals did indeed induce tolerance to K^b in these chimeric mice.

Similarly, Husbands et al. (1992) reported that expression of K^b under the α-lactalbumin promoter in transgenic mice resulted in tolerance. Again, thymic K^b expression was not detectable by histology but only on the mRNA level by S1 nuclease protection analysis. Again, transfer of the transgenic thymus was sufficient to induce tolerance in the nontransgenic chimeric mice. Taken together, establishing a model system for peripheral tolerance one has to take into account that minute amounts of antigen present in the thymus may induce tolerance.

II Parameters controlling peripheral tolerance induction

The various published mouse models also show that the consequences of antigen expression in a tissue-specific manner are hardly predictable. From the following 5 examples it will be apparent that expression of different transgenes under the same rat insulin promoter sequences resulted in either tolerance, ignorance of the self antigen, or induction of autoimmunity. As already mentioned above in RIP-K^b mice thymic tolerance induction was observed (Heath et al., 1992). Mice expressing the influenza hemagglutinin on islet ß cells were also tolerant (Lo et al., 1992) but in this case based on a peripheral event, as the authors concluded. In contrast, expression of the lymphocytic choriomeningitis virus (LCMV) glycoprotein or nucleoprotein under the same promoter did not influence T cell activity at all. Neither tolerance nor autoimmunity was found (Ohashi et al., 1991; Oldstone et al., 1991). Finally, RIP directed expression of the SV40 large T antigen was not ignored by the immune system: expression late during development led to a strong autoimmune response (Adams et al., 1987).

To gain further insight into the processes leading to peripheral tolerance induction we have to consider the important parameters and to develop model systems which allow us to vary the one or few parameters of interest and to keep the others constant. The following parameters may be relevant in determining the consequence of a contact between a T cell and a cell expressing the respective antigen: the affinity of the T cell receptor (TCR), the level of expression of TCR and accessory molecules, and the developmental stage of the T cells encountering the antigen. In contrast to circulating T cells, recent thymic emigrants may be more susceptible to tolerance induction than to activation. This could explain why early expression of antigen on pancreatic cells leads to tolerance, whereas late expression during development leads to autoimmunity (Adams et al., 1987). In addition, the frequency, length and intensity of contacts between T cells and the tolerizing tissue could be important and may depend on the number of antigen bearing cells within the tissue and their accessibility to lymphocytes. Furthermore, the density of the self-antigen and the expression of adhesion molecules on the cell surface of a given cell type should contribute to the intensity of the contact. Finally, the impact of lymhokines released into the microenvironment during an inflamation should also be considered.

III Distinct manifestations of peripheral tolerance.

Based on the considerations above we have generated double transgenic mice to study peripheral tolerance. These mice expressed the major histocompatibility complex (MHC) class I antigen K^b under various tissue specific promoter elements and in addition as a second transgene product a TCR specific for the K^b antigen (Schönrich et al., 1991; Schönrich et al, 1992b). A class I molecule was chosen as a model antigen because a given TCR recognizes only intact MHC molecules and not degradation products which could migrate to the thymus. In addition, we have recently shown that soluble class I molecules do not induce T cell tolerance to the respective cell-bound class I molecule (Arnold et al., 1988). In systems with polyclonal T cell responses tolerance mechanisms could not be studied. Using TCR transgenic mice one has the advantage of having many T cells expressing the same TCR which can be identified by anti-clonotypic antibodies. The fate of these T cells can therefore be followed and the basis of tolerance induction be explored.

The αß TCR genes used were derived from a K^b-specific and CD8-dependent cytotoxic T cell (CTL) clone isolated from B10.BR (H-2^k) mice (Hue et al., 1990). These TCR transgenic mice were crossed with each of the following K^b transgenic mice. We used the albumin (Alb) promoter, the C-reactive protein (CRP) promoter and the keratin IV (2.4 Ker IV)promoter for the liver-specific and epithelium-specific expression of K^b. The glial fibrillary acidic protein (GFAP) promoter was employed to direct expression of K^b to cells of neuroectodermal origin. Expression of the transgene was verified on the mRNA level by conversion into cDNA and multiplication by polymerase chain reaction (PCR) and on the protein level by immunohistology. In all four transgenic mouse lines the K^b antigen could be detected only in extrathymic tissue: either only in hepatocytes (albumin and CRP promoter) or only in epithelial cells in skin, tongue and foot pads (keratin promoter) or only in cell types of neuroectodermal origin (GFAP promoter). Tolerance to K^b was observed in all these mice as well as in their offspring from crosses with the TCR transgenic mice as judged by K^b-positive skin graft acceptance and by growth of K^b-positive tumors (Schönrich et al., 1992b; Ferber et al., unpublished).

Using such an approach to study tolerance induction we were able to keep constant all parameters concerning the T cell and the nature of antigen. In this way we could examine the influence of antigen expression in different anatomical sites. We verified that the observed tolerance was based on a thymic event by the following criteria. Firstly, even by the sensitive reverse transcription PCR assays for the detection of mRNA no expression was found in the thymus. Secondly, using the anti-clonotypic antibody Désiré-1 (Hua et al., 1986) normal maturation of $CD8^+$ clonotype$^+$ thymocytes was observed in the various double transgenic mice. Thirdly, these mature thymocytes could develop into K^b-specific CTL after in vitro stimulation with antigen, and they could proliferate in response to anti-clonotypic antibodies. These responses were identical to those of thymocytes from TCR single-transgenic mice. Finally, CRP-K^b chimeric mice transplanted with a normal thymus and reconstituted after lethal irradiation with bone marrow of TCR transgenic mice were tolerant to K^b as shown by their acceptance of skin grafts (Ferber et al., in preparation). Thus, these data prove that K^b expression on hepatocytes alone is sufficient to induce tolerance.

To clarify peripheral tolerance mechanisms we followed the fate of the $CD8^+$ clonotype$^+$ cells in spleen and lymph nodes of various animals (Schönrich et al., 1992b). In (TCR x GFAP-K^b)F_1 and (TCR x Alb-K^b)F_1 animals, a strong reduction of clonotype$^+$ $CD8^+$ T cells was observed in comparison to the numbers found in TCR single transgenic mice. This reduction was due to downregulation of TCR and CD8 molecules. In (TCR x GFAP-K^b)F_1 animals, downregulation of the TCR could be reversed by antigen stimulation *in vitro*; in (TCR x Alb-K^b)F_1 mice antigenic stimulation failed, but crosslinking of CD2 molecules using immobilized antibodies could induce re-expression of TCR molecules on peripheral T cells. Thus, upregulation of TCR/CD3 complexes in the two types of double transgenic mice required different signals. In (TCR x 2.4 KerIV-K^b)F_1 mice normal numbers of clonotype$^+$ $CD8^+$ cells were found in the lymphoid organs. However, the majority of these cells expressed high levels of Pgp-1 molecules indicating that these cells have indeed seen the K^b antigen (Fowlkes and Pardoll, 1989). Spleen cells from these mice contained the same high frequency of CTL precursors as spleen cells of TCR single transgenic mice. These cells lost their tolerant stage also in vivo when transferred into a neutral environment. Nude mice of H-2^d origin were able to reject K^b-positive skin grafts when grafts were set 10-20 days after injection of spleen cells from (TCR x 2.4 KerIV-K^b)F_1 animals (Alferink et al., in preparation).

Taken together, expression of K^b in different tissues led to different manifestations of peripheral nonresponsiveness. Since tolerant T cells could be distinguished phenotypically and by their capacity for reactivation, depending on the tolerogenic signals, we proposed that T cells can reach different levels of tolerance (Arnold et al., 1993).

IV Peripheral tolerance induction depending on the amount of antigen expressed.

Besides the properties of the tissue on which the self-antigen is expressed the amount of antigen on the individual cells may also be important for the mechanism by which tolerance is induced. To study the influence of this parameter on peripheral tolerance we analyzed (TCR x CRP-K^b)F_1 mice (Ferber et al., in preparation). CRP is an accute phase protein which can be strongly induced within about 12 hours by lipopolysaccharide (LPS) and IL-6 (Ciliberto et al., 1987). Accordingly, K^b was hardly detectable on hepatocytes of these double transgenic animals, but clearly visible after induction with LPS. Despite the very low expression of K^b before induction the mice were tolerant due to downregulation of TCR and CD8 molecules on the respective T cells. Thus, using two different promoter systems (albumin and CRP) to direct K^b expression exclusively to hepatocytes peripheral tolerance induction was achieved by the same mechanism. To study the influence of the amount of K^b expressed on the degree of TCR downregulation, TCR expression on $CD2^+$ $CD4^-$ T cells was measured before and after antigen induction by LPS. Three days after induction the fraction of T cells with normal TCR expression was decreased (30% to 12%). All the other $CD2^+$ $CD4^-$ cells showed only background staining with the anti-TCR antibody indicating complete downregulation of TCR molecules. To exclude an indirect effect of LPS on the immune system potentially causing a higher degree of TCR downregulation, (TCR x Alb-K^b)F_1 mice were also treated with LPS. In these animals no difference in downregulation of TCR molecules could be observed with or without LPS treatment. The complete downregulation of TCR 2.5 days after high expression of K^b cannot be explained by the appearance of fresh thymic emigrants encountering large amounts of K^b on hepatocytes because this event could also be observed in adult thymectomized animals. Therefore, the most reasonable explanation for the rapid change of TCR expression is that contact of the tolerant T cells (expressing already reduced levels of TCR) with the high number of K^b molecules on induced hepatocytes led to a further and more drastic downregulation of the TCR molecules.

V Tolerance induction as a multistep process.

An additional system was established to clarify whether T cells in a mild stage of tolerance could be shifted into a deeper level of anergy or even towards deletion by receiving further tolerogenic signals. The K^{bm1} mutant antigen carries in its $\alpha 2$ domain mutations of aminoacid residues 152, 155 and 156 which affect the peptide binding groove. Therefore the anti-K^b TCR used in the present study displays a lower affinity for K^{bm1} than for K^b wild type molecule (Auphan et al., 1992). In the thymus of bm1 x TCR mice negative selection was still functional but incomplete. Therefore large numbers of $CD8^+$ clonotype$^+$ cells with slightly decreased expression of both types of molecules were found in the periphery (Schönrich et al., submitted). These cells expressed high levels of the T cell activation marker Pgp-1 and were tolerant in vivo because they did not cause rejection of K^b positive skin and spleen grafts. The bm1 x TCR mice were crossed with the Alb-K^b animals. The rational of this experiment was that in these double transgenic mice the T cells would first encounter the K^{bm1} antigen in the thymus and in peripheral lymphoid organs and thereafter the K^b molecule expressed on hepatocytes. If tolerant T cells were inert one should have expected to find mainly peripheral T cells with the phenotype of T cells found in bm1 x TCR mice. In the periphery of these double transgenic mice, however, neither T cells with slightly reduced levels of TCR and CD8 molecules (as in the bm1 x TCR mice) nor T cells with strong TCR and CD8 downregulation (as in TCR x Alb-K^b mice) were found. Apparently T cells which have already been tolerized by the K^{bm1} molecule were deleted by the subsequent encounter with the wild type K^b molecule on hepatocytes.

VI Summary and Conclusions .

The studies described here show that mature T lymphocytes can be rendered tolerant to antigens expressed exclusively outside the thymus. Depending on the tolerogenic signals T cells reach various levels of tolerance either in one or in several consecutive steps. Despite the fact that a detailed knowledge about the mechanisms of peripheral tolerance induction is still lacking the described observations may be of clinical relevance. The finding that different tissues can induce different levels of tolerance could explain why some organs are more susceptible to autoimmune diseases than others. For example, potentially autoreactive T cells in a relatively mild stage of tolerance as described for the (TCR x 2.4 KerIV-K^b)F_1 mice are, of course, dangerous for an individual because tolerance may be easily broken under the influence of an invading pathogen. So far we have discussed tolerance induction to peripheral self antigens constitutively expressed during neonatal life and the possibility that mature T lymphocytes are susceptible to inactivation by antigen for only a short period after exit from the thymus. However, to ensure nonresponsiveness in autoimmune diseases and in allograft transplantation, mature T cells which have been in circulation for a longer time period must be rendered unresponsive. Therefore, strategies to improve immunosuppression in these situations should concentrate on suitable methods to eliminate the circulating T cell pool responsible for graft rejection or onset of autoimmune diseases. Since allograft specific or autoreactive T cells newly exported from the thymus would undergo peripheral tolerance induction, specific immunosuppression could be achieved. Those strategies can, however, only be used after we have learnt how peripheral tolerance to various antigens can predictibly be induced.

References

Adams, T.E., Alpert, S. and Hanahan, D.(1987)Non-tolerance and autoantibodies to a transgenic self antigen expressed in pancreatic b cells. Nature 325:223-228

Arnold, B., Dill, O., Küblbeck, G., Jatsch, L., Simon, M.M., Tucker, J. and Hämmerling, G.J.(1988)Alloreactive immune responses of transgenic mice expressing the foreign transplantation antigen in a soluble form. Proc Natl Acad Sci USA 85:2269-2273

Arnold, B., Schönrich, G. and Hämmerling, G.J.(1993)Multiple levels of peripheral tolerance. Immunol Today 14:12-14

Auphan, N., Jézo-Brémond, A., Schönrich, G., Hämmerling, G.J., Arnold, B., Malissen, B. and Schmitt-Verhulst, A.-M.(1992)Threshold tolerance in H-2Kb-specific TCR transgenic mice expressing mutant H-2Kb: helper-dependent CTL. Int.Immunol 4,12:1419-1428

Ciliberto, G., Arcone, R., Wagner, E.F. and Rüther, U.(1987)Inducible and tissue-specific expression of human C-reactive protein in transgenic mice. EMBO J 6,13:4017-4022

Fowlkes, B.J. and Pardoll, D.M.(1989)Molecular and cellular events of T cell development. Adv Immunol 44:207-264

Hämmerling, G.J., Schönrich, G., Momburg, F., Auphan, N., Malissen, M., Malissen, B., Schmitt-Verhulst, A.-M. and Arnold, B.(1991)Non-deletional mechanisms of peripheral and central tolerance: studies with transgenic mice with tissue specific expression of a foreign MHC class I antigen. Immunol Rev 122:47-67

Heath, W.R., Allison, J., Hoffmann, M.W., Schönrich, G., Hämmerling, G.J., Arnold, B. and Miller, J.F.A.P.(1992)Autoimmune diabetes as a consequence of locally produced interleukin-2. Nature 359:547-549

Hoffmann, M.W., Allison, J. and Miller, J.F.A.P.(1992)Tolerance induction by thymic medullary epithelium. Proc Natl Acad Sci USA 89:2526-2530

Hue, I.., Trucy, J., McCoy, C., Couez, D., Malissen, B. and Malissen, M.(1990)A novel type of aberrant T cell receptor alpha-chain gene rearrangement. Implications for allelic exclusion and the V-J recombination process. J Immunol 144:4410-4419

Husbands, S.D., Schönrich, G., Arnold, B., Chandler, P.R., Simpson, E., Philpott, K.L., Tomlinson, P., OtReilly, L., Cooke, A. and Mellor, A.L.(1992)Expression of major histocompatibility complex class I antigens at low levels in the thymus induces T cell tolerance via a non-deletional mechanism involving downmodulation of CD8 and T cell receptor. Eur J Immunol 22:2655-2661

Knobloch, M., Schönrich, G., Schenkel, J., Malissen, M., Malissen, B., Schmitt-Verhulst, A.-M., Hämmerling, G.J. and Arnold, B.(1992)T cell activation and thymic tolerance induction require different adhesion intensities of the CD8 coreceptor. Int. Immunol 4: 1169-1174

Lo, D., Freedman, J., Hesse, S., Palmiter, R.D., Brinster, R.L. and Sherman, L.A.(1992)Peripheral tolerance to an islet cell-specific hemagglutinin transgene affects both CD4+ and CD8+ T cells. Eur J Immunol 22:1013-1022

Miller, J.F.A.P. and Morahan, G.(1992)Peripheral T cell tolerance. Annu Rev Immunol 10:51-69

Morahan, G., Allison, J. and Miller, J.F.A.P.(1989)Tolerance of class I histocompatibility antigens expressed extrathymically. Nature 339:622-624

Ohashi, P.S., Oehen, S., Buerki, K., Pircher, H., Ohashi, C.T., Odermatt, B., Malissen, B., Zinkernagel, R.M. and Hengartner, H.(1991) Ablation of "tolerance" and induction of diabetes by virus infection in viral antigen transgenic mice. Cell 65:305-317

Oldstone, M.B., Nerenberg, M., Southern, P., Price, J. and Lewicki, H.(1991)Virus infection triggers insulin-dependent diabetes mellitus in a transgenic model: role of anti-self (virus) immune response. Cell 65:319-331

Pircher, H., Hoffmann Rohrer, U., Moskophidis, D., Zinkernagel, R.M. and Hengartner, H.(1991)Lower receptor avidity required for thymic clonal deletion than for effector T cell function. Nature 351:482-485

Ramsdell, F., Lantz, T. and Fowlkes, B.J.(1989)A nondeletional mechanism of thymic self tolerance. Science 246:1038-1041

Salaün, J., Bandeira, A., Khazaal, I., Calman, F., Coltey, M., Coutinho, A. and Le Douarin, N.M.(1990)Thymic epithelium tolerizes for histocompatibility antigens. Science 247:1471-1475

Schönrich, G., Kalinke, U., Momburg, F., Malissen, M., Schmitt-Verhulst, A., Malissen, B., Hämmerling, G.J. and Arnold, B.(1991)Downregulation of T cell receptors on self-reactive T cells as a novel mechanism for extrathymic tolerance induction. Cell 65:293-304

Schönrich, G., Momburg, F., Hämmerling, G.J. and Arnold, B.(1992a)Anergy induced by thymic medullary epithelium. Eur J Immunol 22:1687-1691

Schönrich, G., Momburg, F., Malissen, M., Schmitt-Verhulst, A.-M., Malissen, B., Hämmerling, G.J. and Arnold, B.(1992b)Distinct mechanisms of extrathymic T cell tolerance due to differential expression of self-antigen. Int.Immunol 4,5:581-590

von Boehmer, H.(1990)Developmental biology of T cells in T cell-receptor transgenic mice. Ann Rev Immunol 8:531-556

Transgenic mice carrying the rearranged T cell receptor of a diabetogenic T cell clone

Jonathan D. Katz, Bo Wang, Christophe Benoist, and Diane Mathis
Laboratoire de Génétique Moléculaire des Eukaryotes du CNRS et
Unité 184 - Biologie Moléculaire et Génie Génétique de l'INSERM
Faculté de Médecine
11, rue Humann
67085 Strasbourg Cédéx
France

INTRODUCTION

Insulin-dependent diabetes mellitus (IDDM) is a progressive autoimmune disease that results in destruction of the insulin-producing beta cells in the pancreatic islets of langerhans (reviewed in Castano and Eisenbarth, 1990). The genetic and environmental factors which provoke this disease in humans remain largely unknown. Fortunately, the nonobese diabetic (NOD) mouse develops a spontaneous diabetes remarkably similar to that in humans and has thus greatly facilitated its study (Makino et al, 1980). The disease of NOD mice parallels that of the human in that it first develops as an occult infiltration of lymphocytes into the islets, a process termed insulitis, followed by the destruction of the overwhelming majority of the beta cells, thereby, percipitating an overt diabetes. Autoimmune diabetes in NOD mice is under polygenic control, with the most important single contributor being the major histocompatibility complex (MHC) - H-2^{g7}.

The evidence that T lymphocytes play a major role in the induction and progression of diabetes in NOD mice is now extensive (Miyazaki et al, 1985; Ogawa et al, 1985; Harada and Makino, 1986; Makino et al, 1986; Mori et al, 1986; Koike et al, 1987; Dardenne et al, 1989; Shizuru et al, 1988; Hayward and

NATO ASI Series, Vol. H 80
Autoimmunity: Experimental Aspects
Edited by Moncef Zouali

Schreiber, 1989; Sempé et al, 1991), the most convincing being the transfer of disease by cloned T cell lines (Bendelac et al, 1987; Miller et al, 1988; Haskins et al, 1989; Reich et al, 1989; Haskins and McDuffie, 1990; Nakano et al, 1991; Pankewysz et al, 1991; Yagi et al, 1992; Bradley et al, 1992).

Clearly then, the T lymphocytes in type 1 diabetes are self-reactive. Yet, the principles of self/non-self discrimination hold that these cells should have been illiminated or rendered harmless. How then did these cells avoid clonal deletion or anergy induction in the thymus and periphery? Further, in what immunologic compartment(s) do they normally reside, and in what activation state are they found? In the past, such questions were close to impossible to address given that any one antigen specificity is expressed on fewer than one T cell per thousand. However, recent successes with transgenic mice carrying rearranged T cell receptor (TCR) genes - and therefore dramatically limited T cell repertoires - have rendered these questions more approachable. Earlier TCR transgenic mice studies have lead to a greater understanding of the antigens, cell types, timing and locations of self/non-self discrimination, demonstrating that multiple mechanisms exist to induce T cell tolerance (Kisielow et al, 1988; Sha et al, 1988; Blackman et al, 1990; Ohashi et al, 1991; Rocha and von Boehmer, 1991; Schönrich et al, 1992a; Carlow et al, 1992; Schönrich et al, 1991, 1992b). Therefore, we sought to harness the power of transgenic mouse technology to investigate the T cells involved in type 1 diabetes. To this end, we have generated a strain of mice carrying the rearranged TCR-α and -β genes from a diabetogenic T cell clone.

RESULTS

The $CD4^+$ T cell clone BDC 2.5, derived from a diabetic female NOD mouse, proliferates and releases IL-2 when challenged with islet cells (ICs) from any mouse strain and antigen presenting cells (APCs) from NOD mice, but not by cells from other tissues or APCs from other strains, nor by ICs or APCs alone (Haskins et al, 1988, 1989). Furthermore, when this clone was transferred into 2-4 week old, asymptomatic NOD mice, it rapidly induced insulitis and diabetes (Haskins and McDuffie, 1990). Thus, BDC 2.5 was known to be islet antigen-specific, H-2^{g7}-restricted and diabetogenic; it therefore, seemed a logical source of rearranged TCR genes for constructing transgenic mice.

Generation of BDC 2.5 TCR transgenic mice.

As illustrated in Fig. 1, the Vα-Jα and Vβ-Dβ-Jβ rearrangements from BDC 2.5 were cloned into cosmid vectors previously demonstrated to give rise to efficient and faithful expression of rearranged TCR genes (Uematsu et al, 1988; Blüthmann et al, 1988; Signorelli et al, 1991). Two sets of transgenic mice were generated; those carrying the BDC 2.5 TCR-β were produced by injecting (NOD x C57Bl/6) F1 embryos, while mice carrying both the BDC 2.5 TCR-α and the TCR-β chains were derived from injections into (C57Bl/6 x SJL) F2 eggs. Founder mice were crossed with NOD or C57Bl/6 (B6) mice to generate lines on selecting or non-selecting genetic backgrounds, respectively.

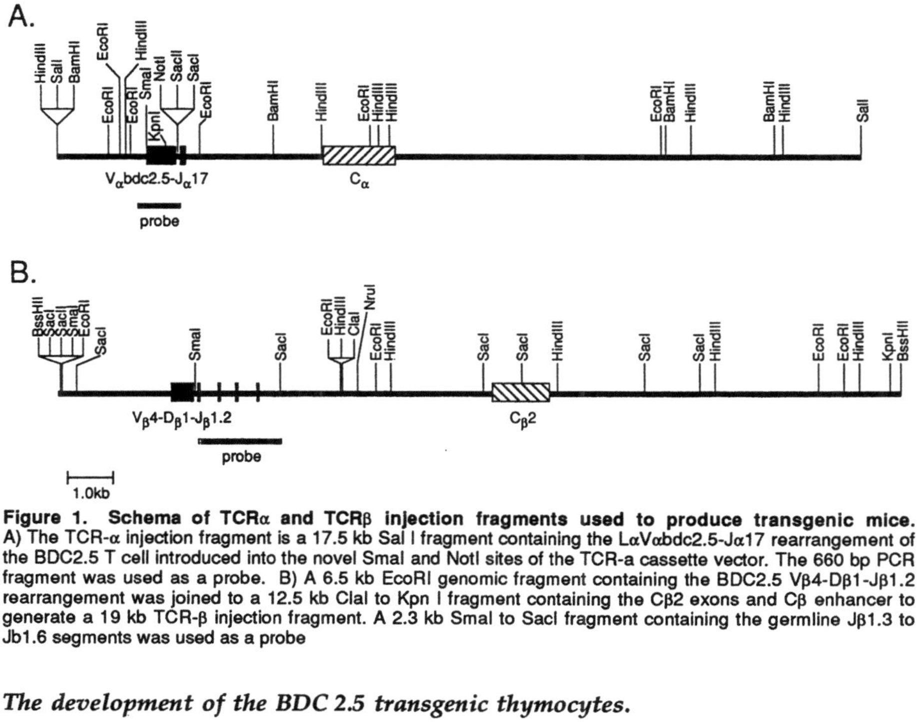

Figure 1. Schema of TCRα and TCRβ injection fragments used to produce transgenic mice. A) The TCR-α injection fragment is a 17.5 kb Sal I fragment containing the LαVαbdc2.5-Jα17 rearrangement of the BDC2.5 T cell introduced into the novel SmaI and NotI sites of the TCR-a cassette vector. The 660 bp PCR fragment was used as a probe. B) A 6.5 kb EcoRI genomic fragment containing the BDC2.5 Vβ4-Dβ1-Jβ1.2 rearrangement was joined to a 12.5 kb ClaI to Kpn I fragment containing the Cβ2 exons and Cβ enhancer to generate a 19 kb TCR-β injection fragment. A 2.3 kb SmaI to SacI fragment containing the germline Jβ1.3 to Jb1.6 segments was used as a probe

The development of the BDC 2.5 transgenic thymocytes.

To evaluate the possible effect of transgenic TCR expression on positive and negative selection of T cells, we compared thymi from TCR-β transgenics, TCR-$\alpha\beta$ transgenics and transgene-negative littermates, produced on either the NOD (H-2^{g7}) or B6 (H-2^{b}) backgrounds. All of the animals were normal by size and weight, and all had normal thymi according to gross examination. Fig. 2A presents cytofluorimetric analysis of thymocytes from H-2$^{g7/g7}$ animals. TCR-β transgenics showed a slight increase in the proportion of CD4^{+}8^{-} at the expense

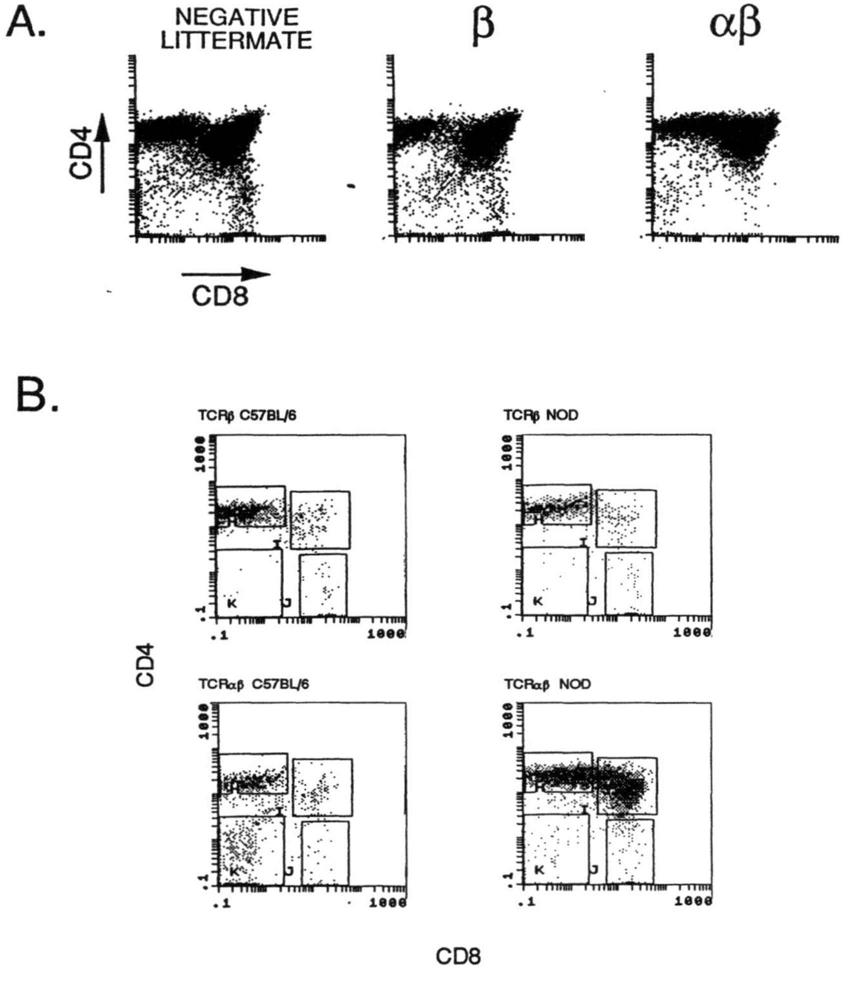

Figure 2. Analyses of BDC2.5 TCR-ab expression in the thymi of transgenic mice. A) Three-color cytofluorometric analysis of negative littermate (left) and transgenic TCR-β (center) or TCR-αβ (right) mice (NOD background) using anti-CD4, anti-CD8. B) Analysis of CD4 and CD8 expression of TCR^{hi} cells from TCR-β (upper panels) and TCR-αβ (lower panels) transgenic thymocytes derived from selective (NOD, right) and non-selective (B6, left) MHC genes.

of $CD4^-8^+$ cells. This skewing was evident in the extreme in the case of TCR-αβ transgenics, which showed an almost complete absence of a $CD4^-8^+$ compartment. Both types of transgenic animals expressed the transgene-encoded Vβ4 segment on >95% of mature T cells and excluded the expression of other Vβ elements (data not shown). Since we lack a monoclonal antibody (mAb) specific for the transgene-encoded Vα1, we could not directly quantitate the proportion of thymocytes displaying the transgenic TCR-α. However, we have indirect evidence that it is highly expressed. The pronounced skewing of mature thymocytes into the CD4+8- compartment appears to signify positive selection of the BDC 2.5 specificity, which occurs only on the selecting NOD but not the non-selecting B6 background. This is illustrated by gating on TCR^{hi} cells and displaying CD4/CD8 profiles, as shown in Fig. 2B.

Therefore, we conclude that the BDC 2.5 specificity is expressed on a significantly large fraction of thymocytes from TCR-αβ transgenic mice, and that these cells undergo positive selection and avoid negative selection within the thymus.

Peripheral T cells in the TCR transgenic mice

To determine whether T cells displaying the BDC 2.5 specificity emigrated from the thymus and seeded and survived in the peripheral lymphoid organs, we performed three-color cytofluorimetric analysis of lymph node cells from the transgenic and control animals described above. While the TCR-β transgenic

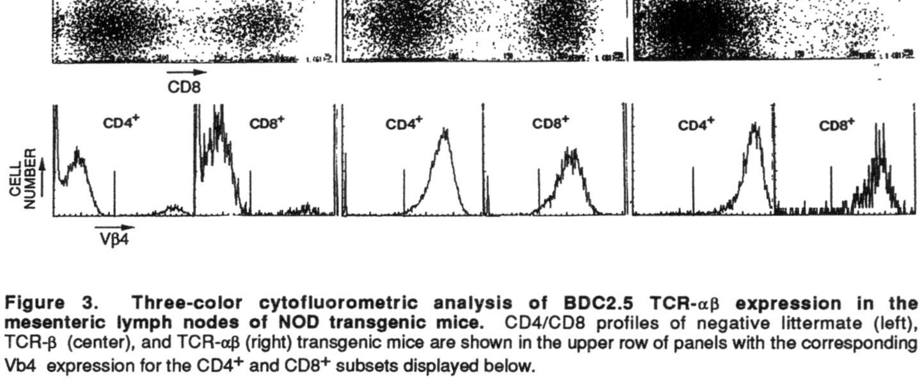

Figure 3. Three-color cytofluorometric analysis of BDC2.5 TCR-αβ expression in the mesenteric lymph nodes of NOD transgenic mice. CD4/CD8 profiles of negative littermate (left), TCR-β (center), and TCR-αβ (right) transgenic mice are shown in the upper row of panels with the corresponding Vb4 expression for the CD4+ and CD8+ subsets displayed below.

mice exhibited a slight increase in the CD4/CD8 ratio compared to the littermate controls, as in the thymus, this skewing became extremely pronounced in the TCR-αβ transgenics, leaving only a dramatically reduced CD8+ T cell compartment.

Both types of transgenic mice demonstrated complete and high-level expression of the BDC 2.5 beta chain, evident from the detection of Vβ4 on >95% of CD4+ and CD8+ T cells (lower panels of Fig. 3), and both transgenic lines exhibited allelic exclusion of the endogenous TCR-β genes, as confirmed by staining with a panel of specific anti-TCR Vβ reagents and by amplification of TCR-β transcripts using the polymerase chain reaction (PCR) (not shown).

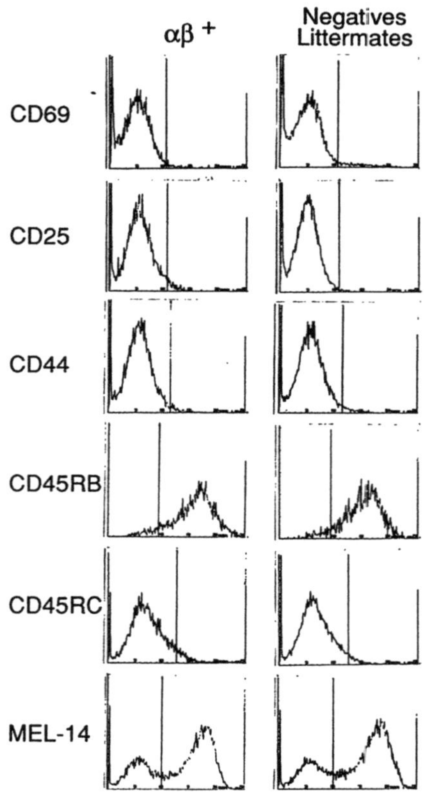

Figure 4. Analysis of activation and memory markers on CD4+ lymphocytes from TCR-αβ transgenic NOD mice. Lymph node cells from 7 week old TCR-αβ transgenic (left-hand column) or negative littermate (right-hand column) mice were labelled simultaneously with mAbs to CD4 (PE), CD8 (Tri-color), CD3 (FITC) and a mAb to the indicated surface protein (TexasRed). Histograms are displayed for $CD4^+CD8^-CD3^+$ for each marker. List-mode data from 6 x 10^4 cells were collected.

It, therefore, became important to assess the activation state of the T cells from the α:β transgenics. Fig. 4 presents four-color cytofluorimetric analysis of lymph node $CD4^+$ cells from a TCR-αβ transgenic mouse and a transgene-negative littermate. Cells were stained with anti-CD4, anti-CD8, an anti-TCR reagent (anti-Vβ4 in the case of the TCR-αβ transgenic or anti-CD3 in the case of the negative littermate), and with a panel of mAbs that recognize activation and memory markers. No differences were observed between these two seven week-old mice.

These data permit us to conclude that T cells expressing the BDC 2.5 specificity immigrate to and seed the periphery of TCR-αβ transgenic mice, and once there, they do not undergo massive clonal deletion nor immediate activation.

Lack of anergy ***<u>in vitro</u>*** ***and*** ***<u>in vivo</u>***

Although T cells displaying the BDC 2.5 specificity were clearly present in the periphery of TCR-αβ transgenic mice, it remained possible that they were

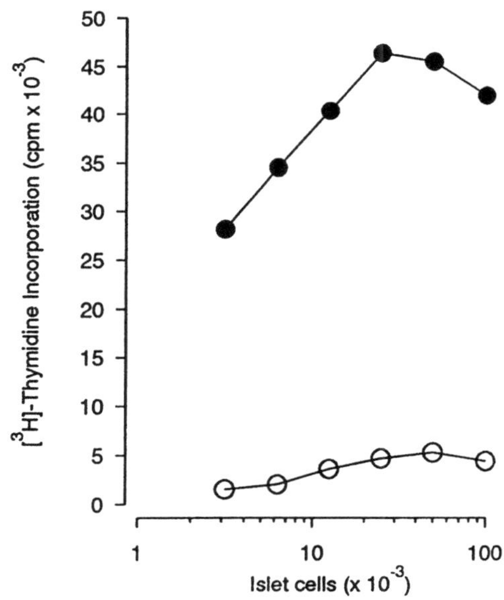

Figure 5. In-vitro proliferation of BDC2.5 TCR-αβ transgenic lymphocytes to islet cell antigen. T cell proliferation was measured by ^{3}H-thymidine (^{3}H-TdR) incorporation. Cultures of 5 x 10^4 transgenic (filled circles) or negative littermate (open circles) lymph node cells were mixed with mitomycin-c treated NOD APCs and various numbers of islet cells, and cultured for 3 days at 37°C. Cultures were pulsed with 1μCi of ^{3}H-TdR for the last 18h, harvested and incorporation assessed by scintillation counting.

functionally anergic. To test this, we cultured lymph node cells from αβ transgenics and transgene-negative littermates with increasing numbers of islet cells in the presence of mitomycin-c treated splenic NOD APCs. After three days, the cultures were harvested and proliferation assessed (Fig. 5). T cells from the αβ transgenics responded energetically (panel A), but neither the littermate control nor β-only transgenic T cells responded (not shown). To rule out that only a few percent of the TCR-αβ T cells were responsive and the rest anergic, we compared the ability of lymph node T cells from αβ transgenics and transgene-negative littermates to respond to anti-CD3 or Con A stimulation, both global activating agents of non-anergic T cells. T cells from the two types of animals responded virtually identically to both stimuli (data not shown). Therefore, at least in vitro, the BDC 2.5 TCR-αβ transgenic peripheral T cells are fully capable of reacting to islet cell antigen.

Yet because of several reports of "split" tolerance - response to a self-antigen in vitro but not in vivo - it was necessary to assess T cell anergy in vivo. This was done by histopathologic evaluation of pancreata from transgenic mice for signs of insulitis. From the second NOD back-cross generation, TCR-αβ

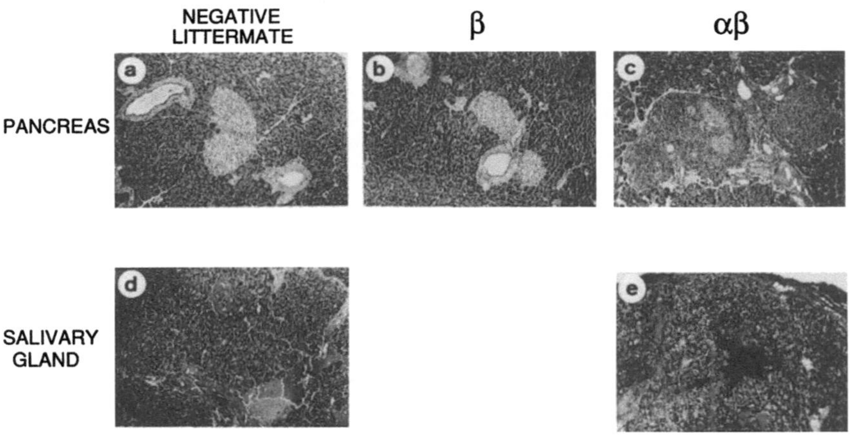

Figure 6. Histology micrographs of pancreas from BDC2.5 TCRαβ transgenic mice. The pancreata from seven week-old mice were removed, fixed in Bouin solution, embedded in parafin, sectioned (5μ) and stained with hematoxylin and eosin. Pancreas sections from a negative littermate (a), a TCR-β transgenic (b) and a TCR-αβ transgenic (c) are displayed. Islets stain lightly against the exocrine pancreas, while lymphoid infiltrating cells stain darkly. Magnification 125x.

transgenics exhibited a severe insulitis from 3-5 weeks of age (Fig. 6c); TCR-β transgenic and control littermates showed absolutely no islet infiltration at this young age (Fig. 6a,b). Furthermore, we demonstrated that the T cells actually involved in the insulitic lesion expressed the BDC2.5 TCR and that the insulitis was active, not just a passive accumulation of anergic T lymphocytes, since the vast majority of infiltrating T cells in 5 to 12 week-old αβ transgenic mice were $CD4^+CD8^-V\beta4^+$ in phenotype as expected, proliferated in the islets (as measured by in situ ^{3}H-thymide-uptake) and were capable of destroying islet beta cells (data not shown). Furthermore, analysis of isolated islets from TCR-αβ mice by PCR showed that all TCR-β transcripts in the pancreata were transgene-encoded and around 40% of TCR-α transcripts were derived from the transgene, a percentage similar to that found for the lymph nodes (data not shown).

We conclude, then, that the T cells from TCR-αβ transgenic mice are not anergic: they respond vigorously to islet cells in vitro and in vivo. Both of these activities require expression of the BDC 2.5 TCR-alpha chain because they were not observed with T cells from three independent lines of TCR-β transgenic mice.

Diabetes induction.

Given the uncontroled lymphocytic infiltration of the pancreatic islets by TCR-αβ transgenic mice, coupled with extensive beta cell destruction, we expected a rapid-onset diabetes. We, therefore, monitored forty transgenic and control mice from the N3 backcross generation (83.75% NOD; all $H\text{-}2^{g7/g7}$) for diabetes by scoring diabetes by the weekly measurement of blood glucose levels. As illustrated in Figure 7, the transgenic animals manifested a slightly accellerated onset of disease as compared to control animals. But it still took a considerable time for the TCR-αβ transgenic mice to develop diabetes given that most had extensive insulitis from 5 weeks yet only started to show overt diabetes from 17 weeks of age. In the end, the disease penetrance was much higher in transgenic animals; the increased incidence was marked when one considers that diabetes in NOD mice is polygenic (Garchon, 1992), and that the experimental animals represent only the N3 generation of NOD backcross. One might have expected only a moderate percentage of animals to contract disease, like the 40% of control animals. The 85% of transgenics that developed diabetes was clearly a significant increase.

Lack of potection from IDDM by the MHC class II E molecule.

NOD mice express no MHC class II E molecules because of a mutation in the Eα promoter region. Kishimoto and colleagues were the first to demonstrate that introduction of a functional Eα transgene onto the NOD genetic background by mating or direct injection leads to protection from insulitis and diabetes (Nishimoto et al, 1989; Uehira et al, 1989; Uno et al, 1991), and this has now been

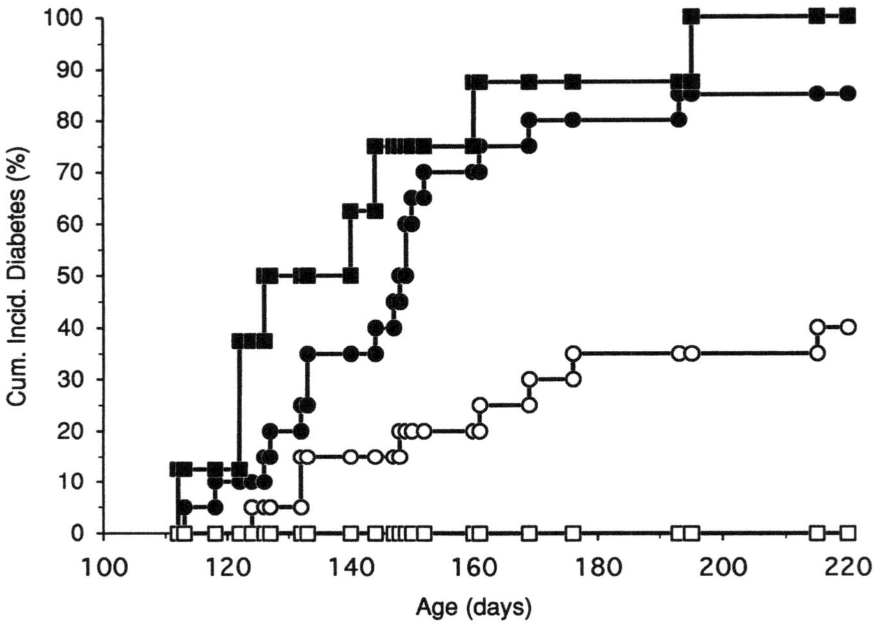

Figure 7. Comparison of TCR-αβ/NOD transgenic and TCR-αβ/NOD-E+ double transgenic mice. Diabetes in TCR-αβ and TCR-αβ/E+ versus negative littermate controls. TCRαβ, filled circles; TCR-αβ/E+, filled squares; NOD negative littermate, open circles; NOD/E+, open squares.

confirmed in several laboratories (Böhme et al, 1990; Lund et al, 1990). To assess the effect of the restored E expression on the phenotype of BDC 2.5 TCR-αβ transgenic mice, we produced double transgenic animals by mating NOD/TCR-αβ transgenics with NOD/Eα transgenics.

The phenotype of the BDC 2.5 TCR-αβ transgenics was unchanged by expression of the class II E molecule. Positive selection of cells with the BDC 2.5 specificity appeared unperturbed in the thymus, and these cells avoided massive clonal deletion, and these cells were exported to and seeded the periphery (not shown). Cells from E-positive TCR-αβ transgenics responded to islet cells <u>in vitro</u> and penetrated the islets <u>in vivo</u> just like those from E-negative TCR transgenics (Data not shown). Finally, in the presence of E molecules, the TCR-αβ transgenics developed a diabetes that paralleled the TCR-αβ transgenics in the absence of E - beginning earlier, showing a more rapid dispersal and manifesting a complete population penetrance (Fig. 7). Note that TCR transgene-negative controls that expressed E molecules showed no diabetes whatsoever, clearly illustrating the protective capacity of this particular Eα transgene.

DISCUSSION

We have generated a line of mice carrying the rearranged TCR-α and -β genes from a diabetogenic T cell clone, BDC 2.5. The transgenic TCR-αβ mice exhibit a striking skewing into the mature $CD4^+$ compartment, vigorous islet cell reactivity in vitro, infiltration into the islets in vivo, and provocation of a high penetrant diabetes. In contrast, all three independent lines of β-only transgenics did not exhibit any of these characteristics, indicating that the phenomena we observe depend on both transgene-encoded TCR chains.

No tolerance

The BDC 2.5 TCR transgenic mice provide a new approach to address questions one might pose about any autoimmune disease. Here we have concentrated on determining how mice "handle" the self-reactive specificity to guard self-tolerance. The answer appears to be quite simple: they don't. These T cells avoided deletion and anergy induction in the thymus and periphery and were fully cognizant of their target antigen on islet cells. This result contrasts with past findings on TCR transgenics with repertoires highly enriched for self-reactive specificities, where deletion, anergy induction, or both were the rule even when the self-antigen was expressed in only very restricted non-thymic locations (reviewed in Ohashi et al, 1989; Miller and Morahan, 1992; also see Schönrich et al, 1992b).

A major difference between BDC 2.5 and the T cell clones used for generating all of the other TCR transgenics is that BDC 2.5 was selected as being not only self-reactive but also pathogenic. In addition, in the BDC 2.5 TCR transgenics, the autoreactive T cells find their target antigen in its natural context, unlike many of the other models. It may be that pathogenic T cell clones are precisely those which have managed to avoid the animal's battery of defenses against autoreactivity and are quiescent only until they encounter their target, an event that might be influenced by genetic or environemental factors, or a combination of the two.

The BDC 2.5 induced disease

The BDC 2.5 TCR-αβ transgenic mice developed a disease that was similar to, but in many ways more aggressive than, diabetes in unmanipulated NOD mice. Insulitis was accelerated, appearing abruptly at 3 weeks and rapidly achieving a severity typical of 12-15 week-old NOD animals. The onset of diabetes was also accelerated - the average age of onset being 144 days, versus 157 days for littermate controls. However, disease progression within an individual was considerably faster and the degree of disease penetrance within the population were also noticeably elevated.

Disease progression in the BDC 2.5 TCR transgenic mice identifies two "check-points" in the development of diabetes. The first involves insulitis. None of the TCR transgenics showed T cell infiltration in the islets at two weeks of age despite the presence of many T cells displaying the transgene-encoded specificity in the lymph nodes and spleen. However, a week later, infiltration was rampant. There exist several possibilities to account for this lag: lack of relevant antigen before the third week after birth, lack of a crucial population of APCs until the third week, inhibition of insulitis by another lymphoid cell is responsible for the delay.

Interestingly, a three week lag period has been noted in other contexts related to diabetes in NOD mice. When T cells from diabetic females were transferred into neonatal NOD hosts, insulitis was detected only after three weeks of age (Bendelac et al, 1987). Similarly, NOD mice expressing the TNF-α gene specifically on islet cells developed a massive insulitis that was detectable at 3 weeks (Higuchi et al, 1993). Finally, neonatal thymectomy protected NOD animals from insulitis while thymectomy at 3 weeks accelerated the disease process (Dardenne et al, 1989). These results, together with ours, suggest that some important change(s) in the immune system or the pancreatic islets of NOD mice may take place at three weeks of age.

The second check-point involves diabetes. Although the TCR-αβ transgenic mice exhibited massive insulitis starting at three weeks of age, and despite the fact that the infiltrating T cells were capable of extensive islet cell destruction, animals did not contract diabetes until much later, at around 4.5 months. This lag is reminiscent of the delay between occult insulitis and overt diabetes known to occur in unmanipulated NOD mice and human patients and may stem from the highly skewed T cell repertoire of these animals. It could be that other effector cells are needed and these take quite some time to accumulate in the transgenics because by far most cells express the transgene-encoded specificity - such as CD8+ T cells which have been implicated in diabetes in several studies (Bendelac et al, 1987; Miller et al, 1988; Nagata et al, 1989; Thivolet et al, 1991; Hayakawa et al, 1991; Nagata and Yoon, 1992; Yagi et al, 1992). Though other cell types may also be important.

MHC class II E molecule and NOD diabetes

NOD mice express only one MHC class II molecule, A^{g7}. Several studies have demonstrated that artificial expression of the E molecule protects NOD mice from insulitis and diabetes (Nishimoto et al, 1989; Uehira et al, 1989; Böhme et al, 1990; Lund et al, 1990; Uno et al, 1991). Although the mechanism of protection remains elusive, diverse possibilities have been suggested: E-mediated clonal deletion (Uehira et al, 1989; Reich et al, 1989), clonal anergy (Uehira et al, 1989), selection of a regulatory T cell (Nishimoto et al, 1989; Böhme et al, 1990), "determinant stealing" (Deng et al, submitted), skewing of the TH1/TH2 balance (unpublished speculations). Insulitis and diabetes in BDC 2.5 TCR transgenic mice, much to our great surprise, were not prevented by the rescue of E molecule expression. TCR-$\alpha\beta$ transgenics which did or did not express an Eα transgene responded to islet cells in vitro, penetrated the islets, and provoked diabetes indistinguishably.

These results argue against certain of the proposed mechanisms of protection. E-mediated clonal deletion would appear to be ruled out, consistent with previous observations (Böhme et al, 1990; Lund et al, 1990) and clonal anergy is rendered very unlikely. Determinant stealing as envisaged as competition between the E and A^{g7} molecules for a critical epitope, while not formally ruled-out, would not seem to be in agreement with the observation, particularly at the level of thymus selection, where no difference was seen between the BDC2.5 TCR on NOD or NOD-E backgrounds. Furthermore, diabetogenic T cells can be stimulated by APCs from both E-positive and E-negative NOD mice (Nakano et al, 1991). Finally, our studies did not address the eventuality that E molecules may perturb the balance of TH1/TH2 phenotypes.

Concluding Remarks

By generating a strain of mice carrying rearranged TCR genes from a diabetogenic T cell clone, we have provided a new approach to address questions of interest to diabetologists and, more generally, to anyone interested in autoimmune diseases. Our first series of analyses has provided new information of NOD diabetes: 1) that diabetogenic T cells can evade all of the so far recognized forms of tolerance induction; 2) that there is a lag before insulitis onset which does not just represent the time required to accumulate a sufficient number of islet-specific T cells, and that the second lag before overt diabetes involves more than inundating the islets with T cells capable of destruction; 3) that the E molecule protects NOD mice from diabetogenic T cells by a mechanism other than clonal deletion, clonal anergy.

ACKNOWLEDGEMENTS

We would like to thank M. Lemeur, C. Marfing, P. Andre, and M. Duval of the transgenic facility for production of the transgenic lines, R. Ceredig, K. Tomanari, H. von Boehmer, B. Malissen, H. Blüthman, M. Steinmetz, T. Shirai and R. Zamoyska for gifts of materials, C. Waltzinger for helpful advice and excellent assistance in flow cytometric analysis, P. Gerber and C. Ebel for technical help, J. Hergueux, P. Michel, N. Zinck and S. Metz for maintaining the mice, J-M. Garnier for technical advice with cDNA and genomic library construction, B. Schubaur and M-A. Laffait for their excellent and tireless assistance with histology, H. Cohen at Hôpital Necker for helpful advice on the isolation of pancreatic islet cells, Dr. H. Kikutani for advice on aldehyde-fushsin staining, K. Signorelli, H.-J. Fëhling, and S. Candéias and V. Kouskoff for advice and materials. This work was supported by institute funds from the INSERM, the CNRS, and the Centre Hospitalo-Univesitaire Régional, and by grants to DM and CB from the Association pour la Recherche sur le Cancer, the Juvenile Diabetes Foundation International. JK was supported by post-doctoral fellowships from the Human Frontiers Science Program and the Juvenile Diabetes Foundation International. BW was supported by a fellowship from the Université Louis Pasteur.

REFERENCES

Bendelac, A., Carnaud, C., Boitard, C., and Bach, J-F. (1987). Syngeneic transfer of autoimmune diabetes from diabetic NOD mice to healthy neonates. J. Exp. Med. 166,823-832.

Blackman, M.A., Gerhard-Burgert, H., Woodland, D.L., Palmer, E., Kappler, J.W., and Marrack, P. (1990). A role for clonal inactivation in T cell tolerance to Mls-1a. Nature 345,540-542.

Blüthmann, H., Kisielow, P., Uematsu, Y., Malissen, M., Krimpenfort, P., Berns, A. (1988). T-cell-specific deletion of T-cell receptor transgenes allows functional rearrangement of endogenous α- and β-genes. Nature 334,156-159.

Böhme, J., Schuhbaur, B., Kanagawa, O., Benoist, C., Mathis, D. (1990). MHC-linked protection from diabetes dissociated from clonal deletion of T cells. Science 249,293-295.

Bradley, B.J., Haskins, K., La Rosa, F.G., Lafferty, K.J. (1992). CD8 T cells are not required for islet destruction induced by a CD4+ islet-specific T-cell clone. Diabetes 41,1603-1608.

Carlow, D.A., Teh, S-J., and Teh, H-S. (1992). Altered thymocyte development resulting from expressing a deleting ligand on selecting thymic epithelium. J. Immunol. 148,2988-2995.

Castano, L., and Eisenbarth, G.S. (1990). Type-I diabetes: A chronic autoimmune disease of human, mouse and rat. Ann. Rev. Immunol. 8,647-679.

Dardenne, M., Lepault, F., Bendelac, A., and Bach, J-F. (1989). Acceleration of the onset of diabetes in NOD mice by thymectomy at weaning. Eur. J. Immunol. 19,889-895.

Deng, H., Apple, R., Clare-Salzler, M., Trembleau, S., Adorini, L., Sercarz, E. (1993). Determinant capture as a possible mechanism of protection afforded by MHC class II molecules in insulin-dependent diabetes mellitus. in press J. Exp. Med.

Garchon, H-J., (1992). Non-MHC-linked genes in autoimmune diseases. Curr. Opin. in Immunol. 4,716-722.

Harada, M., and Makino, S. (1986). Suppression of overt diabetes in NOD mice by anti-thymocyte serum or anti-Thy 1. 2 antibody. Exp. Anim. 35,501-506.

Haskins, K., Portas, M., Bradley, B., Wegmann, D., Lafferty, K.J. (1988). T-lymphocyte clone specific for pancreatic islet antigen. Diabetes 37,1444-1448.

Haskins, K., Portas, M., Bergman, B., Lafferty, K., Bradley B. (1989). Pancreatic islet-specific T-cell clones from nonobese diabetic mice. Proc. Natl. Acad. Sci. USA 86,8000-8004.

Haskins, K., and McDuffie, M. (1990). Acceleration of diabetes in young NOD mice with a CD4+ islet-specific T-cell clone. Science 249,1433-1436.

Schönrich, G., Momburg, F., Hammerling, G.J., and Arnold, B. (1992a). Anergy induced by thymic medullary epithelium. Eur. J. Immunol. 22,1687-1691.

Schönrich, G;, Momburg, F., Malissen M., Schmitt-Verhulst, A-M., Malissen, B., Hämmerling, G.J., and Arnold, B. (1992b). Distinct mechanisms of extrathymic T cell tolerance due to differential expression of self antigen. Int. Immunol. 4,581-590.

Sempé, P., Bédossa, P., Richard, M-F., Villà, M-C., Bach, J-F., and Boitard, C. (1991). Anti-a/b T cell receptor monoclonal antibody provides an efficient therapy for autoimmune diabetes in nonobese diabetic NOD mice. Eur. J. Immunol. 21,1663-1169.

Sha, W.C. Nelson, C.A., Newberry, R.D., Kranz, D.M., Russell, J.H., and Loh, D.Y. (1988). Positive and negative selection of an antigen receptor on T cells in transgenic mice. Nature 336,73-76.

Shizuru, J.A., Taylor-Edwards, C., Banks, B.A., Gregory, A.K., Fathman, C.G. (1988). Immunotherapy of the nonobese diabetic mouse: treatment with an antibody to T-helper lymphocytes. Science 240,659-662.

Signorelli, K., Benoist, C., and Mathis, D. (1992). Why is clonal deletion of neonatal thymocytes defective? Eur. J. Immunol. 22,2487-2493.

Thivolet, C., Bendelac, A., Bedossa, P., Bach, J-F., and Carnaud, C. (1991). $CD8^+$ T cell homing to the pancreas in the nonobese diabetic mouse is $CD4^+$ T cell-dependent. J. Immunol. 146,85-88.

Uehira, M., Uno, M., Kürner, T., Kikutani, H., Mori, K-I., Inomoto, T., Uede, T., Miyazaki, J-I., Nishimoto, H., Kishimoto, T., and Yamamura, K-I. (1989). Development of autoimmune insulitis is prevented in $E\alpha^d$ but not in $A\beta^k$ NOD transgenic mice. Int. Immunol. 1,210-213.

Uematsu, Y., Ryser, S., Dembic, Z., Borgulya, P., Krimpenfort, P., Berns, A., von Boehmer, H., and Steinmetz, M. (1988). In transgenic mice the introduced functional T cell receptor β gene prevents expression of endogenous β genes. Cell 52,831-841.

Uno, M., Miyazaki, T., Uehira, M., Nishimoto, H., Kimoto, M., Miyazaki, J-I., and Yamamura, K-I. (1991). Complete prevention of diabetes in transgenic NOD mice expressing I-E molecules. Immunol. Lett. 31,47-52.

Yagi, H., Matsumoto, M., Kunimoto, K., Kawaguchi, J., Makino, S., and Harada, M. (1992). Analysis of the roles of $CD4^+$ and $CD8^+$ T cells in autoimmune diabetes of NOD mice using transfer to NOD athymic nude mice. Eur. J. Immunol. 22,2387-2393.

Mori, Y., Suko; M., Okudaira, H., Matsuba, I., Tsuruoka, A., Sasaki, A., Yokoyama, H., Tanase, T., Shida, T., Nishimura, M., Terada, E., and Ikeda, Y. (1986). Preventive effects of cyclosporin on diabetes in NOD mice. Diabetologia 29,244-248.

Nagata, M., Yokono, K., Hayakawa, M., Kawase, Y., Hatamori, N., Ogawa, W., Yonezama, K., Shil, K., and Baba, S. (1989). Destruction of pancreatic islet cells by cytotoxic T lymphocytes in nonobese diabetic mice. J. Immunol. 143,1155-1162.

Nagata, M., and Yoon J-W. (1992). Studies on Autoimmunity for T-cell-Mediated β-cell destruction. Diabetes 41,998-1008.

Nakano, N., Kikutani, H., Nishimoto, H., and Kishimoto, T. (1991). T cell receptor V gene usage of islet β cell-reactive T cells is not restricted in non-obese diabetic mice. J. Exp. Med. 173,1091-1097.

Nishimoto, H., Kikutani, H., Yamamura, K-I., and Kishimoto, T. (1989). Prevention of autoimmune insulitis by expression of I-E molecules in NOD mice. Nature 328,432-434.

Ogawa, M., Maruyama, T., Hasegawa, T., Kanaya, T., Kobayashi, F., Tochino, Y., Uda, H. (1985). The inhibitory effect of neonatal thymectomy on the incidence of insulitis in non-obese diabetes (NOD) mice. Biomed. Res. 6,103-105.

Ohashi, P.S., Pircher, H., Mak, T.W., Burki, K., Zinkernagel, R.M., and Hengatner, H. (1989). Ontogeny and selection of the T cell repertoire in transgenic mice. Semin. Immunol. 1,95-104.

Ohashi, P.S., Oehen, S., Buerki, K., Pircher, H., Ohashi, C.T., Odermatt, B., Malissen, B., Zinkernagel, R.M., and Hengartner, H. (1991). Ablation of "tolerance" and induction of diabetes by virus infection in viral antigen transgenic mice. Cell 65,305-317.

Pankewycz, O., Strom, T.R., and Rubin-Kelley, V.E.R. (1991). Islet-infiltrating T cell clones from non-obese diabetic mice that promote or prevent accelerated onset diabetes. Eur. J. Immunol. 21,873-879.

Reich, E-P., Sherwin, R.S., Kanagawa, O., and Janeway, Jr, C.A. (1989). An explanation for the protective effect of the MHC class II I-E molecule in murine diabetes. Nature 341,326-328.

Rocha, B., and von Boehmer, H. (1991). Peripheral selection of the T cell repertoire. Science 251,1225-1228.

Schönrich, G., Kalinke, U., Momburg, F., Malissen, M., Schmitt-Verhulst, A-M., Malissen, B., Hämmerling, G.J., and Arnold, B. (1991). Down-regulation of T cell receptors on self-reactive T cells as a novel mechanism for extrathymic tolerance induction. Cell 65,293-304.

Schönrich, G., Momburg, F., Hammerling, G.J., and Arnold, B. (1992a). Anergy induced by thymic medullary epithelium. Eur. J. Immunol. 22,1687-1691.

Schönrich, G;, Momburg, F., Malissen M., Schmitt-Verhulst, A-M., Malissen, B., Hämmerling, G.J., and Arnold, B. (1992b). Distinct mechanisms of extrathymic T cell tolerance due to differential expression of self antigen. Internat. Immunol. 4,581-590.

Sempé, P., Bédossa, P., Richard, M-F., Villà, M-C., Bach, J-F., and Boitard, C. (1991). Anti-a/b T cell receptor monoclonal antibody provides an efficient therapy for autoimmune diabetes in nonobese diabetic NOD mice. Eur. J. Immunol. 21,1663-1169.

Sha, W.C. Nelson, C.A., Newberry, R.D., Kranz, D.M., Russell, J.H., and Loh, D.Y. (1988). Positive and negative selection of an antigen receptor on T cells in transgenic mice. Nature 336,73-76.

Shizuru, J.A., Taylor-Edwards, C., Banks, B.A., Gregory, A.K., Fathman, C.G. (1988). Immunotherapy of the nonobese diabetic mouse: treatment with an antibody to T-helper lymphocytes. Science 240,659-662.

Signorelli, K., Benoist, C., and Mathis, D. (1992). Why is clonal deletion of neonatal thymocytes defective? Eur. J. Immunol. 22,2487-2493.

Thivolet, C., Bendelac, A., Bedossa, P., Bach, J-F., and Carnaud, C. (1991). CD8+ T cell homing to the pancreas in the nonobese diabetic mouse is CD4+ T cell-dependent. J. Immunol. 146,85-88.

Uehira, M., Uno, M., Kürner, T., Kikutani, H., Mori, K-I., Inomoto, T., Uede, T., Miyazaki, J-I., Nishimoto, H., Kishimoto, T., and Yamamura, K-I. (1989). Development of autoimmune insulitis is prevented in $E\alpha^d$ but not in $A\beta^k$ NOD transgenic mice. Internat. Immunol. 1,210-213.

Uematsu, Y., Ryser, S., Dembic, Z., Borgulya, P., Krimpenfort, P., Berns, A., von Boehmer, H., and Steinmetz, M. (1988). In transgenic mice the introduced functional T cell receptor β gene prevents expression of endogenous β genes. Cell 52,831-841.

Uno, M., Miyazaki, T., Uehira, M., Nishimoto, H., Kimoto, M., Miyazaki, J-I., and Yamamura, K-I. (1991). Complete prevention of diabetes in transgenic NOD mice expressing I-E molecules. Immunol. Lett. 31,47-52.

Yagi, H., Matsumoto, M., Kunimoto, K., Kawaguchi, J., Makino, S., and Harada, M. (1992). Analysis of the roles of CD4+ and CD8+ T cells in autoimmune diabetes of NOD mice using transfer to NOD athymic nude mice. Eur. J. Immunol. 22,2387-2393.

The maintenance and reestablishment of self-tolerance: T cell receptor (TCR) peptide-specific regulatory T cells

Vipin Kumar and Eli Sergarz
Department of Microbiology and Molecular Genetics
405 Hilgard Avenue,University of California
Los Angeles, CA 90024-1489,U.S.A.

While we know that mature peripheral T cells are generally unresponsive to dominant determinants on self-antigens, the mechanism inducing this self tolerance is still unclear. Clonal deletion of self-reactive T cells during the thymic selection process has been described as one of the major mechanisms for inducing central tolerance (Kappler et al., 1987; von Boehmer, 1990). The presence of autoaggressive cells in healthy individuals as well as the demonstration of physiological presentation of self-antigens in vivo, suggest involvement of other mechanisms in maintenance of peripheral tolerance, for example anergy or suppressive regulation (Schwartz, 1990; Kumar and Sercarz, 1991). It has also been hypothesized that a deficiency of regulatory T cells may lead to autoimmunity (Sakaguchi and Sakaguchi, 1990; Powrie and Mason, 1990). Many studies suggest that potentially pathogenic autoimmune cells are not tolerized at the outset, but rather become activated normally and then controlled by regulatory T cells. Thus the equilibrium between autoreactive and regulatory T cells may determine the expression of autoimmunity.

We will be primarily discussing our work on the regulation of the immune response to self-myelin basic protein (MBP) that results in an autoimmune condition, experimental allergic encephalomyelitis (EAE). EAE is a T cell-mediated demyelinating disease of the central nervous system and is generally self-limiting in B10.PL mice. A majority of MBP-induced T cells recognize N-terminal fragment Ac1-9, and predominantly utilize TCR Vβ8.2 and Vα2.3 variable (V) gene segments (Urban et al. 1988; Kumar et al, 1989). We have addressed the issue of whether the TCR itself can be processed and presented in the context of either class I or class II molecules, in vivo, in a physiological manner. If so, do T cells activated by TCR

NATO ASI Series, Vol. H 80
Autoimmunity: Experimental Aspects
Edited by Moncef Zouali

peptides become involved in the regulation of T cell responses to MBP normally as well as during the course of the disease?

Experiments employing a T cell vaccination approach by Cohen and his co-workers suggested that motifs on the T cell receptors themselves may be engaged in regulating the cells on which they are displayed (Ben-Nun et al, 1981). Recently it has been shown that immunization with particular V region peptides induces regulation and protects rats from EAE (Howell et al, 1989; Vandenbark et al, 1989). To map out all the potential determinants within the V regions of one prototypic T cell receptor (Vß8.2 and Vα2.3) expressed on a majority of Ac1-9-reactive T cells in B10.PL mice (Urban et al, 1988), we have synthesized five 25-30-mer overlapping peptides from each TCR chain, encompassing the entire variable region. In testing their capacity to induce T cell proliferation in B10.PL mice, we have found that immunogenic self-determinants are prevalent in both the ß- (three) and α- (two) chains. These determinants are spread along the V region and are not from the previously reported CDR II region that induced regulation of EAE in rats (Vandenbark et al, 1989).

Thus we found that three Vß-chain peptides B2 (21-50), B4 (61-90) and B5 (76-101) were immunogenic when injected into B10.PL mice as peptides. However, based on the proliferative recall responses after immunization with T cells expressing the receptor, there appears to be only a single immunodominant TCR determinant, B5, on the ß-chain (Kumar et al 1993). Thus, B5 appears dominant, while the determinants on B2 and B4 can be considered cryptic. Since, antigen processing and the structural context of determinants influence dominance, it is important to discuss some of the issues regarding the presentation of B5 in vivo. Importantly, we found that both cloned $CD4^+$ T cells reactive to B5 as well as the peptide itself are capable of protecting mice from MBP-induced EAE, indicating an important role for presentation of B5 in the natural recovery from the disease.

Self-TCR peptide B5 can induce regulatory T cells that specifically down-regulate Ac1-9 responses. When injected prior to MBP immunization, the immunodominant ß-chain determinant B5 is specifically able to suppress the Ac1-9 proliferative response in both lymph nodes and spleens of B10.PL, PL/J and (SJL/JxB10.PL) F1 mice (Kumar et al 1993). B5 immunization also significantly protects these mice from antigen-induced EAE (Kumar et al, 1993). Interestingly, other TCR determinants like B2 or B4, although strongly immunogenic, do not down-regulate Ac.1-9 responses, suggesting that two distinct functional sets of T cells are involved. As a control throughout these studies, we have examined the response to other determinants of guinea pig MBP (gpMBP), not present on mouse MBP (mMBP): responses to these unique gpMBP determinants, which presumably do not utilize Vß8 genes, are not affected (Kumar et al, 1993).

We have generated T cell lines and clones reactive to the immunodominant TCR-peptide B5. These T cells express the CD4 coreceptor and are restricted to the I-A^u class II molecule. Upon adoptive transfer in vivo in relatively small numbers ($1x10^5$ activated T cells), the TCR-peptide B5-specific T cell clone B5.2 is able to down-regulate Ac1-9 responses in recipient B10.PL mice and protect mice from antigen -induced EAE (Kumar and Sercarz, 1993). These regulatory T cells (Treg) utilize a very limited set of TCR Vß gene segments (Vß14 and Vß3), and a large majority of cell clones utilize the Vß14 gene segment (Kumar and Sercarz, 1993).

Do mice recover from EAE because of a TCR-based T-T regulatory network which becomes operative during the course of the disease and suppresses the initial encephalitogenic response? To address this crucial question directly, we examined the B10.PL mouse after its initial episode of MBP-induced EAE and asked whether peripheral T cells specific for the immunodominant TCR ß-chain peptide B5 had been induced via antigenic stimulation. Indeed, without any previous exogenous challenge with the TCR peptide, B5-specific Treg become primed in vivo during the initial challenge with antigen (Kumar and Sercarz, 1993). Also, preliminary experiments suggest that the deletion of $Vß14^+$ T cells from the peripheral T cell repertoire indeed leads to severe EAE and poor recovery from EAE in (B10.PL X SJL)F1 mice. These data further support our view that the TCR $Vß14^+$ T cells constitute a crucial part of a regulatory T cell network responsible for recovery from EAE.

How do B5-specific regulatory T cells regulate responses to MBP? CD8 deletion experiments suggest that once activated, the regulatory $CD4^+$ T cells recruit $CD8^+$ T cells which ultimately down-regulate encephalitogenic T cells (Kumar and Sercarz, 1993). Which are the structures that could serve as targets for regulation by $CD8^+$ T cells? One of the important motifs is a framework sequence within the variable region of the T cell receptor, which is exposed on the molecule (Pullen et al, 1990) and therefore could possibly be recognized directly by idiotypic T cells (TCR-TCR interaction). More likely, determinants from nascent TCR chains, subject to degradation within the T cell, enter the Class I pathway and are presented on the cell surface continually in an MHC class I context (Jiang et al, 1991). Recognition of MBP-specific effector cells by TCR-idiopeptide-specific $CD8^+$ T cells may thus be one of the major pathways of regulation. It is most interesting that $CD4^+$ T cells appear to be essential in the activation of the $CD8^+$ T cells. It is likely that TCR-peptide-specific $CD4^+$ T cells provide help for the priming of $CD8^+$ T cells in vivo as has been shown for a viral peptide recently (Fayolle et al, 1991). One key prediction is that mice lacking regulatory $CD8^+$ T cells should not be able to recover well from EAE and also should be susceptible to repeated relapses. Experiments in mice devoid of $CD8^+$ T cells either by anti-CD8 treatment or CD8 gene targeting confirm this prediction (Jiang et al, 1992; Koh et al, 1992).

It is likely that certain dominant TCR regulatory peptides are most readily presented to T cells following antigen processing, as is the pattern in class II responses against conventional antigens. It could also be true that the favored determinant lies within a framework region appearing in different V gene families, which would give rise to very broad regulatory families. There could be a hierarchy of broadly public as well as very private TCR-idiopeptides that would permit both types of regulatory cells to gain access to a single target T cell. Accordingly, a network of V domains could be structured in which many of the cellular participants would be regulatory T cells.

Should regulation be specific for each private idiotope on every antigen-specific T cell? It seems likely that evolution would have favored an immune system spending a minimal amount of its resources on specific regulation. A focus on particular dominant public motifs of certain V regions of the TCR can provide coordinate and effective regulation, economically. It is possible that these dominant TCR motifs are presented by activated antigen-specific B cells, although TCR-idiotope-specific B cells must first be activated. Interestingly, significant levels of anti-TCR-Ab response are induced in animals recovering from MBP or Ac1-9-induced EAE (Kumar, unpublished). Activation of such anti-TCR specific B cells may be an important factor in recruiting a cascade of regulatory players in vivo. Therefore, in the mouse system, processing and presentation of TCR peptides by professional APCs may be a determining event in the induction of TCR-based regulation.

Human T cells expressing class II MHC molecules have been shown to efficiently process and present antigen (Lanzavecchia et al, 1988). Mouse T cells do not generally display class II MHC molecules and they may be unable to present self TCR peptides (Benoist and Mathis, 1990; Lorber et al, 1982), although this point is disputable, especially in activated T cells. How then do TCR-peptide-reactive $CD4^+$ T cells become primed in vivo. It is possible that the TCR-specific B cells mentioned previously could pick up TCRs at the inflammatory site and present TCR-peptides in the context of class II MHC molecules to $CD4^+$ regulatory T cells. Since B cells which express antibody specific for a particular antigen are highly efficient APC for that antigen, capable of activating T cells when provided with 1/1000th the antigen concentration required of non-specific APC (Casten et al, 1988), a minimal amount of TCR molecule should be sufficient to turn on regulatory T cells in vivo. Another possibility is that dendritic cells, which are very efficient APC may be involved in processing and presentation of TCR-peptides.

Interestingly, we have found that spleen cells from naive animals are capable of stimulating TCR-peptide-reactive T cells in vitro in the absence of exogenous peptide (Kumar et al, 1993). There are two explanations for this observation. (i) APCs constitutively process and potentially present B5 in vivo. In view of the prevalence of the TCR Vβ8 family in peripheral T cells (15-30%), one may imagine that TCR Vβ8 peptides could be generated that are capable of binding to MHC molecules in vivo. TCR-specific B cells would be the most parsimonious solution to this requirement for presentation, since they require a minimal quantity of specific antigen. (ii) Another possible interpretation is that the B5 determinant could be cross-reacting (Bhardwaj et al, 1993) with some other molecules on the surface of splenic APC. We are currently characterizing the peptide isolated from these APC.

There is evidence that MHC molecules themselves may guide processing by binding peptides as they are generated and protecting them from further enzyme degradation (Donermeyer et al, 1989; Sercarz et al, 1986). We find that the minimal core of the determinant on B5 is confined to residues from the TCR variable framework region III, and does not contain any N, D or J residues. Residues at this site have been shown to be involved in binding to superantigens (Pullen et al, 1990) and therefore are thought to be exposed. Also, human autoantibodies have been shown to bind to TCR framework region III peptides (Marchalonis et al, 1992). It is probable that this region becomes easily accessible for binding to the MHC molecule and therefore dominant.

Typically, dominant determinants on self-antigens have induced tolerance (Gammon and Sercarz, 1989). Thus, why is the T cell repertoire directed against dominant TCR peptide B5 is still intact! Are these TCR-peptide/MHC complexes not present in the thymus or do they fail to mediate negative selection? Preliminary data suggest that the TCR-peptide B5-specific T cell repertoire is not tolerized even after neonatal treatment of mice with B5. It is tempting to speculate that T cell regulatory networks may be operative through the recognition of some dominant TCR-peptides, for example, B5 (Kumar et al, 1991, 1993), that are part of a "homuncular network" (Cohen and Young, 1991). One of the necessities in such a network is that regulatory cells not themselves be susceptible to removal through tolerance. Therefore, understanding the processing and the presentation of the self-TCR-peptides in vivo is extremely important to explore mechanistic rules for TCR-based network regulation.

Interestingly, B5-reactive T cells are not revealed before the onset of EAE in mice immunized with MBP/CFA and pertussis toxin (PT). There are at least two explanations for this phenomenon: (i) regulatory TCR Vß14$^+$ T cells are only primed after the expansion of MBP-reactive Vß8.2$^+$ T cells which induce EAE, and subsequently only upon optimum priming and expansion of B5-reactive T cells which results in the down-regulation of encephalitogenic Ac1-9-reactive T cells. (ii) immunization with MBP in CFA and pertussis toxin somehow inhibits the normal initial expansion of Vß14$^+$ T cells temporarily; upon release of inhibition, these cells expand leading to the down-regulation of the Vß8.2$^+$ effector T cells. Preliminary data suggest that B5-reactive T cells appear to be present normally but in an inhibited state. In conclusion, it is likely that the TCR Vß14$^+$ regulatory T cells are an essential part of the physiological T cell regulatory network that maintains and reestablishes tolerance to self MBP. The organism can withstand the perturbations and flare-ups that probably characterize the delicate balance inherent in the immune system. It appears that disruption of this cellular circuitry can lead to a failure to maintain an overt equilibrium, resulting in autoimmunity.

REFERENCES

Bhardwaj, V., Kumar, V., Geysen, H.M. and Sercarz, E. (1993) Degenerate recognition of a dissimilar antigenic peptide by MBP-reactive T cells: implications for thymic education and autoimmunity. J. Immunol. in press.

Ben-Nun, A., Wekerle, H., and Cohen, I.R. (1981) Vaccination against autoimmune encephalomyelitis using attenuated cells of a T lymphocyte line reactive against myelin basic protein. Nature 292: 60-61.

Benoist, C. and Mathis, D. (1990) Regulation of major histocompatibility class-II genes: X, Y, and other letters of the alphabet. Ann. Rev. Immunol. 8: 681-715.

Casten, L.A., Kaumaya, P. and Pierce, S.K. (1988) Enhanced T cell responses to antigenic peptides targeted to B cell surface Ig, Ia, or class I molecules. J. Exp. Med. 168: 171-180.

Cohen I.R., and Young D.B. (1991) Autoimmunity, microbial immunity and the immunological homunculus. Immunol Today 12:105-110.

Donermeyer, D.L., and Allen, P.M. (1989) Binding to Ia protects an immunogenic peptide from proteolytic degradation. J. Immunol. 142: 1063-1068.

Fayolle, C., Deriaud, E. and Leclerc, C. (1991) In vivo induction of cytotoxic T cell response by a free synthetic peptide requires $CD4^+$ T cell help. J. Immunol. 147: 4069-4073.

Gammon, G. and Sercarz, E. (1989) How some T cells escape tolerance induction. Nature 342: 183-185

Howell, M.D., Winters, S.T., Olee, T., Powell, H.C., Carlo, D.J. and Brostoff, S.W. (1989) Vaccination against experimental allergic encephalomyelitis with T cell receptor peptides. Science 246:668-670.

Jiang, H., Sercarz, E., Nitecki, D. and Pernis, B. (1991) The problem of presentation of T cell receptor peptides by activated T cells. Annals of the New York Academy of Sciences. 636: 28-32.

Jiang, H., Zhang S., and Pernis, B. (1992) Role of $CD8^+$ T cells in experimental allergic encephalomyelitis. Science 256: 1213-1215.

Kappler, J., Roehm, N. and Marrack, P. (1987) T cell tolerance by clonal elimination in the thymus. Cell 49: 273-280.

Kumar, V., Kono, D.H., Urban, J.L., and Hood. L. (1989) T-cell receptor repertoire and autoimmune diseases. Ann. Rev. Immunol. 7: 657-682.

Kumar, V. and Sercarz, E. (1991) Regulation of autoimmunity. Current Opinion in Immunology 3:888-895.

Kumar, V. and Sercarz, E. (1992) T cell regulatory circuitry: antigen-specific and TCR-idiopeptide-specific T cell interactions in EAE. Intl. Rev. Immunol 9: 269-279.

Kumar, V., Tabibiazar, R. and Sercarz, E. (1993) A homuncular T cell network exists in the B10.PL mouse controlling autoimmunity and based on TCR-peptide recognition by regulatory T cells. J. Immunol. 150: II, 174A.

Kumar, V. and Sercarz, E. (1993) Involvement of TCR-peptide-reactive $CD4^+$ T cells in recovery from antigen-induced EAE. J. Exp. Med. in press.

Koh, D-R., Fung-Leung, W-P., Ho, A., Gray, D., Acha-Orbea, H., and Mak, T-W. (1992) Less mortality but more relapses in experimental allergic encephalomyelitis in $CD8^{-/-}$ mice. Science. 256: 1210-1213.

Lanzavecchia, A., Roosnek, E., Gregory, T., Berman, P. and Abrignani, S. (1988) T cells can present antigens such as HIV gp120 targeted to their own surface molecules. Nature 334: 530-532.

Lorber, M. L., Locken, R., Stall, A.M. and Fitch, F.W. (1982) I-A antigen on cloned alloreactive murine T lymphocytes are acquired passively. J. Immunol. 128: 2798-2801.

Marchalonis, J.J., Kaymaj, H., Dedeoglu, F., Schluter, S.F., Yokum, D.E. and Edmundson, A.B. (1992) Human autoantibodies reactive with synthetic autoantigens from T-cell receptor ß chain. Proc. Natl. Acad. Sci. U.S.A. 89: 3325-3329.

Powrie, F. and Mason, D. (1990) $OX\text{-}22^{high}$ $CD4^+$ T cells induce wasting disease with multiple organ pathology: prevention by the $OX\text{-}22^{low}$ subset. J. Exp. Med. 172: 1701-1708

Pullen, A. N., Wade, T., Marrack, P. and Kappler, J.W. (1990) Identification of the region of T cell receptor ß chain that interacts with the self-superantigen $Mls\text{-}1^a$. Cell 61: 1365-1375.

Sakaguchi, S. and Sakaguchi, N. (1990) Thymus and autoimmunity: Capacity of normal thymus to produce pathogenic self-reactive T cells and conditions required for their induction of autoimmune disease. J. Exp. Med. 172: 537-545.

Schwartz, R. (1990) A cell culture model for T lymphocyte clonal anergy. Science 248: 1349-1356

Sercarz, E., Wilbur, s., Sadegh-Nasseri, s., Miller, A., Manca, F., Gammon, G. and Shastri, N. (1986) The molecular context of a determinant influences its dominant expression in a T cell response heirarchy through fine processing. Prog. Immunol. 6: 227-237.

Urban J.L., Kumar V., Kono D.H., Gomez C., Horvath S.J., Clayton J., Ando D.G., Sercarz E.E. and Hood, L. (1988) Restricted use of T cell receptor V genes in murine autoimmune encephalomyelitis raises possibilities for antibody therapy. Cell 54:577-592.

Vandenbark, A.A., Hashim, G. and Offner, H. (1989) Immunization with a synthetic T-cell receptor V-region peptide protects against experimental autoimmune encephalomyelitis. Nature 341:541-544.

Von Boehmer, H.(1991) Positive and negative selection of the αß T cell repertoire in vivo. Curr. Opinion. Immunol. 3: 210-215

Prevention of gastric autoimmunity by transgenic expression of a proton pump subunit in the thymus

Ban-Hock Toh, Frank Alderuccio, Seong-Seng Tan, Paul A. Gleeson, and Ian R. Van Driel
Department of Pathology and Immunology,
Monash University Medical School, Alfred Hospital,
Melbourne, Victoria, 3181,
Australia

INTRODUCTION

Autoimmune gastritis is an excellent example of the organ-specific autoimmune diseases. The gastric lesion, characterized by loss of parietal and chief cells from the mucosa and submucosal lymphocytic infiltration, is the underlying basis for pernicious anaemia, the most common cause of vitamin B12 deficiency in whites of northern European origin (Gleeson and Toh, 1991; Toh et al, 1992). The gastric parietal cell appears to be the principal cell targeted since autoantibodies to parietal cells and to intrinsic factor (a parietal cell secretory product) are found in most patients with pernicious anaemia and parietal cell autoantibody titres correlate with severity of gastritis. These parietal cell autoantibodies are directed against the catalytic α and the glycoprotein β-subunit of the gastric H^+/K^+-ATPase (proton pump), the enzyme responsible for gastric acid secretion (Karlsson et al,

NATO ASI Series, Vol. H 80
Autoimmunity: Experimental Aspects
Edited by Moncef Zouali

1988; Goldkorn et al, 1989; Toh et al, 1990). To investigate the immunopathogenesis of autoimmune gastritis, we have adopted an experimental model of autoimmune gastritis induced by neonatal thymectomy carried out 2-4 days after birth in BALB/c mice. The murine disease is an ideal model since it displays the major features of the human disease with loss of parietal and chief cells accompanied by submucosal immunocytic infiltration and circulating parietal cell autoantibodies (Fukuma et al, 1989). These parietal cell autoantibodies also target the α and β-subunits of the proton pump (Jones et al, 1991). The murine disease appears to be cell-mediated, since the disease can be transferred by $CD4^+$ T cells but not by autoantibodies (Sakaguchi et al, 1985).

To investigate the antigen specificity of the pathogenic T cells, we have produced transgenic mice that express the β-subunit of the H/K ATPase driven by the MHC class II I-E^k promoter (βH/K-transgenic), resulting in expression of the β-subunit in major histocompatibility complex (MHC) class II-expressing cells, including those in the thymus. Transgenic expression of the β-subunit prevented the development of neonatal-thymectomy induced autoimmune gastritis. Further, thymocytes from βH/K-transgenic mice, in contrast to thymocytes from normal BALB/c and non-transgenic littermates, failed to transfer the disease to BALB/c nude mice, suggesting that tolerance was induced within the thymus (Alderuccio, et al, 1993).

RESULTS

The construct bearing the MHC class II (I-E^k) promoter and the β-subunit transgene was injected into (BALB/c x C57BL/6)F_1 X BALB/c oocytes, generating 4 lines (lines 25, 50, 57 and 59) of transgenic mice. Polymerase chain reaction showed expression of H/K ATPase β-subunit mRNA only in the stomach of non-transgenic littermates while βH/K-transgenic mice expressed β-subunit mRNA in many tissues including the thymus. Littermates from line 25 were thymectomised at day 3. None (0/23) of the transgenic mice developed parietal cell autoantibodies to the α or β-subunits of the H/K ATPase or immunocytic infiltrates in the gastric mucosa. Thymectomised non-transgenic littermates developed anti-H/K ATPase antibodies and gastric immunocytic infiltrates with a frequency of 45% (9/20), an incidence consistent with previous reports (Fig. 1). Anti-H/K ATPase antibodies were also not found in transgenic littermates of the other three βH/K-transgenic lines (lines 50, 57 and 59) following neonatal thymectomy. Therefore, disease prevention was associated with transgene expression and not to a chance transgene insertion into the mouse genome. Prevention of autoimmune gastritis in the transgenic mice was also accompanied by the unexpected absence of autoantibodies to the α-subunit. The α-subunit may not contain MHC class II restricted-T cell epitopes. In which case, the B cell response to the α-subunit may be driven by β-subunit specific T cell help. Alternatively, the α-subunit may carry hidden determinants which become immunogenic following an immune response to the β-subunit.

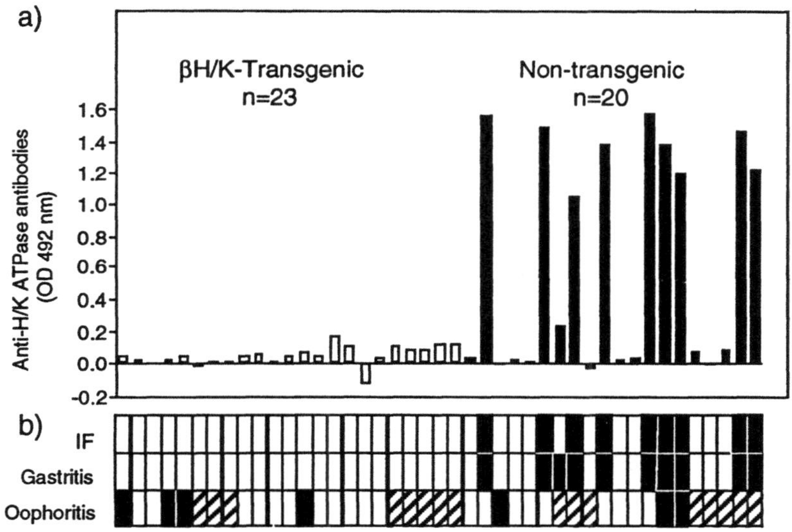

Figure 1. H/K ATPase autoantibodies, gastritis and oophoritis in neonatally thymectomised mice. A) Mice were thymectomised at day 3 and transgenic status determined at weaning. Sera were collected at 3-5 months and mice were killed for histological examination. Anti-H/K ATPase antibodies were detected by ELISA. B) Immunofluorescence (IF) for anti-parietal cell autoantibodies and histological examination of stomachs and ovaries of mice from *a*. Mice are represented in the same order as *a* and are shown directly below corresponding ELISA readings. Presence of parietal cell autoantibodies, gastritis or oophoritis is indicated by a black box. Striped boxes indicate male mice excluded from analysis of oophoritis. Reprinted from Alderuccio et al. J. Exp. Med. 178: 419-426 1993.

As neonatally thymectomised female BALB/c mice also developed autoimmune oophoritis (Tung et al, 1987), we examined mice for this disease. The incidence of oophoritis was identical in βH/K-transgenic (4/16, 25%) and non-transgenic (3/12, 25%) mice (Fig. 1), indicating that the prevention of autoimmune disease was gastric-specific.

The prevention of experimental autoimmune gastritis in the transgenic mice suggests that the autoreactive T cells which cause disease in normal BALB/c or non-transgenic littermates have been tolerized. Thymocytes transferred from adult BALB/c mice to BALB/c nude mice induce autoimmune gastritis (Sakaguchi and Sakaguchi, 1990). Therefore, to assess the presence of autoreactive T cells within the thymus of transgenic and non-transgenic littermates, thymocytes were transferred to BALB/c nude recipients. Autoimmune gastritis was induced in the nude mice by thymocyte transfer from normal BALB/c mice (5/6) and non-transgenic littermates (5/7), but not from transgenic mice (0/7), suggesting that the induction of tolerance of pathogenic T cells in the transgenic mice occured within the thymus.

CONCLUSION

We conclude that a T cell response to the H/K ATPase β-subunit is an absolute requirement for the development of autoimmune gastritis. Our findings have general implications for tolerance to peripheral, self-antigens as it suggests that tolerance to the H/K ATPase β-subunit in normal animals must occur at an extra-thymic location. The ability to identify the causative antigens in autoimmune diseases has obvious implications for developing rational immunotherapeutic strategies. This approach directed towards the identification of T cell targets in autoimmune gastritis is applicable to other autoimmune diseases.

ACKNOWLEDGMENTS

This study is supported by grants from the National Health and Medical Research Council of Australia, the William Buckland Foundation and the Alfred Group of Hospitals.

REFERENCES

Alderuccio, F., B.H. Toh, S.S. Tan, P.A. Gleeson, and I.R. van Driel. (1993). An autoimmune disease with multiple molecular targets abrogated by the thymic expression of a single autoantigen. J. Exp. Med. 178, 419-426.

Fukuma K, Sakaguchi S, Kuribayashi K, Chen WL, Morishita R, Sekita K, Uchino H and Masuda T. (1988). Immunologic and clinical studies on murine experimental autoimmune gastritis induced by neonatal thymectomy. Gastroenterology 94: 274-283.

Gleeson PA and Toh BH. (1991). Molecular targets in pernicious anaemia.Immunol. Today 12: 233-238.

Goldkorn I, Gleeson PA and Toh BH. (1989). Gastric parietal cell antigens of 60-90 kDa, 92 kDa and 100-120 kDa associated with autoimmune gastritis and pernicious anaemia. Role of N-glycans in the structure and antigenicity of the 60-90 kDa component. J. Biol.Chem. 264: 18768-18774.

Jones C M, Callaghan JM, Gleeson PA , Mori Y, Masuda T and Toh BH. (1991). The parietal cell autoantigens recognised by neonatal thymectomy-induced murine gastritis are the a and b subunits of the gastric proton pump. Gastroenterology 101: 287-294.

Karlsson FA, Burman P, Loof L, and Mardh S. (1988). Major parietal cell antigen in autoimmune gastritis with pernicious anaemia is the acid-producing H+, K+ adenosine triphosphatase of the stomach. J. Clin. Invest. 81: 475-479.

Sakaguchi S, Fukuma K, Kuribayashi K and Masuda T. (1985). Organ-specific autoimmune diseases induced in mice by elimination of T cell subset. J. Exp. Med. 161: 72-87.

Sakaguchi S and Sakaguchi N. (1990). Thymus and autoimmunity: capacity of the normal thymus to produce pathogenic self-reactive T cells and conditions required for their induction of autoimmune disease. J. Exp. Med.; 172: 537-545.

Toh BH, Gleeson PA, Simpson RJ, Moritz, RL, Callaghan J, Goldkorn I, Claerwen M Jones, Martinelli TM, Mu FT, Humphris DC, Pettitt JM, Mori Y, Masud T., Sobieszczuk P, Weinstock J, Mantamadiotis T, and Baldwin GS. (1990). The 60-90 kDa parietal cell autoantigen associated with autoimmune gastritis is a β-subunit of the gastric H^+/K^+-ATPase (proton pump). Proc. Nat. Acad. Sci. USA 87: 6418-6422.

Toh BH, Ian R Van Driel, and Gleeson PA (1992). Autoimmune gastritis: tolerance and autoimmunity to the gastric H/K ATPase (proton pump). Autoimmunity 13: 165-172.

Tung KSK, Smith S, Teuscher C, Cook C and Anderson RE. (1987) Murine autoimmune oophoritis, epididymoorchitis and gastritis induced by day 3 thymectomy. Am. J. Pathol. 126: 293-302.

The Anti-Self B Cell Repertoire

Robert S. Schwartz
Hematology-Oncology Division
New England Medical Center
Boston MA 02111
USA

A notable feature of B cells that has been observed recently in several laboratories is that a relatively small group of V_H segments predominates in the expressed V_H gene repertoire of normal humans. The results of a number of different experiments suggest that the phenomenon of repertoire restriction is due to clonal selection at the level of both mature and immature B cells. For instance, in examining sequences of the over-represented V_H genes we found examples of somatic mutations that were typical of an antigen-driven immune response, and which were manifestations of selection of mature B cell clones (Huang, et al., 1992). But we also found cases of over-represented V_H genes that had germline sequences. Each of these unmutated V_H segments was linked to a clonally unique and unmutated (D-J_H) CDR3 (Stewart, et al., 1992). These latter findings suggest a process of positive selection that is linked to the heavy chain in B cells that do not mutate their V_H genes. These and related results have led us and others to propose that clonal selection occurs not only in mature B cells but also in pre-B cells (Schwartz, 1993; Nishimoto, et al., 1991; Decker, et al., 1991). We propose that positive selection of pre-B cells differs from clonal selection of mature B cells in two ways: (a) pre-B cells lack a mechanism for V gene somatic mutation; and (b) positively selected pre-B cells have a differentiation advantage, whereas antigen-driven mature B cells gain a proliferative advantage. Analogous types of clonal selection are well-established in T cells. Mature T cells undergo

NATO ASI Series, Vol. H 80
Autoimmunity: Experimental Aspects
Edited by Moncef Zouali

clonal expansion when stimulated by an antigen, whereas immature T cells differentiate but do not proliferate when positively selected (Kisielow, et al., 1988; Macdonald, 1989; Mueller, et al., 1989; Schwartz, 1989; Tomonari, 1990; Von Boehmer, et al., 1990).

This hypothesis has another parallel with known properties of T cells in that it links B cell differentiation to the anti-self binding properties of the pre-B cell's antigen receptor. Positive, or negative, selection of immature T cells requires binding of the T cell receptor to a MHC-peptide complex displayed by an antigen-presenting cell in the thymic cortex (reviewed by Von Boehmer, 1990). In the normal fetus, self proteins are the sole source of such peptides, except for any maternal proteins that might leak through the placenta. Indeed, chemical analysis has shown that the peptides that are constitutively bound to MHC molecules originate from intracellular self-proteins (Buus, et al., 1988; Chicz, et al., 1992; Hunt, et al., 1992; Madden, eet al., 1991; Murphy, et al., 1992; Rudensky, et al., 1991; Srinivasan, et al., 1991). Whether the immature T cell undergoes positive or negative selection probably depends on the affinity of its receptor for the (MHC) plus (self-peptide) complex displayed in the thymic cortex. It has been proposed that a high-affinity receptor-MHC interaction kills the immature T cell, whereas low-affinity binding promotes its differentiation (Sprent, et al., 1990). The consequence of this arrangement is that endogenous peptides and self-reactive T cell receptors mold the naive T cell repertoire (Pullen, et al., 1988).

It is likely that all IgM antibodies encoded by germline variable region segments are low-affinity polyreactive autoantibodies (reviewed by Stewart, et al., 1992; Schwartz, 1987; Schwartz, 1988). We propose that pre-B cells at the surface μ^+/light chain$^-$ stage complete their differentiation if their surface Ig heavy chains bind to self-antigens. It is likely that the surface μ chain transduces signals that instruct the pre-B cells to continue their differentiation. For example, expression of a functional μ chain stimulates differentiation in IL-7-dependent immature mouse B cells; when the μ^+, IL-7-dependent B cells express a functional light chain they lose their dependence on IL-7 (Era, et al., 1991). Indeed, targeted disruption of the membrane exon of the μ chain gene resulted in a severe deficiency of mature B cells, although there were normal numbers of pre-B cells in the bone marrow (Kitamura, et al., 1991). Furthermore, clonal expansion of marrow pre-B cells in the mouse probably depends on expression of the heavy chain (Decker, et al., 1991).

In our hypothesis, a pre-B cell that expresses a polyreactive anti-self heavy chain would have an advantage, whereas a pre-B cell that expresses a μ heavy chain lacking those properties would have but a small possibility of moving to the next step in differentiation, and may, as a result, die. Certain V_H genes may therefore appear frequently in the expressed pre-immune B cell repertoire because the heavy chain region they encode allows binding to multiple autoantigens.

The light chain has an important role in specifying the antigen-binding properties of the immunoglobulin molecule. The light chain may thus also influence clonal selection, and hence the representation of a V_H gene in the

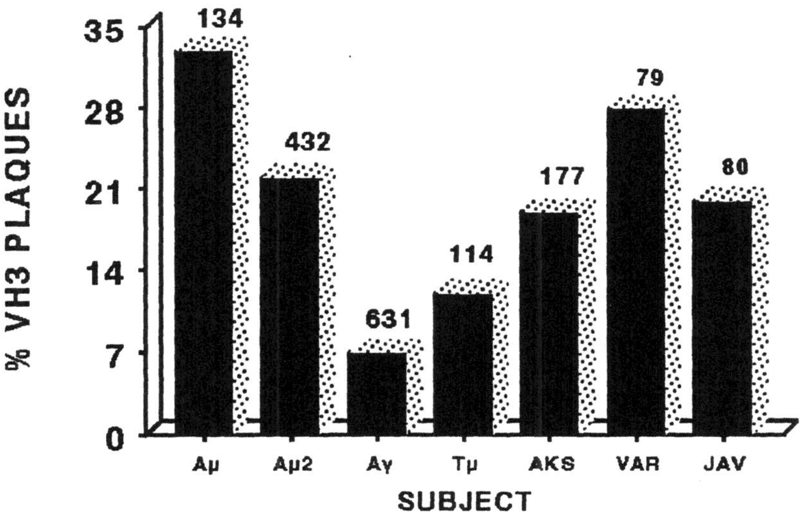

Figure 1: Representation of a single VH gene (*VH 18/2*) in mature B cells from the peripheral blood of normal adults (From Stewart et al, J.Exp.Med. 1992,177: 1227).

expressed B cell repertoire. Our studies have not shown any linkage of *V_H18/2*, an over-represented V_H gene, to a particular light chain (Young, et al., 1990). However, possible influences of the light chain on the frequency of a V_H gene in the expressed repertoire has not been systematically investigated.

REFERENCES

Buus, S., Sette, A., and Colon, S. M. (1988) Autologous peptides constitutively occupy the antigen binding site of Ia. *Science* **242**:1045-1047.

Chicz, R. M., Urban, R. G., Lane, W. S., Gorga, J. C., Stern, L. J., Vignali, D. A. A. et al. (1992) Predominant Naturally Processed Peptides Bound to HLA-DR1 Are Derived from MHC-Related Molecules and Are Heterogeneous in Size. *Nature* **358**:764-768.

Decker, D. J., Boyle, N. E., Koziol, J. A., and Klinman, N. R. (1991) The Expression of the Ig H-Chain Repertoire in Developing Bone Marrow B-Lineage Cells. *J. Immunol.* **146**:350- 361.

Era, T., Ogawa, M., Nishikawa, S. I., Okamoto, M., Honjo, T., Akagi, K. et al. (1991) Differentiation of growth signal requirement of B lymphocyte precursor is directed by expression of immunoglobulin. *EMBO J.* **10**:337-342.

Huang, C. C., Stewart, A. K., Schwartz, R. S., and Stollar, B. D. (1992) Immunoglobulin heavy chain gene expression in peripheral blood B-lymphocytes. *J. Clin. Invest.* **89**:1331-1343.

Hunt, D. F., Henderson, R. A., Shabanowitz, J., Sakaguchi, K., Michel, H., Sevilir, N. et al. (1992) Characterization of peptides bound to the class-I MHC molecule HLA-A2.1 by mass spectrometry. *Science* **255**:1261-1263.

Kisielow, P., Bluthmann, H., Staerz, U. E., and Steinmetz, M. (1988) Tolerance in T-cell-receptor transgenic mice involves deletion of nonmature CD4+CD8+ thymocytes. *Nature* **333**: 742-746.

Kitamura, D., Roes, J., Kuhn, R., and Rajewsky, K. (1991) A B cell-deficient mouse by targeted disruption of the membrane exon of the immunoglobulin μ chain. *Nature* **350**:423- 426.

MacDonald, H. R. (1989) Mechanisms of immunological tolerance. *Science* **246:** 982-983.

Madden, D. R., Gorga, J. C., Strominger, J. L., and Wiley, D. C. (1991) The structure of HLA-B27 reveals nonamer self- peptides bound in an extended conformation. *Nature* **353**:321-325.

Mueller, D. L., Jenkins, M. K., and Schwartz, R. H. (1989) Clonal expansion versus functional clonal inactivation. *Ann. Rev. Immunol.***7**:45-80.

Murphy, D. B., Rath, S., Pizzo, E., Rudensky, A. Y., George, A., Larson, J. K. et al. (1992) Monoclonal Antibody Detection a Major Self Peptide: MHC Class-II Complex. *J. Immunol.* **148**:3483-3491.

Nishimoto, N., Kubagawa, H., Ohno, T., Gartland, G. L., Stankovic, A. K., and Cooper, M. D. (1991) Normal pre-B cells express a receptor complex of μ heavy chains and surrogate light-chain proteins. *Proc. Natl. Acad. Sci. (USA)* **88**:6284-88.

Pullen, A. M., Marrack, P., and Kasppler, J. W. (1988) The T-cell repertoire is heavily influenced by tolerance to polymorphic self-antigens. *Nature* **335**:796-801.

Rudensky, A. Y., Prestonhurlburt, P., Hong, S. C., Barlow, A., and Janeway, C. A. (1991) Sequence Analysis of Peptides Bound to MHC Class-II Molecules. *Nature* **353**:622-627.

Schwartz, R. H. (1989) Acquisition of immunologic self- tolerance. *Cell* **57**:1073-1081.

Schwartz, R. S. (1993) The human B cell repertoire: Is repertoire bias a manifestation of life saver heavy chain genes? *Festschrifft For Professor Ivan Roitt* **In press**:

Schwartz, R. S. (1987) Autoantibodies and normal antibodies: Two sides of the same coin. *Harvey Lect.* **81**:53.

Schwartz, R. S. (1988) Polyvalent anti-DNA autoantibodies: Immunochemical and biological significance. *Int. Rev. Immunol.* **3**:97-115.

Sprent, J., Gao, E. K., and Webb, S. R. (1990) T-cell reactivity to MHC molecules - Immunity versus tolerance. *Science* **248**:1357-1363.

Srinivasan, M., Marsh, E. W., and Pierce, S. K. (1991) Characterization of Naturally Processed Antigen Bound to Major Histocompatibility Complex Class-II Molecules. *Proc. Natl. Acad. Sci. (USA)* **88**:7928-7932.

Stewart, K., Huang, C. C., Stollar, B. D., and Schwartz, R. S. (1992) High frequency representation of a single VH gene in the rearranged and expressed human B cell repertoire. *J. Exp. Med.* **177**:409-418.

Stewart, A. K., Huang, C., Long, A. A., Stollar, B. D., and Schwartz, R. S. (1992) VH gene representation in autoantibodies reflects the normal human B cell repertoire. *Immunol. Rev.* **128**: 101-122

Tomonari, K. (1990) Tolerance In vivo and In vitro. *Immun. Rev.* **116**:139-157.

Von Boehmer, H., and Kisielow, P. (1990) Self-nonself discrimination by T-cells. *Science* **248**:1369-1373.

Von Boehmer, H. (1990) Developmental biology of T cells in cell receptor transgenic mice. *Ann. Rev. Immunol.* **8**:531-556.

Young, F., Tucker, L., Rubinstein, D., Guillaume, T. A. ,. S. J., Barrett, K. J., Schwartz, R. S. et al. (1990) Molecular analysis of a germ line encoded idiotypic marker of pathogenic human lupus autoantibodies. *J. Immunol.* **145**:2545-53.

In vivo suppression of IgG2a^b production accomplished by CD4$^+$8$^-$ Th1 cell clones specific for a γ2a^b allopeptide determinant

Kristian Bartnes and Kristian Hannestad
Department of Immunology,
Institute of Medical Biology,
University of Tromsø,
School of Medicine,
N-9037 Tromsø,
Norway.

INTRODUCTION

As autoantigens, immunoglobulins (Ig) are of particular interest for several reasons: recognition of autologous Ig variable region determinants has been implicated in idiotypic network immune regulation (Jerne 1974), in the pathogenesis of autoimmune diseases (Hannestad and Kristoffersen, 1988; Hannestad et al, 1992), and immune surveillance of B cell neoplasms (Lynch et al, 1972; Jørgensen et al, 1980). Furthermore, IgG Fc regions are the targets for rheumatoid factor-producing B cells.

Previous work from this laboratory had established that normal B cells can process and present an idiopeptide from CDR3 of the light chain of an endocytosed IgA,λ2 myeloma protein to T helper cells (Th) (Hannestad et al, 1986). Since this idiopeptide contained three mutated amino acids that were essential for recognition by polyclonal T cells, the prevalence of normal B cells producing the idiopeptide is probably extremely low. To bypass the problem of low prevalence of B cells that express a given idiotype we established an experimental model for Th/B interactions across T cell receptor (TcR)/Ig peptide/major histocompatibility complex (MHC) class II bridges based on BALB/c (IgCH^a (Ig heavy chain constant region allotype a), H-2^d) CD4$^+$ T cell clones that specifically recognize Fc of IgG2a^b in an IgCH allotype, MHC class II (I-A^d) restricted fashion. When transferred to newborn IgCH allogeneic, MHC syngeneic (BALB/c x B10.D2)F$_1$ mice (IgCH$^{a/b}$, H-2^d) the T cells induced a

NATO ASI Series, Vol. H 80
Autoimmunity: Experimental Aspects
Edited by Moncef Zouali

profound suppression of IgG2a^b production (Bartnes and Hannestad, 1991). Because the suppression was IgG2a^b iso- and allotype-specific and the T cells recognized a processed $\gamma 2a^b$ peptide (Bartnes et al, 1993), it probably resulted from cognate interactions between the T cells and B cells that presented endogenous $\gamma 2a^b$ as allopeptides complexed to I-A^d. Interestingly, the clones crossreacted significantly with the corresponding peptide of autologous $\gamma 2a^a$, suggesting that Th recognizing autologous Ig C-region peptides may influence B cell differentiation.

RESULTS

An Fc$\gamma 2a^b$/I-A^d-specific T cell line (TL.1) was isolated from a BALB/c mouse immunized with purified IgG2a^b. From this line four $CD4^{+}8^{-}$, TcRα/β^{+} clones (A6, B2, B4 and B5) were derived. Upon activation they secreted IL-2, TNF-α, IFN-γ, but no IL-4, assigning them to the Th1 subset. To probe the outcome of their interactions with B cells TL.1 was cultured with T-depleted spleen cells from (BALB/c x B10.D2)F_1 mice. This strain was chosen as B cell donor because it expresses I-A^d and the target b as well as the autologous a allotype of IgG2a. In these experiments, TL.1 induced a five-fold increase in IgG2a^a levels, while IgG2a^b was hardly detectable (Table 1).

Table 1. Effect of the IgG2a^b-specific T cell line TL.1 on IgG2a^b secretion by (BALB/c x B10.D2)F_1 B cells *in vitro*

Culture	[IgG2a^a]	[IgG2a^b]
F_1 spleen cells alone	250±50	35±7
F_1 spleen cells + TL.1	1200±230	24±4

Spleen cells from (BALB/c x B10.D2)F_1 mice were depleted of T cells by treatment with Thy-1.2-, CD4- and CD8-specific monoclonal antibodies and complement before culturing 5 x 10^5 cells/well alone or together with 2 x 10^4 TL.1 cells. Supernatants were harvested at day 9 and concentrations of IgG2a of the a and b allotypes were determined in ELISA. Values represent mean concentration (ng/ml) ± SD of five parallell cultures. (Reproduced from Bartnes and Hannestad, 1991, with permission from Eur. J. Immunol.)

When adoptively transferred to newborn (BALB/c x B10.D2)F_1 TL.1 and the four clones derived from it induced a striking suppression of IgG2a^b: in 12 out of 15 6-week-old recipients IgG2a^b levels were < 5% of those found in the age-matched controls (Fig. 1). The suppressive effect was isotype and allotype specific, since IgG1^b and IgG2a^a levels were unaffected.

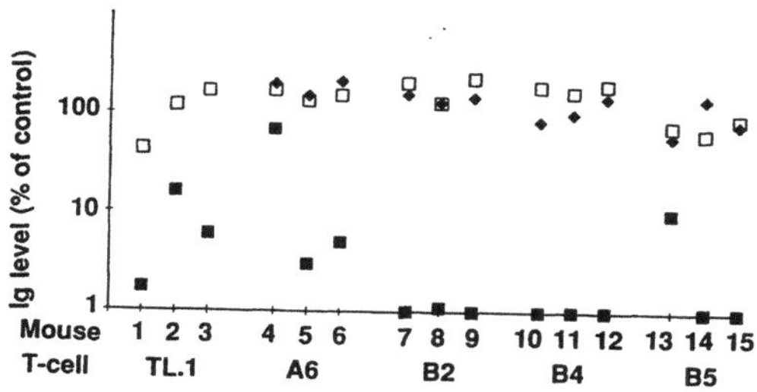

Figure 1. Effect of adoptively transferred Fc γ2a^{b}-specific BALB/c T cells on Ig levels in MHC-syngeneic Ig$^{a/b}$ mice *in vivo*.
0.5-2.0 x 10^{6} TL.1, A6, B2, B4 and B5 T cells were transferred to (BALB/c x B10.D2)F_1 mice by intraperitoneal injections on day 3 and 10 after birth. At 6 weeks of age, serum concentrations of IgG2a^{a} (open squares), IgG2a^{b} (filled squares) and IgG1^{b} (diamonds) were determined in ELISA and compared to levels in age-matched, sham-injected controls. (Reproduced from Bartnes and Hannestad, 1991, with permission from Eur. J. Immunol.)

By testing responses of hybridomas from clones A6 and B5 to HPLC-purified CNBr cleavage fragments of Fcγ2a^{b} the antigenic determinant was mapped to γ2a^{b} 438-478. Subsequent challenge with a panel of synthetic peptides encompassing this sequence demonstrated that γ2a^{b} 435-451 contained the epitope(s) (Bartnes et al, 1993). Truncation variants of antigenic peptides revealed that γ2a^{b} 440-446 (KLRVQKS) which comprises an I-A^{d} allele specific motif (Sette et al, 1988) represented the determinant core. This sequence exhibits a high degree of homology to the corresponding regions of the autologous γ1^{a} and γ2a^{a} (Fig. 2).

	SEQUENCE																
	435					440					445					450	
γ2a^{b}	Y	F	M	Y	S	K	L	R	V	Q	K	S	T	W	E	R	G
γ2a^{a}	-	-	-	-	-	-	-	-	-	E	-	K	N	-	V	E	R
γ1^{a}	-	-	V	-	-	-	-	N	-	-	-	-	N	-	-	A	-

Figure 2. Amino acid sequence of the γ2a^{b} T cell allo-determinant compared with the corresponding regions of the autologous γ2a^{a} and γ1^{a} (Kabat et al, 1991).
Dashes indicate homology with γ2a^{b}. The determinant core is framed. (Reproduced from Bartnes et al, 1993, with permission from Eur. J. Immunol.)

To assess the influence of those core residues that distinguish $\gamma 2a^b$ from $\gamma 1^a$ and $\gamma 2a^a$, i.e. R^{442}, Q^{444} and S^{446}, the hybridomas were challenged with synthetic peptides comprising residues 435-451 of $\gamma 1^a$ and $\gamma 2a^a$ (Fig. 3). Both clones cross-reacted with the $\gamma 2a^a$ peptide with a 6000-fold potency difference in favor of the allopeptide. $\gamma 1^a$ 435-451 was non-stimulatory.

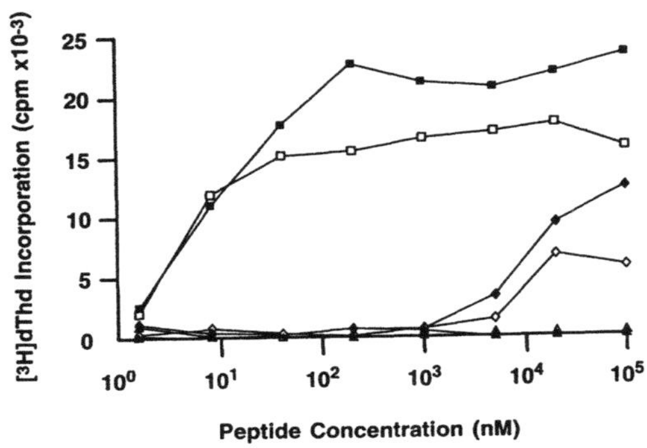

Figure 3. Discrimination between the allopeptide $\gamma 2a^b$ 435-447 and peptides encompassing residues 435-451 of the auto-antigens $\gamma 1^a$ and $\gamma 2a^a$.
T cell hybridomas A6 (filled symbols) and B5 (open symbols) were co-cultured with irradiated BALB/c spleen cells and challenged with various concentrations of $\gamma 2a^b$ 435-447 (squares), $\gamma 1^a$ 435-451 (triangles) and $\gamma 2a^a$ 435-451 (diamonds), and the IL-2 activity in 5% 24-h supernatants was determined. Values represent [^{3}H]thymidine incorporation by IL-2-responsive HT-2 cells (mean of triplicates). (Reproduced from Bartnes et al, 1993, with permission from Eur. J. Immunol.)

The role of individual amino acids in the formation of the determinant was explored by challenge with analogs of the $\gamma 2a^b$ 435-447 peptide carrying single amino acid substitutions in the core. This experiment revealed that residues K^{440}, R^{442} and K^{445} which are shared by the autologous and allogeneic $\gamma 2a$, as well as residues Q^{444} and S^{446} which represent allogeneic differences, are critical for recognition (Fig. 4).

In subsequent experiments (Bartnes et al, 1993), analogs with single alanine substitutions in positions 445 and 446 efficiently inhibited presentation of the parent peptide, indicating that K^{445} and S^{446} served to anchor the peptide to IA^d. A requirement for processing of IgG2a^b prior to T cell recognition was inferred from the findings that $\gamma 2a^b$ 436-451, but not IgG2a^b was effectively presented by aldehyde-fixed antigen presenting cells.

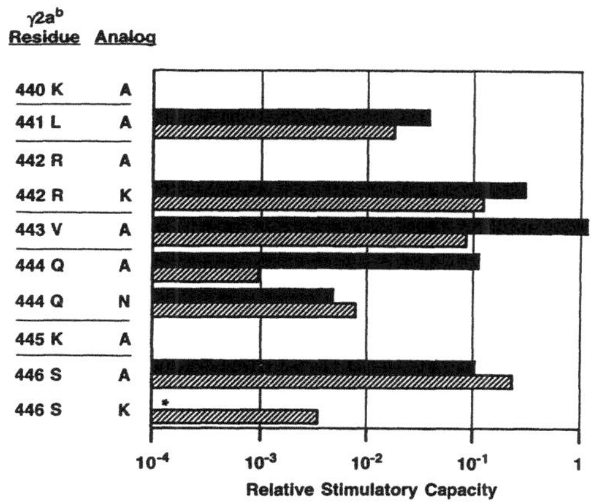

Figure 4. Relative antigenicity of γ2a^b 435-447 analogs carrying single amino acid substitutions in the determinant core.
Hybridomas A6 (filled bars) and B5 (hatched bars) were co-cultured with irradiated BALB/c spleen cells and challenged with various concentrations of peptide. The relative stimulatory capacity of a given analog represents the molar concentration of the parent peptide 435-447 which gave 50% of maximal antigen-induced hybridoma IL-2-response divided by the molar concentration of the analog required to obtain the same level of stimulation. * The maximal response of A6 to γ2a^b 435-447(K^{446}) was 29% of maximal Ag-induced hybridoma IL-2-response at 118,000 nM, corresponding to a relative stimulatory capacity of approximately 1.6×10^{-4}. (Reproduced from Bartnes et al, 1993, with permission from Eur. J. Immunol.)

DISCUSSION

We have established a BALB/c (Ig^a, $H\text{-}2^d$) T cell line and clones of the Th1 subset that recognize an IgG2a^b H chain allopeptide together with an autologous MHC class II molecule ($I\text{-}A^d$). Hybridomas of clones A6 and B5 recognized a processed $\gamma 2a^b$ allopeptide containing a core sequence spanning $K^{440}LRVQKS^{446}$. Their influence on (BALB/c x B10.D2)F_1 B cells was investigated because these T/B interactions were assumed to resemble those based on Th1 recognition of self Ig idiopeptides.

TL.1 cells induced production of $IgG2a^a$, but not $IgG2a^b$ *in vitro*. We interpret this as bystander stimulation of $mIgG2a^{a+}$ B cells (probably via IFN-γ) following activation of the T cells on $mIgG2a^{b+}$ B cells and/or professional splenic APC. Intrigued by the observation that levels of $IgG2a^b$, the cognate antigen for the T cells, were not enhanced, we turned to *in vivo* experiments. When adoptively transferred to newborn (BALB/c x B10.D2)F_1 the T cells selectively suppressed the target $IgG2a^b$ allotype. Taken together, our data indicate that the suppression resulted from cognate interactions between the $\gamma 2a^b$-specific T cells and $mIgG2a^{b+}$ B cells that presented $I\text{-}A^d$-associated endogenous Ig peptides. As B lymphocytes can process and present endogenous Ig on MHC class II (Bikoff et al, 1988; Bikoff and Eckhardt, 1989; Rudensky and Yurin, 1989; Weiss and Bogen, 1989), and Ig-peptides have been eluted from MHC class II molecules of B cells (Rudensky et al, 1992), we suggest that idiopeptide-specific Th1 cells may suppress normal and possibly malignant B cells in a manner similar to the allotype suppression reported here.

Both clones A6 and B5 crossreacted, although weakly, with the 435-451 peptide of the autologous $\gamma 2a^a$. This opens the possibility that self-peptides of Ig C regions can target B cells for regulatory interactions with autologous Th that escape negative selection due to low affinity. For example, under pathological conditions that favor class II-associated presentation of cryptic Cγ-region determinants, Th cells responding to Cγ peptides might promote differentiation of rheumatoid factor-producing B cells.

ACKNOWLEDGEMENTS
We thank Dr. J.-P. Briand and Ø. Rekdal for synthesizing peptides, Dr. K. Sletten for HPLC purification of CNBr cleavage fragments, and Ms. B. Harbak for technical assistance.

REFERENCES

Bartnes K and Hannestad K (1991) Igh-1b-specific $CD4^+8^-$ T cell clones of the T_h1 subset selectively suppress the Igh-1b allotype *in vivo*. Eur J Immunol 21: 2365-2371

Bartnes K, Rekdal Ø, Briand J-P and Hannestad K (1993) Th1 clones that suppress $IgG2a^b$ specifically recognize an allopeptide determinant comprising residues 435-451 of $\gamma 2a^b$. Eur J Immunol 23 (in press)

Bikoff EK, Yu H, Eckhardt LA (1988) T cell recognition of endogenous IgG_{2a} expressed in B lymphoma cells. Eur J Immunol 18: 341-348

Bikoff EK, Eckhardt LA (1989) Presentation of IgG2a antigens to class II-restricted T cells by stably transfected B lymphoma cells. Eur J Immunol 19: 1903-1909

Hannestad K, Kristoffersen G, Briand J-P (1986) The T lymphocyte response to syngeneic λ2 light chain idiotopes. Significance of individual amino acids revealed by variant λ2 chains and idiotope-mimicking chemically synthesized peptides. Eur J Immunol 16: 889-893 and correction (1986) 16: 1470

Hannestad K, Kristoffersen G (1988) Are idiotypic peptides from variable domains critical for T-dependent production of rheumatiod factors? Scand J Rheum Supplement 75: 97-101

Hannestad K, Andreassen K, Kristoffersen G (1992) Anti-idiotypic immune responses against adjuvant-free isologous IgM monoclonal antibodies and their augmentation by complex-formation between IgM and albumin in bovine serum. Eur J Immunol 22: 321-327

Jerne NK (1974) Towards a network theory of the immune system. Ann Immunol (Paris) 125C: 373-389

Jørgensen T, Gaudernack, G and Hannestad K (1980) Immunization with the light chain and the V_L domain of the isologous myeloma protein 315 inhibits growth of mouse plasmacytoma MOPC 315. Scand J Immunol 11: 20-35

Kabat EA, Wu TT, Perry HM, Gottesmann KS, Foeller C (1991) Sequences of proteins of immunological interest. 5th Edn, p. 692

Lynch RG, Graff RJ, Sirisinha S, Simms E, Eisen HN (1972) Myeloma proteins as tumor-specific transplantation antigens. Proc Natl Acad Sci USA 69: 1540-1544

Rudensky AY, Yurin VL (1989) Immunoglobulin-specific T-B cell interaction. I. Presentation of self immunoglobulin determinants by B lymphocytes. Eur J Immunol 19: 1677-1683

Rudensky AY, Preston-Hurlburt P, Al-Ramadi BK, Rothbard J, Janeway CA (1992) Truncation variants of peptides isolated from MHC class II molecules suggest sequence motifs. Nature 359: 429-431

Sette A, Buus S, Colon S, Miles C, Grey HM (1988) I-A^d-binding peptides derived from unrelated protein antigens share a common structural motif. J Immunol 141: 45-48

Weiss S and Bogen B (1989) B-lymphoma cells process and present their endogenous immunoglobulin to major histocompatibility complex-restricted T cells. Proc Natl Acad Sci USA 86: 282-286

Expression cloning using antibodies from a patient with rheumatoid arthritis of an autoantigen homologous to the Drosophila splicing regulator, *suppressor-of-white-apricot*

Robert Lafyatis and André Capron
Centre d'Immunologie et de Biologie Parasitaire
Institut Pasteur de Lille, B.P. 245
1, rue P. Calmette
59019 Lille cedex
France

INTRODUCTION

Although the primary stimulus of inflammation in rheumatoid arthritis (RA) is unknown, data strongly suggest that this inflammation is mediated by activated T-cells, including the prevalence of T-cells in the pannus, the association of the disease with certain MHC class II isotypes (Todd et al, 1988), and the dramatic clinical response to immunosuppressive agents including Cyclosporin A (Dougados et al, 1988; Yocum et al, 1988). Further, several antigen-induced and infectious animal models of arthritis (Collins et al, 1940; Cromartie et al, 1977; Decker et al, 1976; Pearson, 1956; Stuart et al, 1984), as well as chronic Lyme arthritis (Steere et al, 1980) are similar to rheumatoid arthritis, suggesting that unidentified antigens might be responsible for stimulating the T-cells causing this disease. This pathogenic mechanism is also supported by the observation that in several of these animal models, antigen stimulated T-cells and antigen specific T-cell clones derived from animals with arthritis are able to induce a similar illness in naive animals (Holoshitz et al, 1983; Stuart et al, 1984).

Although autoantigens potentially involved in disease pathogenesis have been identified and cloned in a variety of autoimmune diseases, antigens in RA are less well defined. Antibodies against immunoglobulins, i.e. rheumatoid factors (RFs), are strongly associated with RA, but it remains unclear whether immunoglobulins are a primary immunologic stimulus of this disease for several reasons: RFs appear in other illnesses associated with prolonged activation of the immune system; they are produced in response to stimuli of polyclonal B-cell

NATO ASI Series, Vol. H 80
Autoimmunity: Experimental Aspects
Edited by Moncef Zouali

activation, including aggregated IgG and immune complexes; and their role may be physiologic - to enhance antibody affinity through cross-linking (Carson, 1985). On the other hand, in patients with RA the presence of somatic mutations in at least some RFs suggests B-cell antigen stimulation (Victor et al, 1991). Stimulation of somatic mutation in RF^+ B-cells might occur after antigen-antibody complexes, due to unidentified antigen(s), stimulate T-cell dependent B-cell activation through surface (RF) immunoglobulin (Carson et al, 1991; Roosnek et al, 1991).

Other proteins have been suggested to be important antigens in RA including collagen II (Rowley et al, 1986; Trentham et al, 1980), the 65 kD heat-shock protein (Holoshitz et al, 1986), and the RA-associated nuclear antigen (RANA; Alspaugh et al, 1976). The first two of these proteins have been clearly implicated in causing or mediating experimental models of arthritis in rats (collagen and adjuvant arthritis, respectively; Stuart et al, 1984; van Eden et al, 1988); however, the importance of these antigens as the primary stimuli of the immune response in RA remains uncertain, in part because antibodies against these antigens are also seen frequently in other arthritic disorders (Holoshitz et al, 1986; Mottonen et al, 1988). Antibodies to both RANA and the Epstein-Barr nuclear antigen-1 (EBNA-1), seen frequently in patients with RA, but also in patients early after infection with Epstein-Barr virus (EBV; Alspaugh et al, 1978; Catalano et al, 1980; Catalano et al, 1979), are likely present at increased levels in patients with RA due to the presence of increased numbers of circulating B-cells infected with EBV in patients with RA; this is possibly due to a disease associated T-cell defect rather than to a discrete autoimmune dysfunction against EBV (Tosato et al, 1984). Alternative roles for EBV in the pathogenesis of RA have been suggested in other studies showing that an antigen in rheumatoid synovial tissue cross-reacts with antibodies to EBNA-1 (Fox et al, 1986), and that there is a shared epitope between the HLA-DR4 region associated with RA and the EBV glycoprotein gp110 (Roudier et al, 1989), but the importance of these findings in disease pathogenesis has not yet been defined.

Other autoantibodies are seen occasionally in patients with RA including antinuclear antibodies (ANA; Harris, 1985), and antineutrophil cytoplasmic antibodies (ANCA; Savige et al, 1991), however, antibodies against these antigens are primarily associated with other autoimmune diseases (systemic lupus erythematosus (SLE) and Wegener's granulomatosus, respectively). In contrast, several groups, including our group, have described autoantibodies primarily associated with RA (Hassfeld et al, 1989; Hoet et al, 1991; Lafyatis et al, 1992). By immunoblotting we have found autoantibodies in synovial fluids of patients with RA that react against several proteins in synovial tissue. These antibodies were generally seen primarily in patients with RA, and most of the corresponding antigens, although originally detected in rheumatoid synovium, were also detected in normal synovial, dermal and liver tissues (Lafyatis et al, 1992). On the basis of these studies, we decided to attempt to directly clone protein antigens from a λgt11 expression library using RA synovial fluid as the source of antibodies for immunologic screening of the library. Using this method we have cloned an autoantigen recognized by antibodies in the synovial fluid of a patient with RA that is highly homologous to the Drosophila splicing regulator, *suppressor-of-white-apricot* (*su*(w^a)). This autoantigen is not only recognized by antibodies from patients with RA, but also by patients with SLE.

RESULTS

Screening of an expression library for autoantigens

In our previous study, we found that the autoantigens identified in synovial tissue, were also a found in other tissues (Lafyatis et al, 1992). We, therefore, probed a λgt11 placenta cDNA library using synovial fluid as the source of primary antibodies. Although we had some information regarding potential autoantigens that might be cloned based on these previous immunoblotting results, we chose not to limit the search to specific proteins seen on immunoblot for several reasons. First, although immunoblotting is quite sensitive, it is possible that important antigens could be missed for any of several technical reasons (bad transfer, low level expression in the tissue, blocking of autoantigen by another highly expressed protein of the same molecular weight). Second, immunoblotting might tend to accentuate relatively low-affinity interactions between antibodies and prevalent proteins due to the relatively higher protein concentration in those bands; although similar considerations must be taken into account in screening an expression library, it seems likely that the relative levels of protein expression between different clones falls within a more narrow range than tissue proteins seen by immunoblotting. For these two reasons one might anticipate identifying autoantigens binding the highest affinity antibodies by direct screening of an expression library.

Initial screenings were carried out using a variety of rheumatoid synovial fluids that had been absorbed extensively with *E. coli* acetone powder. Absorption was absolutely necessary to permit identification of positive clones over background staining of *E. coli* proteins, but was not uniformly successful. Seven synovial fluids were chosen, which gave low background. Positive clones were found using only one of the synovial fluids. Results from the third screening of a typical positive clone are shown in Figure 1, background staining of negative clones can also be seen.

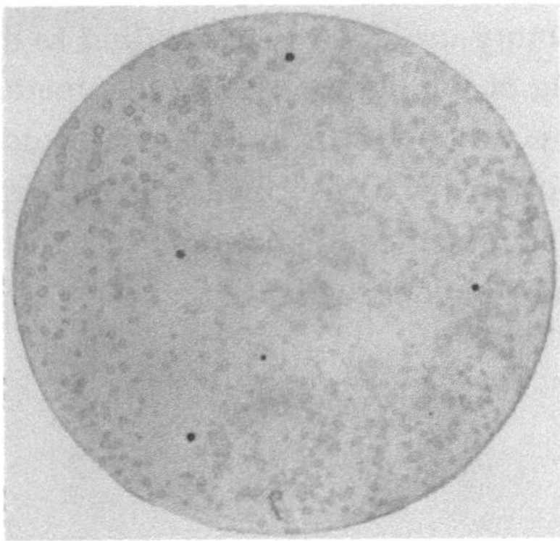

Figure 1. Positive clones from screening of a λgt11 cDNA expression library. Results from the third round of plaque purification are shown. Positive colonies stain darkly as open or filled circular regions, while negative colonies stain lightly due to antibodies against *E. coli* and/or viral proteins.

Analysis, subcloning, and sequencing of clones.

Multiple screenings of the λgt11 library resulted in only two positive clones from one of the seven patients, patient B6; none of the other six patient synovial fluids stained any positive clones. To determine whether these clones represented the same or different cDNAs, the insert of each clone was amplified by PCR using primers to each side of the insert DNA, analyzed on an agarose gel, and a probe prepared from one clone hybridized to a Southern Blot of the amplified cDNAs from both clones. Amplification of the cDNAs revealed inserts of the same size, which cross-hybridized, indicating that the two clones were from the identical DNA.

Amplified fragment of one the clones was ligated into pBluescript. This clone was completely sequenced and was shown to consist of one large open reading frame (ORF) covering the entire clone. The translation product of this ORF was used to search GenBank for possible homologous sequences, revealing that the clone was homologous to a Drosophila alternative splicing regulatory gene called *suppressor-of-white-apricot* (*su(w^{a})*; Chou et al, 1987; Zachar, 1987), and we therefore refer to this antigen as HsSWAP (Denhez et al, submitted).

Construction of plasmids for protein expression and analysis of synovial fluid foı antibodies against the cloned, expressed protein

In order to confirm that the SWAP protein represents an autoantigen and to determine whether other patients produce anti-SWAP antibodies, the SWAP cDNA was cloned into the pGEX bacterial expression vector. Proteins expressed in this vector are fusion proteins with the 27.5 kD glutathione S-transferase (GST) protein of *Shistosoma japonicum* (Smith et al, 1988). The GST tail of the protein permits one step affinity-purification of the recombinant fusion protein using glutathione-agarose.

Bacteria containing the SWAP cDNA in the GST expression vector, were treated with isopropylthio-β-D-galactoside to induce expression of the fusion protein and the recombinant fusion protein purified, and analyzed by PAGE. Staining of the gel revealed a single, highly purified protein (Figure 4). This protein was immunoblotted and probed using synovial fluid from patient B6, originally used to clone this cDNA, revealing strong antibody reactivity against this protein (Figure 4). The size of the fusion protein was only 44 kD - markedly smaller than the calculated size of the fusion protein (27.5 kD GST + 59.5 kD SWAP), indicating that the fusion protein had been partially degraded during

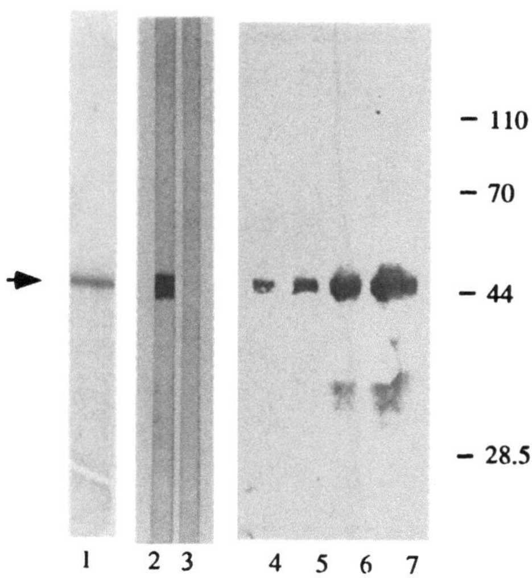

Figure 2. Expression anc anti ıı·dy binding to recombinant SWAP fusion protein. Purified SWAP/GST fusion protein (f6,1) is s:.own after staining of the gel (lane 1), or after electrophoresis and blotting of 1 ug (lanes 2,3,5), 0.5 ı·σ (lane 4), 2 ug (lane 6), or 4 ug (lane 7) of protein and incubation with either synovial fluid B6 (lanes 2 and 4-7), or normal serum (lane 3) followed by incubation with goat anti-human IgG/alkaline phosphatase and development.

expression. The calculated molecular weight of the SWAP fragment would therefore be about 16.5 kD, assuming that the GST part of the protein is largely intact. Reactivity of the B6 antibodies suggests that at least one recognized B-cell epitope on the SWAP protein is rather proximal to the 5' end of this clone (as diagrammed in Figure 5). Control antibody staining against the GST fusion partner did not result in positive staining (not shown).

Analysis of patients with RA and SLE for antibodies against the recombinant SWAP protein.

Since the SWAP protein has not been previously characterized as an autoantigen, we determined whether antibodies against this protein could be found in other patients with autoimmune disease using immunobloting of the recombinant protein. Of 14 different synovial fluids from patients with RA, three were positive for antibodies against the SWAP protein (not shown).

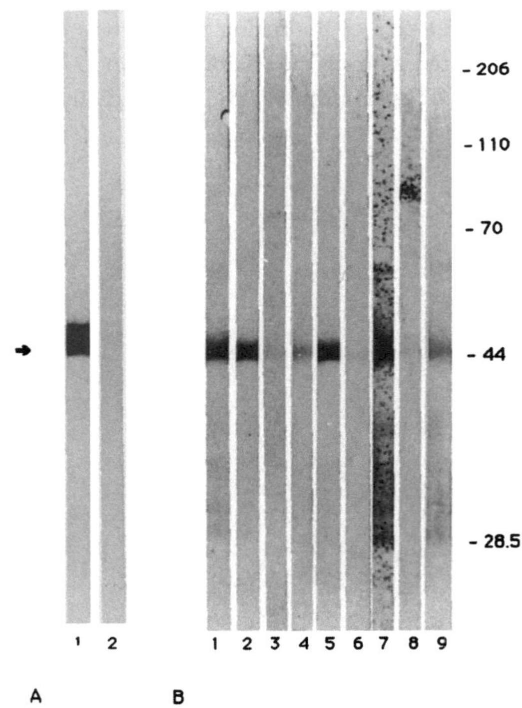

Figure 3: Analysis of SLE serum samples for reactivity to fusion proteins. Fusion protein f6.1 was immunoblotted and probed using serum or synovial fluid. Filter strips were incubated with: A) a 1:100 dilution of the RA synovial liquid B6 (lane 1), or a 1:100 dilution of normal human serum (lane 2), or B) 1:100 dilutions of serum from nine different patients with SLE (lanes 1-9).

Autoantibodies found in patients with other autoimmune illnesses are also occasionally found in patients with RA; therefore, it seemed possible that antibodies against SWAP might also be found in patients with other autoimmune diseases, such as SLE. To begin to address this question, we studied 9 patients with SLE using the same immunoblot procedure as above (Figure 5). These studies revealed that antibodies against the SWAP autoantigen are quite common in patients with SLE (6/9 patients studied), at least as common as in patients with RA. None of the synovial fluids or sera used in these studies reacted against the GST fusion partner alone (not shown).

Epitope mapping of the SWAP protein

Epitope mapping of several autoantigens, for example Ro and the U1 70K snRNP, by other groups has shown that autoantibodies frequently react against multiple epitopes on these autoantigens. To begin to more clearly define the

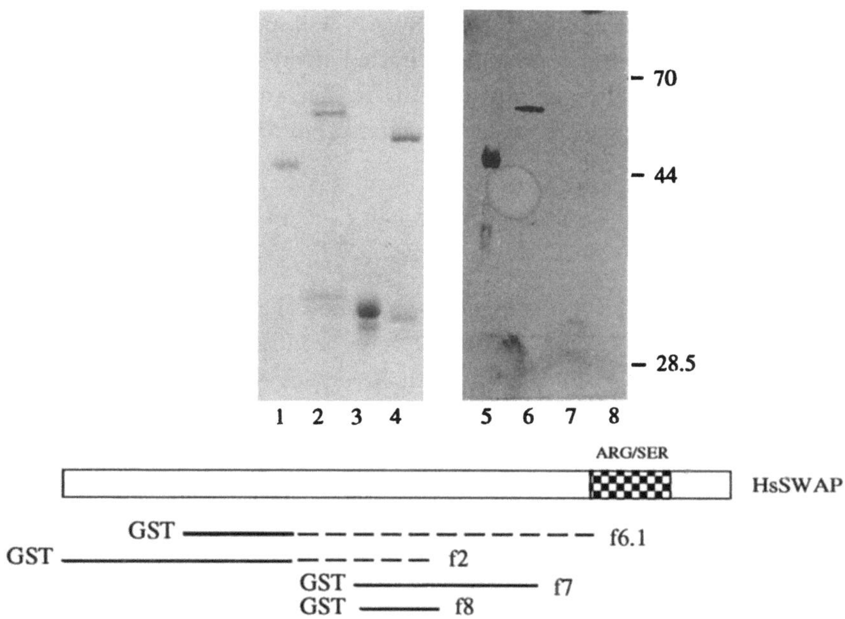

Figure 4. Epitope mapping of the SWAP autoantigen. GST fusion proteins were constructed containing different parts of the SWAP protein. **Bottom panel:** cloned portions (solid and dotted lines) and apparent regions stably produced as fusion proteins based on the molecular weight of the fusion proteins (solid line only) are shown below a diagram of the complete protein. **Top panel:** fusion proteins produced by plasmids pX6.1 (lanes 1 and 5), pXf2 (lanes 2 and 6), pXf8 (lanes 3 and 7) or pXf7 (lanes 4 and 8) were stained after PAGE (lanes 1,2,3, and 4) or immunoblotted and incubated with the B6 synovial fluid as the source of primary antibody (lanes 5,6,7, and 8).

epitope(s) responsible for reactivity of the B6 synovial fluid against the SWAP autoantigen, several expression plasmids were constructed containing different regions of the SWAP cDNA. These fusion proteins were purified and stained, or immunoblotted as above. cDNAs cloned for expression were overlapping fragments of either the carboxy half of the protein (pXf2), the central region of the protein (pXf7), or a short region corresponding to the 5' region of the pXf7 insert (pXf8, Figure 6).

The fusion protein (f2) produced by the amino terminal cDNA (pXf2) appeared truncated as had been seen for the original fusion protein (f6.1, produced by pX6.1). The other two fusion proteins (f7 and f8, produced by respectively plasmids pXf7 and pXf8) appeared to be roughly of the anticipated size based on the calculated molecular weights of the encoded peptide regions. The B6 synovial fluid reacted against f2, which overlaps with f6.1, but not with f7 or f8. These results do not allow precise epitope mapping, but do show that large parts of the protein are not targeted - at least by this patient serum. Further studies using other patients with antibodies against SWAP will permit an evaluation of whether reactivity against this protein is directed always against the same region. The contruction of smaller fusion proteins in the aminoterminal region will permit a more complete evaluation of the targeted epitope(s).

DISCUSSION

We have cloned an autoantigen using synovial fluids from patients with RA homologous to the Drosophila $su(w^a)$ gene product (DmSWAP) by directly screening a cDNA expression library with RA synovial fluids. Although using this technique "natural" autoantibodies could have theoretically bound to proteins not necessarily associated with an autoimmune response. Natural autoantibodies are generally low-affinity, IgM antibodies, whereas we specifically screened only antigens recognized by IgG antibodies. Further, this screening procedure requires antibodies of high-affinity to succeed (Sambrook et al, 1989). Therefore, the strong binding of antibodies of the IgG class to this protein by immunoblot, along with the clear structural and functional relationship between this protein with other autoantigens, indicates that this protein represents an authentic autoantigen and is not simply a target of natural antibodies.

The very low number of highly restricted clones obtained through this screening procedure is of note. The total number of cDNA clones screened for each synovial liquid was $1x10^6$. Since only one of six cDNAs would be expected to be in the correct orientation and reading frame in a given phage clone, a more accurate estimate of the number of proteins screened is $1x10^6/6 = 1.6x10^5$. Further, without doubt many proteins are represented by more than one clone. Given this overrepresentation, it is still likely that between 10^4 and 10^5 different proteins were screened using each synovial fluid. The fact that only two positive identical clones were obtained after screening of seven different synovial fluids indicates that high-affinity autoantibodies in RA synovial fluids are reactive against a highly selective number of autoantigens. Therefore, the autoantibody response in patients with RA appears to be quite restricted. This conclusion, has essentially already been reached in other autoimmune diseases such as polymyositis, SLE and Sjogren's syndrome with their apparently highly restricted antibody response against antigens such as Jo-1, or Ro and La. Our data suggest that the same type of phenomenon may also exists in patients with RA. As in the case with other autoantigens, one does not yet understand the relationship between disease pathogenesis and the production of these antibodies, nor, indeed, why certain autoantibodies are associated with specific autoimmune disorders; however, increasing evidence for antibody maturation in autoimmune illnesses suggests that the production of autoantibodies is in some manner directly related to a process of autoimmunization.

A comparison of this autoantigen with the U1 70K snRNP seems particularly relevant given several features shared between these two autoantigens. First, both of these autoantigens are proteins implicated in splicing and splicing regulation. The U1 70K snRNP is one of the snRNP proteins associated uniquely with the U1 snRNA. This protein has also been shown to potentially influence splicing regulation. The Drosophila SWAP analogue, on

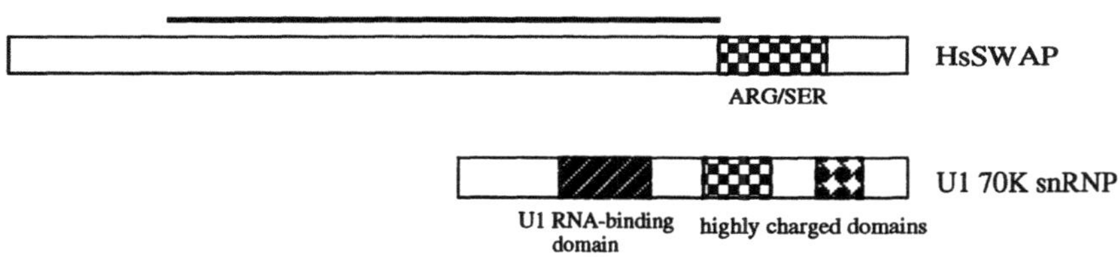

Figure 5. Comparison of the SWAP and U1 70K snRNP autoantigens. HsSWAP has a R/S rich region in the carboxyterminus, while the U1 70K snRNP protein has an RNA-binding motif in its aminoterminus and two highly charged regions in the center and carboxyterminus, one of which is structurally highly similar to the R/S module in HsSWAP (marked with a checkered motif). The bar above the HsSWAP protein marks the region cloned by screening of the expression library.

the other hand, is defined primarily as a splicing regulator; it has not been clearly shown to be associated with a snRNP, but it likely influences regulated splicing through interactions with snRNP proteins.

The primary structural parallel between these two proteins is in the highly charged carboxytermini of each of these proteins. In the U1 70K snRNP protein this charged region is divided into two smaller regions: one which is characterized by a high frequency of aspartic and glutamic acid residues frequently alternating with arginine residues; the other is characterized by a high frequency of basic and hydroxylated amino acids with multiple stretches of arginine/serine repeats (R/S motif) (Query et al, 1989; Theissen et al, 1986). This second motif has been found in a variety of proteins shown to be important in splicing and splicing regulation. These proteins include a family of proteins designated SR proteins (Mayeda et al, 1992), which include splicing factors SF2/ASF (Ge et al, 1991; Krainer et al, 1991) and SC35 (Fu et al, 1992); U2AF (Zamore et al, 1992); and two other Drosophila splicing regulators, *transformer* and *transformer-2* (Amrein et al, 1988; Boggs et al, 1987; Goralski et al, 1989). Therefore the motif shared between these two autoantigens likely reflects a common function. Apparently the R/S motif is important in targeting DmSWAP and *tra* to a subnuclear speckled compartment (Li et al, 1991), and is also directly involved in the functional splicing activity of this family of proteins (Valcarcel et al, 1993; Zamore et al, 1992).

This R/S region was not the target of the autoantibodies reactive against the original SWAP clones, since this region was not present in these clones (see Figure 3). Rather, the anti-SWAP antibodies from patient B6 clearly react with a region in the aminoterminus of the protein. Reactivity against epitopes of the U1 70K snRNP protein have been studied in some detail, showing that multiple regions within this protein can be targets of autoantibodies (Cram et al, 1990; Takeda et al, 1991). In both these studies fusion proteins containing part of the R/S region were studied, although in the study by Takeda et al. only a small part of the R/S region was studied in the context of a much larger region of the U1 70K snRNP. In the study of Cram et al, fusion proteins containing different parts of the U1 70K snRNP R/S region were apparently recognized relatively frequently by autoimmune sera. However, the R/S region was split between two different fusion proteins in this study and in both of these two fusion proteins contained other regions of the U1 70K snRNP. Therefore, it remains unclear how commonly the R/S region is a target of the autoantibody response against the U1 70K snRNP autoantigen. The identification of the SWAP autoantigen with a similar region suggests that this region might in some way be important in the autoantibody response against each of these proteins, perhaps by breaking T-cell tolerance presumably required for the production of autoantibodies. This could occur through several mechanisms. For example, the highly charged residues within this region are capable of more easily breaking T-cell tolerance than other peptides, leading to autoantibody formation. Another possibility is that these two proteins undergo similar intracellular degradation pathways and might be associated with surface MHC molecules more frequently, potentially leading to a break in tolerance in patients with autoimmune disease.

In SLE much controversy has existed regarding the relative importance of antigen-driven versus polyclonal B-cell activation, apparently both mechanisms may be important in SLE (Klinman et al, 1991). In RA this issue is complicated by the data cited in the introduction, which indicate that T-cells, rather than B-cells may be primarily responsible for stimulating synovial inflammation. In this regard, although antibodies reactive against autoantigen may not have a direct pathologic role in RA, nevertheless, if RA is indeed antigen-driven, it seems likely - given the important connections between B-cell and T-cell activation, including cognate T-cell help - that activated T-cells in RA might be directed against at least some of the same antigens as activated B-cells. A more complete evaluation of the autoantibody response against the SWAP autoantigen should give insight into the potential importance of this autoantigen in patients with RA and SLE, and further study of the relation between SWAP and the U1 70K snRNP might permit more general insights into the formation of autoantibodies in patients with RA and other autoimmune disorders.

ACKNOWLEDGEMENTS

We would like to thank R.-M. Flipo, B. Duquesnoy and J.-P. Dessaint for help obtaining clinical materials and helpful discussions. R. Lafyatis was supported in part by a Nord/Pas-de-Calais regional grant and in part by INSERM. Support for the work in this manuscript was also provided by the Association de Recherche sur la Polyarthrite rhumatoide (A.R.P.), and the Commission mixte de recherche hospitalo-universitaire (C.H.R.U., Lille).

REFERENCES

Alspaugh, M.A., Jensen, F.C., Rabin, H., and Tan, E.M. (1978). Lymphocytes transformed by Epstein-Barr virus: Induction of nuclear antigen reactive with antibody in rheumatoid arthritis. J. Exp. Med. 147,1018-1027.

Alspaugh, M.A., and Tan, E.M. (1976). Serum antibody in rheumatoid arthritis reactive with a cell-associated antigen. Arthritis Rheum. 19,711-719.

Amrein, H., Gorman, M., and Nothiger, R. (1988). The sex-determining gene *tra-2* of Drosohila encodes a putative RNA binding protein. Cell 55,1025-1035.

Boggs, R.T., Gregor, P., Idriss, S., Belote, J.M., and McKeown, M. (1987). Regulation of sexual differentiation in D. melanogaster via alternative splicing from the transformer gene. Cell 50,739-747.

Carson, D.A. (1985) Rheumatoid factor. In Textbook of Rheumatology, Edited by W. N. Kelley, E. D. Harris, S. Ruddy and C. B. Sledge. W.B. Saunders, Philadelphia, 664-680.

Carson, D.A., Chen, P.P., and Kipps, T.J. (1991). New roles for rheumatoid factor. J. Clin. Invest. 87,379-383.

Catalano, M.A., Carson, D.A., Neiderman, J.C., Feorino, P., and Vaughan, J.H. (1980). Antibody to the rheumatoid arthritis nuclear antigen. J. Clin. Invest. 65,1238-1242.

Catalano, M.A., Carson, D.A., Slovin, S.F., Richman, D.D., and Vaughan, J.H. (1979). Antibodies to the Epstein-Barr virus-determined antigens in normal subjects and in patients with rheumatoid arthritis. Proc. Natl. Acad. Sci. 11,5825-5828.

Chou, T.-B., Zachar, Z., and Bingham, P.M. (1987). Developmental expression of a regulatory gene is programmed at the level of splicing. EMBO J. 6,4095-4104.

Collins, D.H., and Goldie, W. (1940). Observations on polyarthritis and on experimental erysipelothrix infection of swine. Pathol Bacteriol. 50,323-353.

Cram, D.S., Fisicaro, N., Coppel, R.L., Whittingham, S., and Harrison, L.C. (1990). Mapping of mulitple B-cell epitope on the 70-kilodalton autoantigen on the U1 ribonucleoprotein complex. J. Immunol. 145,630-635.

Cromartie, W.J., Craddock, J.G., Schwab, J.H., Anderle, S.K., and Yang, C.H. (1977). Arthritis in rats after systemic injection of streptococcal cells or cell walls. J. Exp. Med. 146,1585-1602.

Decker, J.L., and Barden, J.A. (1976) Mycoplasma arthritis in the pig. In Infection and Immunology in the Rheumatic Diseases, Edited by D. C. Dumond. Blackwell Scientific Publications, Oxford, 191-193.

Dougados, M., Awada, H., and Amor, B. (1988). Cyclosporin in rheumatoid arthritis: a double blind placebo controlled study in 52 patients. Ann. Rheum. Dis. 47,127-133.

Fox, R., Sportsman, R., Rhodes, G., Luka, J., Pearson, G., and Vaughan, J. (1986). Rheumatoid arthritis synovial membrane contains a 62,000-molecular-weight protein that shares an antigenic epitope with the Epstein-Barr-virus-encoded associated nuclear antigen. J. Clin. Invest. 77,1539-1547.

Fu, X.-D., and Maniatis, T. (1992). Isolation of a complementary DNA that encodes the mammalian splicing factor SC35. Science 256,535-538.

Ge, H., Zuo, P., and Manley, J.L. (1991). Primary structure of the human splicing factor ASF reveals similarities with Drosophila regulators. Cell 66,373-382.

Goralski, T.J., Edstrom, J.-E., and Baker, B.S. (1989). The sex determination locus *transformer-2* of drosophila encodes a polypeptide with similarity to RNA binding proteins. Cell 56,1011-1018.

Harris, E.D. (1985). Antinuclear antibodies. In Textbook of Rheumatology, Edited by W. N. Kelley, E. D. Harris, S. Ruddy and C. B. Sledge. W.B. Saunders, Philadelphia, 690-704.

Hassfeld, W., Steiner, G., Hartmuth, K., Kolarz, G., Scherak, O., Graninger, W., Thumb, N., and Smolen, J.S. (1989). Demonstration of a new antinuclear antibody (anti-RA33) that is highly specific for rheumatoid arthritis. Arthritis Rheum. 32,1515-1520.

Hoet, R.M.A., Boerbooms, A.M.T., Arends, M., Ruiter, D.J., and van Venrooij, W.J. (1991). Antiperinuclear factor, a marker autoantibody for rheumatoid arthritis: colocalisation of the perinuclear factor and profilaggrin. Ann. Rheum. Dis. 50,611-618.

Holoshitz, J., Klajman, A., Drucker, I., Lapidot, Z., Yaretzky, A., Frenkel, A., Eden, W.v., and Cohen, I.R. (1986). T lymphocytes of rheumatoid arthritis patients show augmented reactivity to a fraction of mycobacteria cross-reactive with cartilage. Lancet ii,305-309.

Holoshitz, J., Naparstek, Y., Ben-Nun, A., and Cohen, I.R. (1983). Lines of T lymphocytes induce or vaccinate against autoimmune arthritis. Science 219,56-58.

Klinman, D., Shirai, A., Ishigatsubo, Y., Conover, J., and Steinberg, A. (1991). Quantitation of IgM- and IgG-secreting B cells in the peripheral blood of patients with systemic lupus erythematosus. Arthritis Rheum. 34,1404-1410.

Krainer, A.R., Mayeda, A., Kozak, D., and Binns, G. (1991). Functional expression of cloned human splicing factor SF2: homology to RNA-binding proteins, U1 70K, and drosophila splicing regulators. Cell 66,383-394.

Lafyatis, R., Flipo, R., Duquesnoy, B., and Capron, A. (1992). Antibodies in rheumatoid synovial fluids bind to a restricted series of proteins in rheumatoid synovial tissue. Arthritis Rheum. 35,1016-1027.

Li, H., and Bingham, P.M. (1991). Arginine/serine-rich domains of the *su(wa)* and *tra* processing regulators target proteins to a subnuclear compartment implicated in splicing. Cell 67,335-342.

Mayeda, A., Zahler, A.M., Krainer, A.R., and Roth M.B. (1992). Two members of a conserved family of nuclear phosphoproteins are involved in pre-mRNA splicing. Proc. Natl. Acad. Sci. USA 89,1301-1304.

Mottonen, T., Hannonen, P., Oka, M., Rautiainen, J., Jokinen, I., Arvilommi, H., Palosuo, T., and Aho, K. (1988). Antibodies against native type II collagen do not precede the clinical onset of rheumatoid arthritis. Arthritis Rheum. 31,776-779.

Pearson, C.M. (1956). Development of arthritis, periarthritis and periostitis in rats given adjuvant. Proc. Soc. Exp. Biol. Med. 91,95-101.

Query, C.C., Bentley, R.C., and Keene, J.D. (1989). A common RNA recognition motif identified within a defined RNA binding domain of the 70K U1 snRNP protein. Cell 57,89-101.

Roosnek, E., and Lanzavecchia, A. (1991). Efficient and selective presentation of antigen-antibody complexes by rheumatoid factor B cells. J. Exp. Med. 173,487-489.

Roudier, J., Petersen, J., Rhodes, G.H., Luka, J., and Carson, D.A. (1989). Susceptibility to rheumatoid arthritis maps to a T-cell epitope shared by the HLA-Dw4 DR beta-1 chain and the Epstein-Barr virus glycoprotein gp110. Proc. Natl. Acad. Sci. USA 86,5104-5108.

Rowley, M., Tait, B., Mackay, I.R., Cunningham, T., and Phillips, B. (1986). Collagen antibodies in rheumatoid arthritis. Arthritis Rheum. 29,174-184.

Sambrook, J., Fritsch, E.F., and Maniatis, T. (1989) In Molecular Cloning, 2nd edition, Cold Spring Harbor Laboratory, New York, 12.1-12.44.

Savige, J.A., Gallicchio, M.C., Stockman, A., Cunningham, T.J., Rowley, M.J., Georgiou, T., and Davies, D. (1991). Anti-neutrophil cytoplasmic antibodies in rheumatoid arthritis. Clin. Exp. Immunol. 86,92-98.

Smith, B., and Johnson, K. (1988). Single-step purification of polypeptides expressed in Escheria coli as fusions with glutathione S-transferase. Gene 67,31-40.

Steere, A.C., Brinckerhoff, C.E., Miller, D.J., Drinker, H., Harris, E.D., and Malawista, S.E. (1980). Elevated levels of collagenase and prostaglandin E2 associated with erosion of cartilage and bone in a patient with chronic Lyme arthritis. Arthritis Rheum. 23,591-598.

Stuart, J.M., Townes, A.S., and Kang, A.H. (1984). Collagen autoimmune arthritis. Annu. Rev. Immunol. 2,199-218.

Takeda, Y., Nyman, U., Winkler, A., Wise, K.S., Hoch, S.O., Pettersson, I., Anderson, S.K., Wang, R.J., Wang, G.S., and Sharp, G.C. (1991). Antigenic domains on the U1 small nuclear riboncleoprotein-associated 70K polypeptide: a comparison of regions selectively recognized by human and mouse autoantibodies and by monoclonal antibodies. Clin. Immunol. Immunopathol. 61,55-68.

Theissen, H., Etzerodt, M., Reuter, R., Schneider, C., Lottspeich, F., Argos, P., Luhrmann, R., and Philipson, L. (1986). Cloning of the human cDNA for the U1 RNA-associated 70K protein. EMBO J. 5,3209-3217.

Todd, J.A., Acha-Orbea, H., Bell, J.I., Chao, N., Fronek, Z., Jacob, C.O., McDermott, M., Sinha, A.A., Timmermann, L., Steinman, L., and McDevitt, H.O. (1988). A molecular basis for MHC class II associated autoimmunity. Science 240,1003-1009.

Tosato, G., Steinberg, A.D., Yarchoan, R., Heilman, C.A., Pike, S.E., De Seau, V., and Blaese, R.M. (1984). Abnormally elevated frequency of Epstein-Barr virus-infected B cells in the blood of patients with rheumatoid arthritis. J. Clin. Invest. 73,1789-1795.

Trentham, D.E., Dynesius, M.S., Rocklin, R.E., and David, J.R. (1980). Cellular sensitivity to collagen in rheumatoid arthritis. N. Engl. J. Med. 299,327-332.

Valcarcel, J., Singh, R., Zamore, P.D., and Green, M.R. (1993). The protein Sex-lethal antagonizes the splicing factor U2AF to regulate alternative splicing of *transformer* pre-mRNA. Nature 362,171-175.

van Eden, W., Thole, J.E.R., Zee, R.v.d., Noordzij, A., van Embden, J.D.A., Hensen, E.J., and Cohen, I.R. (1988). Cloning of the mycobacterial epitope recognized by T lymphocytes in adjuvant arthritis. Nature 331,171-173.

Victor, K.D., Randen, I., Thompson, K., Forre, O., Natvig, J.B., Fu, S.M., and Capra, J.D. (1991). Rheumatoid factors isolated from patients with autoimmune disorders are derived from germline genes distinct from those encoding the Wa, Po, and Bla cross-reactive idiotypes. J. Clin. Invest. 87,1603-1613.

Yocum, D.E., Klippel, J.H., Wilder, R.L., Gerber, N.L., Austin, H.A., Wahl, S.M., Lesko, L., Minor, J.R., Preuss, H.G., Yarboro, C., Berkebile, C., and Dougherty, S. (1988). Cyclosporin A in severe treatment-refractory rheumatoid arthritis: a randomized study. Ann. Intern. Med. 109,863-869.

Zachar, Z., Chou, T.-B., and Bingham, P.M. (1987). Evidence that a regulatory gene autoregulates splicing of its transcript. EMBO J. 6,4105-4111.

Zamore, P.D., Patton, J.G., and Green, M.R. (1992). Cloning and domain structure of the mammalian splicing factor U2AF. Nature 355,609-614.

Use of synthetic peptides for the analysis of B-cell epitopes in autoantigens

Sylviane Muller
UPR 9021 Immunochimie des Peptides et des Virus
Institut de Biologie Moléculaire et Cellulaire, CNRS
15 rue René Descartes
67000 Strasbourg
FRANCE

INTRODUCTION

In systemic autoimmune diseases, patients develop antibodies to a variety of cellular components. Like the majority of conventionally induced antibodies, autoantibodies recognize discontinuous epitopes which are made up of amino acid residues that are not contiguous in sequence but are brought together by the folding of the polypeptide chain. Since, short synthetic peptides can, to a certain extent, mimic such conformational epitopes and they can be used in solid-phase assays to replace complex antigens that are often difficult to purify and prepare in large quantities. In recent years the development of improved methods of peptide synthesis and the availability of peptides have stimulated a number of investigations using sera from autoimmune patients that react with various self-proteins such as histones, ubiquitin, polypeptides B/B′ and D of the Sm antigen, polypeptide A of the U1 snRNP antigen, 60 KDa SSA/Ro antigen, phosphorylated ribosomal proteins and E2 (dihydrolipoamide acetyltransferase) mitochondrial antigen. Assays based on the use of synthetic peptides corresponding to sequences within these proteins have proved useful for the detection and in some cases for the discrimination of antibodies present in related systemic and organ-specific autoimmune diseases. In addition to increasing the specificity of immunoassays, the use of peptides offers the possibility of introducing during synthesis modified residues or cofactors (e.g. lipoic acid in E2 antigen) which are normally present in natural molecules. This approach allows one to examine the effects of post-translational events (e.g. acetylation, methylation, phosphorylation or

NATO ASI Series, Vol. H 80
Autoimmunity: Experimental Aspects
Edited by Moncef Zouali

ubiquitination) or the influence of substrates and cofactors on the antigenicity of peptides and thus permits the identification of unique conformational epitopes. This article reviews the findings of epitope analysis based on the use of synthetic peptides from four antigens recognized by antibodies of autoimmune patients which have been investigated in our laboratory, namely histones, ubiquitin, poly(ADP-ribose) polymerase and E2 antigen. An underlying goal in the characterization of epitopes recognized by autoantibodies is the anticipation that such information might help to understand the etiological and pathogenic mechanisms that lead to autoimmunity.

HISTONES

Although core histones are considered to be weak immunogens, a feature which has been ascribed to the fact that they are remarkably conserved, anti-histone antibodies (AHA) are commonly found in the serum of patients suffering from various autoimmune diseases [Monestier and Kotzin, 1992; Muller and Van Regenmortel, 1993]. AHA have been reported in systemic lupus erythematosus (SLE), drug-induced lupus (DIL), rheumatoid arthritis (RA), Felty's syndrome and certain autoimmune disorders of the liver. They are also present in dogs with lupus and in various murine models of SLE.

The distribution of the epitopes on core histones has been studied by several groups by using natural and synthetic histone fragments both in ELISA and Western blot assay [Hardin and Thomas, 1983; Gohill et al., 1985; Costa and Monier, 1986; Portanova et al., 1987, 1990; Craft et al., 1987; Muller et al., 1989]. In SLE, the epitopes recognized in histones H2A, H2B, H3 and H4 are located in the N- and C-terminal domains which are exposed at the surface of nucleosomes. For example, many sera from lupus patients contain antibodies to four terminal peptides (residues 1-21 and 130-135 of H3, 1-29 of H4 and 1-25 of H2B) [Muller et al., 1989]. Additionally, monoclonal antibodies from MRL-*lpr/lpr* and NZB x NZW/F1 autoimmune mice also frequently recognize the basic N-terminal domain of core histones [Muller et al., 1985; Portanova et al., 1990; Monestier et al., 1993]. Based on these results it has been argued that in an autoimmune context, structures equivalent to nucleosome particles rather than histones alone may represent the immunologic stimulus that gives rise to autoantibody production. Investigation of immunogenic properties of nucleosome particles has shown that the histone domains recognized by SLE sera are very similar to those recognized by antibodies raised in rabbits against nucleosomes and against a total histone fraction complexed to RNA in the presence or absence of adjuvant [Einck et al., 1982; Muller et al., 1989; Atanassov et al., 1991].

Peptides in the internal domains of core histones are also recognized by human or murine autoantibodies. For example, antibodies from lupus patients react with peptides 65-85 of H2A and 40-55 of H3 and a murine hybridoma to synthetic peptide 30-45 of H3 has been described [Muller et al., 1989; Monestier et al., 1993]. Recently the crystal structure of the core histone octamer has been determined at 3.1 Å resolution [Arents et al., 1991]. In the three-dimensional structure, the regions encompassing residues 65-85 of H2A and 30-55 of H3 contain loops which are presumed to be exposed at the surface of the histone octamer (Figure 1).

The binding specificity of anti-H1 and anti-H5 autoantibodies was also studied with natural or synthetic fragments [Hardin and Thomas, 1983; Costa et al.,1986; Gohill and Fritzler, 1987; Pauls et al., 1990; Creemers et al., 1992]. Most of the studies agree that the major antigenic domain recognized in ELISA and immunoblotting by autoantibodies from SLE and DIL patients is present in the basic C-terminus of the molecule. Using the multi-peg technology and concurrent overlapping hexameric peptides, linear epitopes were also identified in the globular domain of H5 [Pauls et al., 1993]. Recently two anti-H1 monoclonal antibodies derived from splenic lymphocytes of patients with idiopathic thrombocytopenia and SLE have been studied with H1 peptides and also sequenced. One of these antibodies reacted with the C-terminal region of human H1b (residues 204-218), the other reacted with both the N- and C-terminal regions of human H1b (residues 1-16 and 204-218). The latter antibody presumably recognizes an epitope constituted by residues of both the N- and C-termini of H1 clustered together at the surface of the molecule, as has been described for other monoclonals [Altschuh et al., 1985].

UBIQUITIN AND UBIQUITINATED HISTONE H2A

Antibodies reacting with the low molecular weight stress protein ubiquitin have been found in human and murine lupus [Muller et al., 1988, 1990; Muller, 1992]. The specificity was detectable in ELISA with commercial ubiquitin and ubiquitin synthesized by the solid-phase method of Merrifield [Briand et al., 1989].

Figure 1: STRUCTURE OF HISTONE H3 WITHIN HISTONE OCTAMER AND AS [H3 H4]$_2$ TETRAMER

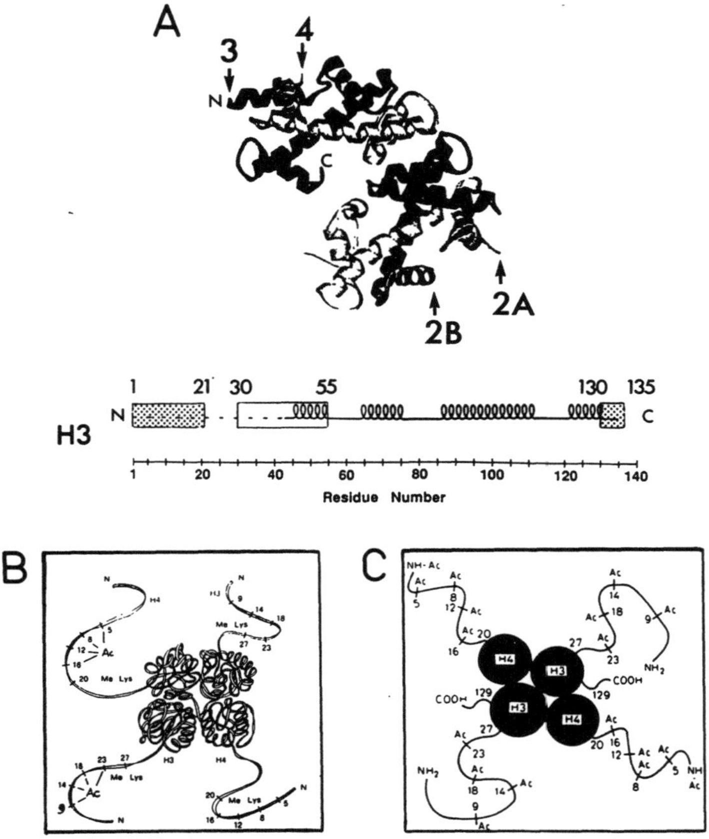

A. Model of the H2A-H2B and H3-H4 dimer domains within the histone octamer proposed by Arents et al. [1991]. The structure of the nucleosomal core histone octamer has been determined at 3.1 Å resolution and shows that histone molecules within the octamer contain several a-helices. The amino end of each histone in the model is marked by an arrow. For reasons of clarity, only one of each histone pair is presented. Domains of H3 recognized by autoantibodies from SLE patients are shown along the extended chain. Residue numbers have been assigned from the structure proposed by Arents et al. (Photographs, E.N. Moudrianakis).
B. Model of the [H3 H4]$_2$ tetramer built from results obtained by NMR studies [Bradbury, 1983]. The NMR results are consistent with a model consisting of a central structured region having the basic N-terminal domain free and mobile. Sites for post-translational acetylation (Ac) and methylation (Me) are shown.
C. Schematic representation of the [H3 H4]$_2$ tetramer deduced from the determination of protease-sensitive and protected domains of H3 and H4 associated as a tetramer [Böhm et al., 1981]. The C-terminal domain 129-135 of H3 is accessible in this model [Figures 1 A-C with permission].

Almost all patient sera reacting with ubiquitin also reacted with a synthetic peptide corresponding to residues 22-45. This peptide inhibited the binding of anti-ubiquitin antibodies to ubiquitin. Ubiquitin antibodies were found essentially to occur in SLE. Less than 16% of patients with other rheumatic autoimmune diseases (e.g. RA or scleroderma) and 3% of normal individuals possessed antibodies to ubiquitin. Serendipitously, a strong reactivity was found by immunoblotting with SLE sera and ubiquitinated complexes containing one or several molecules of either histone H2A or both H2A and H2B which contaminated the commercial ubiquitin preparation [Plaué et al., 1989]. In the nucleus, ubiquitin is enzymatically conjugated via an isopeptide bond between its C-terminal glycine residue (α-COOH group) and the ε-amino groups of lysine residues 119 of H2A and 120 in H2B, respectively, thus forming branched molecules (U-H2A and U-H2B). The reactivity of sera from autoimmune patients was studied by means of synthetic peptides corresponding to the branched moiety of U-H2A. When a branched peptide consisting of four residues of ubiquitin and four residues of H2A in an isopeptide linkage was tested in ELISA, 96% of the SLE sera that recognized ubiquitin also reacted with the octapeptide. None of the sera from patients with other rheumatic diseases such as RA contained antibodies able to bind this branched peptide. The use of such unique synthetic peptides allowed us to characterize antibodies reacting only with U-H2A and not with ubiquitin nor with H2A alone. Parallel experiments showed that antibodies induced against H2A or against ubiquitin conjugated to KLH do not react with the U-H2A branched peptide.

POLY(ADP-RIBOSE) POLYMERASE

Poly(ADP-ribose) polymerase (PARP) is a chromatin-bound, DNA-dependent enzyme that catalyses the ADP-ribosylation of nuclear acceptor proteins by using NAD^+ as a substrate. Once nuclear enzymes are poly(ADP-ribosyl)ated, their activity is generally inhibited and this modification has thus been correlated with a variety of regulatory functions, including DNA repair, gene expression and cell differentiation. Nuclear structural proteins such as histones and high mobility group proteins 1 and 2 are also well-known acceptors for poly(ADP-ribose) and this unique change represents the most drastic post-translational protein modification that can affect chromatin structure. Poly(ADP-ribosyl)ation of histones H1 and H2B induces rearrangements of the nucleosome topography which can be visualized by antibodies against histones. Such modifications in the rearrangement of histones within nucleosomes are also observed when free poly(ADP-ribose) interacts with nucleosomes [Huletsky et al., 1989; Thibeault et al., 1992].

Antibodies to PARP were detected in the serum of patients with rheumatic diseases [Yamanaka et al., 1989]. By using natural fragments of the enzyme, they were shown to react in Western blot with the 54-KDa NAD^{+}-binding domain (C-terminal half of the molecule), but not with the 40-KDa fragment contained in the 54-KDa domain which represents the minimal active domain of the enzyme (Figure 2). We have recently examined a series of sera from patients with a variety of systemic autoimmune diseases, including SLE and primary (pSS) and secondary (sSS) Sjögren's syndrome, with two functional domains of the enzyme, namely the recombinant human 40-KDa fragment and a synthetic peptide encompassing the second zinc finger (F2) of the enzyme. The latter fragment has been shown previously to be essential for binding of the enzyme to DNA breaks [De Murcia et al., 1991] and thus plays an important role in DNA repair. We demonstrated that peptide F2 was effectively able to bind ^{65}Zn.

Figure 2: Schematic structure of poly(ADP-ribose) polymerase (PARP).

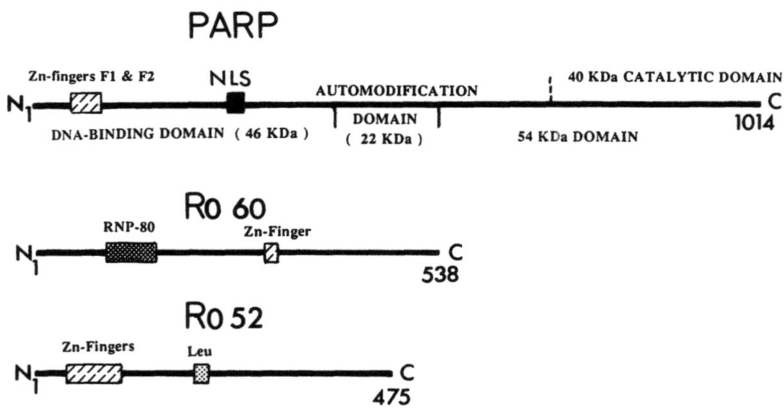

The three functional domains of PARP are shown, namely the 46-KDa DNA binding domain containing the two zinc fingers F1 and F2 and a nuclear localization signal (NLS), the 22-KDa automodification domain and the 54-KDa domain containing the 40-KDa catalytic domain. The position of finger motifs contained in two proteins of the SSA/Ro antigen (Ro 60 and Ro 52) are also shown. Peptides corresponding to Ro 60 and Ro 52 zinc fingers are not recognized by patient antibodies binding peptide F2.

In agreement with the results of Yamanaka et al. [1989], a rather low reactivity and frequency of antibody binding was observed with the natural enzyme purified from calf thymus as well as with the recombinant human PARP. Likewise, very few sera reacted with the recombinant 40-KDa fragment. In contrast, high levels of IgG antibodies reacting specifically with the zinc finger peptide F2 but not with unrelated zinc finger peptides of other self proteins (in 52-KDa and 60-KDa SSA/Ro protein; Figure 2), were found in SLE and Sjögren's syndrome. These antibodies were rarely detected in other autoimmune sera.

The observation that in autoimmune patients antibodies binding the PARP domain which is directly involved in the recognition of DNA breaks are produced suggests a possible link between DNA damage and autoimmunity. In this regard, increased DNA strand breaks were found in mononuclear cells from patients with RA [Bhusate et al., 1992] and poly(ADP-ribose) metabolism may be altered in autoimmune patients and mice [Sibley et al., 1989; Haug et al., 1987]. Following extensive DNA strand breaks, NAD and ATP levels are decreased to such an extent that programmed-cell death may take place [Carson et al., 1986]. This event could result in the release of a variety of cellular components into the circulation, the consequences of which may be immune dysfunction since an excess of these molecules would act as a stimulus for the production of specific antibodies or for inducing polyclonal B cell proliferation [Bell and Morrison, 1991; Mohan et al., 1993].

DIHYDROLIPOAMIDE ACETYLTRANSFERASE, THE E2 COMPONENT OF THE PYRUVATE DEHYDROGENASE COMPLEX (PDH)

The specific marker for serologic diagnosis of primary biliary cirrhosis (PBC) is the presence of autoantibodies reacting with a mitochondrial antigen termed M2 [Berg and Klein, 1990; Gershwin and Mackay, 1991]. These antibodies react with at least three polypeptides with approximate Mw of 70, 45 and 39 KDa, as defined by immunoblottin experiments. Up to 95% of PBC patients possess antibodies to the 70-KDa protein. The two other proteins are less frequently recognized by PBC sera. Gershwin et al. [1987] first isolated and sequenced a cDNA clone that expressed the 70-KDa M2 autoantigen recognized by PBC sera. This protein was then identified as the dihydrolipoamide acetyltransferase (E2) of the PDH complex [Yeaman et al., 1988; Coppel et al., 1988]. In eukaryotic cells, the PDH complex is located within the inner membrane-matrix compartment and catalyses the conversion of pyruvate to acetyl-CoA. Human enzyme E2 contains four structural domains, two of which possess a site for lipoic acid (LA) and are called outer and inner lipoyl domains (in residues 1-98 and 128-229 respectively). These two domains show 92% homology between residues 40-52 and 167-179.

Several reports from Gershwin's group [Van de Water et al., 1988 ; Surh et al., 1990] have indicated that the dominant epitope in E2 was localized in the inner lipoyl domain and that the outer lipoyl domain had only weak reactivity. These authors have used six synthetic peptides encompassing residues 150-169, 167-186, 191-200, 207-226, 274-289 and 287-300 in human E2. Van de Water et al. [1988] showed that among these peptides, the fragment 167-186 adsorbed nearly all ELISA reactivity from PBC sera when the latter were diluted at least 1/80 000. Analysis of recombinant fusion proteins expressed from restriction cDNA fragments encoding the inner lipoyl domain showed a requirement of 75 amino acid residues (fragment 146-221) for detectable autoantibody binding and 93 residues (fragment 128-221) for strong binding [Gershwin and Mackay, 1991]. The authors concluded that the requirement for such a large peptide region indicates that a conformational epitope was involved. Recently we have shown that IgG antibodies in PBC sera indeed recognized a unique conformational epitope in E2 which could be mimicked by both the 18-mer peptide 167-184 associated with lipoic acid covalently attached to the lysine residue 173 and the 13-mer peptide 40-52 with lipoic acid attached to lysine residue 46 [Tuaillon et al., 1992; Briand et al., 1992]. In our case, and in contrast to the results described by Gershwin and coworkers, peptides without LA were not recognized by PBC antibodies. The free peptide 167-184 with LA but not the peptide 167-184 without LA was found to totally absorb M2 antibody reactivity from PBC sera tested by indirect immunofluorescence with rat kidney, stomach and liver cells or with human HEp-2 cells as substrate. Because the two peptides LA^+ are recognized by a large number of PBC sera (85% and 67% for peptides 40-52 LA^+ and 167-184 LA^+ respectively in our study) and very rarely by antibodies from other autoimmune sera (in particular from patients with other liver diseases), they could be used as valuable probes for PBC diagnosis. The whole enzyme in contrast is recognized by antibodies in several autoimmune diseases including liver and rheumatic diseases [Briand et al., 1992 ; Zurgil et al., 1992].

CONCLUSION

Many immunologic abnormalities have been described in patients with autoimmune rheumatic diseases, including T-lymphocyte suppressor cell deficiency and polyclonal B-cell activation. It is well accepted, however, that in view of the disease specificity of most autoantibodies present in systemic diseases, such non-specific mechanisms alone cannot account for the production of antibodies characterized in the serum of autoimmune patients [Steinberg et al., 1990; Boitard, 1992; Drake and Kotzin, 1992]. A number of studies have thus been devoted to identification of cellular structures which may induce the specific production of antibodies. Several previous reports have shown that certain autoantibody subsets recognize molecular complexes or proteins in their native form whereas others react preferentially with the denatured forms of proteins [Huff et al., 1990; Itoh and Reichlin, 1992]. The finding that antibodies react with conformational determinants suggests that native epitopes may directly drive the autoimmune response. Thus, particles such as the nucleosome, the U1 snRNP particle, the ribosome and components of the nucleolus are examples of potential subcellular complexes which may trigger autoimmune response [Hardin, 1986; Tan, 1991]. Although it has been shown that some antibodies to native complexes are able to inhibit function, it is not known whether these antibodies play a direct role in the pathogenesis of systemic diseases, and if so, whether they play a greater role than antibodies recognizing rather other states, somewhat denaturated, of the polypeptides as they may be presented, for example, at the cell surface.

The fine specificity of antibodies in systemic autoimmune diseases has been investigated by many groups using patient sera or monoclonal antibodies from lupus patients and experimental animals, and recombinant or synthetic fragments [Petterson, 1992]. A number of epitopes on various nuclear, cytoplasmic, nucleolar and mitochondrial proteins have been identified. Certain discrepancies have appeared in some studies and have been discussed previously [Horsfall et al., 1991; Elkon, 1992; Guldner, 1992; Rokeach et al., 1992]. They result in part from the fact that long as well as short peptides usually adopt different conformations when they are adsorbed to the plastic of a microtiter plate, coupled on polyethylene pins, chemically conjugated to a protein carrier or present in the free form in solution. Therefore different assay formats based on different types of peptide presentation should be tested before concluding that a peptide is not recognized by the antibodies tested and, consequently, that the sequence does not contain an epitope for these antibodies. Furthermore since the "true" epitope recognized by a particular antibody is extremely difficult to mimic, in particular in the case of autoantibodies which are induced by an unknown but possibly complex antigen, it is difficult to conceive that a single test will be sufficient to characterize the fine specificity of antibodies. The type of assay to apply will depend on the peptide tested (i.e. composition, length, ability to fold) and on the antibodies, the fine specificity of which, even for a dominant epitope, may differ slightly from one patient to another. Thus the results obtained in different assays should be examined in an additive rather than exclusive manner [Muller et al., 1986].

The study of epitopes in intracellular antigens that are recognized by autoantibodies has revealed striking features which may be relevant to the mechanisms involved in the stimulation of autoreactive B-cells. Several authors have proposed that molecular mimicry between host antigens and exogenous (i.e. bacterial or viral) proteins may play an essential role in the specific stimulation of autoreactive B-cell clones and the production of certain autoantibodies [Horsfall, 1992; Jones et al., 1993; Baum, 1993]. However it has also to be pointed out that autoantibodies frequently bind sites in proteins that are post-translationally modified or portions of proteins that are associated with a substrate or a cofactor. Examples of such reactivities include antibodies reacting with phosphorylated RNA polymerase I, ubiquitinated histone H2A and mitochondrial antigens containing LA [Stetler and Jacob, 1984; Plaué et al., 1989; Tuaillon et al., 1992]. It is possible that in predisposed patients, under certain metabolism conditions (for example in case of excessive DNA damage or tissue injury), abnormal levels of such modifications will occur and lead to a change of the antigenicity of "hypermodified" proteins. The resulting antibodies may interact specifically with modified molecules. Unexpectedly, however, we have shown in the case of histone H4 that antibodies induced by the acetylated histone reacted *only* with the *non*-modified parent protein and peptides [Muller et al., 1987]. In this regard, the presence of a dominant autoepitope in the C-terminus of ribosomal P proteins which naturally contain an important site of phosphorylation is also striking [Hasler et al., 1991].

Although it remains important to carefully study the native structures recognized by autoantibodies in macromolecular complexes, identification of B-cell epitopes provides useful information permitting the replacement, in diagnostic immunoassays, of natural purified proteins by recombinant or synthetic fragments which may increase the sensitivity and specificity of the tests [Van Regenmortel, 1993]. Such investigations may bring us closer to the elucidation of the mechanisms involved in the specific antigen-driven activation of B-cells in autoimmunity and may result in therapeutic applications.

ACKNOWLEDGEMENTS

The author wishes to thank Dr. M.H.V. Van Regenmortel for his constant support and many colleagues for helpful discussions. This work has been made possible by support provided by CNRS, ARC (Association pour la Recherche sur le Cancer) and NATO (900515).

REFERENCES

Altschuh D, Al Moudallal Z, Briand JP and Van Regenmortel MHV (1985) Immunochemical studies of tobacco mosaic virus. VI. Attempts to localize viral epitopes with monoclonal antibodies. Mol. Immunol. 22: 329-337.

Arents G, Burlingame RW, Wang BC, Love WE and Moudrianakis EN (1991) The nucleosomal core histone octamer at 3.1 Å resolution: a tripartite protein assembly and a left-handed superhelix. Proc. Natl. Acad. Sci. USA 88: 10148-10152.

Atanassov C, Briand JP, Bonnier D, Van Regenmortel MHV and Muller S (1991) New Zealand white rabbits immunized with RNA-complexed total histones develop an autoimmune-like response. Clin. exp. Immunol. 86: 124-133.

Baum H, Butler P, Davies H, Sternberg MJE and Burroughs AK (1993) Autoimmune disease and molecular mimicry: an hypothesis. Trends Biochem. Sci. 18:140-144.

Bell DA and Morrison B (1991) The spontaneous apoptotic cell death of normal human lymphocytes *in vitro* : the release of, and immunoproliferative response to, nucleosomes *in vitro*. Clin. Immunol. Immunopathol. 60: 13-26.

Berg PA and Klein R (1990) Autoantibodies in primary biliary cirrhosis. Springer Semin. Immunopathol. 12: 85-99.

Bhusate LL, Herbert KE, Scott DL and Perret D (1992) Increased DNA strand breaks in mononuclear cells from patients with rheumatoid arthritis. Ann. Rheum. Dis. 51: 8-1.

Böhm L, Briand G, Sautière P and Crane-Robinson C (1981) Proteolytic digestion studies of chromatin core-histone structure. Identification of the limit peptides of histones H3 and H4. Eur. J. Biochem. 119: 67-74.

Boitard C (1992) B-cell responses in autoimmune diseases. Curr. Opinion Immunol. 4: 741-747.

Bradbury EM (1983) Flexibility in chromosomal proteins. In : Mobility and Recognition in Cell Biology (Sund and Veeger, eds.), Walter de Gruyter, Berlin, pp. 173-194.

Briand JP, Van Dorrselaer A, Raboy B and Muller S (1989) Total chemical synthesis of ubiquitin using BOP reagent: biochemical and immunochemical properties of the purified synthetic product. Peptide Res. 2: 381-388.

Briand JP, André C, Tuaillon N, Hervé L, Neimark J and Muller S (1992) Multiple autoepitope presentation for specific detection of antibodies in primary biliary cirrhosis. Hepathology 16: 1395-1403.

Carson DA, Seto S, Wasson DB, Carrera CJ (1986) DNA strand breaks, NAD metabolism and programmed cell death. Exp. Cell. Res. 164: 273-281.

Coppel RL, McNeilage LJ, Surh CD, Van de Water J, Spithill TW, Whittingham S and Gershwin ME (1988) Primary structure of the human M2 mitochondrial autoantigen of primary biliary cirrhosis: dihydrolipoamide acetyltransferase. Proc. Natl. Acad. Sci. USA 85: 7317-7321.

Costa O and Monier JC (1986) Antihistone antibodies detected by ELISA and immunoblotting in systemic lupus erythematosus and rheumatoid arthritis. J. Rheumatol. 13: 722-725.

Costa O, Tchouatcha-Tchouassom JC, Roux B and Monier JC (1986) Anti-H1 histone antibodies in systemic lupus erythematosus: epitope localization after immunoblotting of chymotrypsin-digested H1. Clin. exp. Immunol. 63: 608-613.

Craft JE, Radding JA, Harding MW, Bernstein RM and Hardin JA (1987) Autoantigenic histone epitopes: a comparison between procainamide- and hydralazine-induced lupus. Arthritis Rheum. 30: 689-694.

Creemers P, Monestier M and Böhm L (1992) Epitope recognition in histone H1 by SLE autoantibodies in the presence of a DNA-ligand. Autoimmunity 12: 167-174.

De Murcia G, Ménissier-de Murcia J and Schreiber V (1991) Poly(ADP-ribose) polymerase: molecular biological aspects. BioEssays 13: 455-462.

Drake CG and Kotzin BL (1992) Genetic and immunological mechanisms in the pathogenesis of systemic lupus erythematosus. Curr. Opinion Immunol. 4: 733-740.

Einck L, Dibble R, Frado L-LY and Woodcock CLF (1988) Nucleosomes as antigens. Exp. Cell Res. 139: 101-110.

Elkon KB (1992) Use of synthetic peptides for the detection and quantification of autoantibodies. Mol. Biol. Rep. 16: 207-212.

Gershwin ME and Mackay IR (1991) Primary biliary cirrhosis: paradigm or paradox for autoimmunity. Gastroenterology 100: 822-833.

Gershwin ME, Mackay IR, Sturgess A and Coppel RL (1987) Identification and specificity of a cDNA encoding the 70 Kd mitochondrial antigen recognized in primary biliary cirrhosis. J. Immunol. 138: 3525-3531.

Gohill J, Cary PD, Couppez M and Fritzler MJ (1985) Antibodies from patients with drug-induced and idiopathic lupus erythematosus react with epitopes restricted to the amino and carboxyl termini of histone. J. Immunol. 135: 3116-3121.

Gohill J and Fritzler MJ (1987) Antibodies in procainamide-induced and systemic lupus erythematosus bind the C-terminus of histone 1 (H1). Mol. Immunol. 24: 275-285.

Guldner HH (1992) Mapping of epitopes recognized by anti-(U1)RNP autoantibodies. Mol. Biol. Rep. 16: 155-164.

Hardin JA (1986) The lupus autoantigens and the pathogenesis of systemic lupus erythematosus. Arthritis Rheum. 29: 457-460.

Hardin JA and Thomas JO (1983) Antibodies to histones in systemic lupus erythematosus: localization of prominent autoantigens on histones H1 and H2B. Proc. Natl. Acad. Sci. USA 80: 7410-7414.

Hasler P, Brot N, Weissbach H, Parnassa AP and Elkon KB (1991) Ribosomal proteins P0, P1 and P2 are phosphorylated by casein kinase II at their conserved carboxyl termini. J. Biol. Chem. 266: 13815-13820.

Haug BL, Sibley JT and Lee JS (1987) Increased synthesis of poly(ADP-ribose) in isolated liver nuclei from autoimmune NZB/NZW mice. FEBS Lett. 215: 252-256.

Horsfall AC, Hay FC, Soltys AJ and Jones MG (1991) Epitope mapping. Immunol. Today 12: 211-213.

Horsfall AC (1992) Molecular mimicry and autoantigens in connective tissue diseases. Mol. Biol. Rep. 16: 139-147.

Huff JP, Roos G, Peebles CL, Houghten R, Sullivan KF, Tan EM (1990) Insights into native epitopes of proliferating cell nuclear antigen (PCNA), using recombinant DNA protein products. J. Exp. Med. 172: 419-429.

Huletsky A, de Murcia G, Muller S, Hengartner M, Ménard L, Lamarre D and Poirier GG (1989) The effect of poly(ADP-ribosyl)ation on native and H1-depleted chromatin. J. Biol. Chem. 264: 8878-8886.

Jones DB, Coulson AFW and Duff GW (1993) Sequence homologies between hsp60 and autoantigens. Immunol. Today 14: 115-118.

Itoh Y and Reichlin M (1992) Autoantibodies to the Ro/SSA antigen are conformation dependent. I: Anti-60 kD antibodies are mainly directed to the native protein; anti-52 Kd antibodies are mainly directed to the denatured protein. Autoimmunity 14: 57-65.

Mohan C, Adams S, Stanik V and Datta SK (1993) Nucleosome: a major immunogen for pathogenic autoantibody-inducing T cells of lupus. J. Exp. Med. 177:1367-1381.

Monestier M and Kotzin BL (1992) Antibodies to histones in systemic lupus erythematosus and drug-induced lupus syndromes. Rheum. Dis. Clin. North Am. 18: 415-436.

Monestier M, Fasy TM, Losman MJ, Novick KE and Muller S (1993) Structure and binding properties of monoclonal antibodies to core histones from autoimmune mice. Mol. Immunol. in press.

Muller S (1992) Place et rôle de l'ubiquitine dans le développement de l'autoimmunité au cours de la maladie lupique. Médecine/ Sciences 8: 223-232.

Muller S and Van Regenmortel MHV (1993) Histones. In: Structure of Antigens (Van Regenmortel MHV, Ed.) Vol. 2. CRC Press, Boca Raton, pp 149-178.

Muller S, Barakat S, Watts R, Joubaud P and Isenberg D (1990) Longitudinal analysis of antibodies to histones, Sm-D peptides and ubiquitin in the serum of patients with systemic lupus erythematosus, rheumatoid arthritis and tuberculosis. Clin. Exp. Rheum. 8: 445-453.

Muller S, Bonnier D, Thiry M and Van Regenmortel MHV (1989) Reactivity of autoantibodies in systemic lupus erythematosus with systemic core histone peptides. Int. Arch. Allergy Appl. Immunol. 89: 288-296.

Muller S, Briand JP and Van Regenmortel MHV (1988) Presence of antibodies to ubiquitin dduring the autoimmune response associated with systemic lupus erythematosus. Proc. Natl. Acad. Sci. USA 85: 8176-8180.

Muller S, Isabey A, Couppez M, Plaué S, Sommermeyer G and Van Regenmortel MHV (1987) Specificity of antibodies raised against triacetylated histone H4. Mol. Immunol. 24: 779-789.

Muller S, Plaué S, Couppez M and Van Regenmortel MHV (1986) Comparison of different methods for localizing antigenic regions in histone H2A. Mol. Immunol. 23: 593-601.

Muller S, Couppez M, Briand JP, Gordon J, Sautière P and Van Regenmortel MHV (1985) Antigenic structure of histone H2B. Biochim Biophys. Acta 827: 235-246.

Pauls JD, Gohill J and Fritzler MJ (1990) Antibodies from patients with systemic lupus erythematosus and drug-induced lupus bind determinants on histone 5 (H5). Mol. Immunol. 27: 701-711.

Pauls JD, Edworthy SM and Fritzler MJ (1993) Epitope mapping of histone 5 (H5) with systemic lupus erythematosus, procainamide-induced lupus and hydralazine-induced lupus sera. Mol. Immunol. 30: 709-719.

Pettersson I (1992) Methods of epitope mapping. Mol. Biol. Rep. 16: 149-153.

Plaué S, Muller S and Van Regenmortel MHV (1989) A branched, synthetic octapeptide of ubiquitinated histone H2A as target of autoantibodies. J. Exp. Med. 169: 1607-1617.

Portanova JP, Arndt RE, Tan EM and Kotzin BL (1987) Anti-histone antibodies in idiopathic and drug-induced lupus recognize distinct intrahistone regions. J. Immunol. 138: 446-451.

Portanova JP, Cheronis JC, Blodgett JK and Kotzin BL (1990) Histone autoantigens in murine lupus. Definition of a major epitope within an accessible region of chromatin. J. Immunol. 144: 4633-4640.

Rokeach LA and Hoch SO (1992) B-cell epitopes of Sm autoantigens. Mol. Biol. Rep. 16: 165-174.

Sibley JT, Haug BL and Lee JS (1989) Altered metabolism of poly(ADP-ribose) in the peripheral blood lymphocytes of patients with systemic lupus erythematosus. Arthritis Rheum. 32: 1045-1049.

Steinberg AD, Krieg AM, Gourley MF and Klinman DM (1990) Theoretical and experimental approaches to generalized autoimmunity. Immunol. Rev. 118: 129-163.

Stetler DA and Jacob ST (1984) Phosphorylation of RNA polymerase I augments its interaction with autoantibodies of systemic lupus erythematosus patients. J. Biol. Chem. 259: 13629-13632.

Surh CD, Coppel RL and Gershwin ME (1990) Structural requirement for autoreactivity on human pyruvate dehydrogenase E2, the major autoantigen of primary biliary cirrhosis. Implication for a conformational autoepitope. J. Immunol. 144: 3367-3374.

Tan EM (1991) Autoantibodies in pathology and cell biology. Cell 67: 841-842.

Thibeault L, Hengartner M, Lagueux J, Poirier GG and Muller S (1992) Rearrangements of the nucleosome structure in chromatin by poly(ADP-ribose). Biochim. Biophys. Acta 1121: 317-324.

Tuaillon N, André C, Briand JP, Penner E and Muller S (1992) A lipoyl synthetic octapeptide of dihydrolipoamide acetyltransferase specifically recognized by anti-M2 autoantibodies in primary biliary cirrhosis. J. Immunol. 148: 445-450.

Van de Water J, Gershwin, ME , Leung P, Ansari A and Coppel RL (1988) The autoepitope of the 74 Kd mitochondrial autoantigen of primary biliary cirrhosis corresponds to the functional site of dihydrolipoamide acetyltransferase. J. Exp. Med. 167: 1791-1199.

Van Regenmortel MHV (1993) Synthetic peptides versus natural antigens in immunoassays. Ann. Biol. Clin. 51: 39-41.

Yamanaka H, Willis EH and Carson DA (1989) Human autoantibodies to poly (adenosine diphosphate-ribose) polymerase recognize cross-reactive epitopes associated with the catalytic site of the enzyme. J. Clin. Invest. 83: 180-186.

Yeaman SJ, Danner DJ, Mutimer DJ, Fussey SPM, James OFW and Bassendine MF (1988) Primary biliary cirrhosis: identification of two major M2 mitochondrial autoantigens. Lancet 1: 1067-1070.

Zurgil N, Bakimer R, Moutsopoulos HM, Tzioufas AG, Youinou P, Isenberg DA,

Scheinberg M, Kveder T, Rozman B, Luderschmidt C, Meroni PL, Kaplan MM and Shoenfeld Y (1992) Antimitochondrial (pyruvate dehydrogenase) autoantibodies in autoimmune rheumatic diseases. J. Clin. Immunol. 12: 201-209.

Molecular mimicry as a mechanism of induction of autoantibodies

Paola Migliorini, Barbara Marchini, Maria Pia Dolcher, Alessandra Sabbatini and Antonio Puccetti
Clinical Immunology Unit
University of Pisa
Via Roma 67
56123 Pisa
Italy

Introduction

Antibodies against small nuclear ribonucleoproteins (snRNP) are frequently produced in systemic autoimmune disorders and different mechanisms have been proposed to account for their production. Molecular mimicry is a well-established phenomenon which can involve exogenous antigens as well as autoantigens. An example of an exogenous antigen, i.e. a viral protein, was reported by Guldner et al. (1990). A 5-amino acid sequence of the p68 protein, a component of U1 snRNP, is shared with the matrix protein of the influenza B virus. A subset of patients with autoimmune disorders, whose sera contain anti-p68 antibodies, react with this shared sequence and with the entire matrix protein. The same p68 protein bears in another region a strong sequence-homology with the third complementarity-determining region (CDR3) of an immunoglobulin light chain. Immunization of normal animals with the light chain induces the production of anti-p68 antibodies (Puccetti et al, 1990). In this case, the sequence mimicking the ribonucleoprotein is contained in the variable region of the immunoglobulin, where the idiotope is located.

The SmD antigen and Epstein-Barr virus

In the model we studied, a protein encoded by the Epstein-Barr virus (EBV) mimicks a ribonucleoprotein, SmD, which is a target of autoantibodies produced

NATO ASI Series, Vol. H 80
Autoimmunity: Experimental Aspects
Edited by Moncef Zouali

in systemic lupus (SLE). Infection with EBV, an ubiquitous DNA virus, is associated with production of autoantibodies. The virus infects B cells and triggers their proliferation (Klein, 1989). Antibody secretion by these infected B cells is the source of autoantibodies produced in acute EBV infection (infectious mononucleosis) and in other conditions associated with polyclonal B cell activation. The specificities of these autoantibodies are those frequently represented in the preimmune repertoire, such as rheumatoid factors, anti-DNA and anti-cytoskeleton antibodies (Dighiero et al, 1985; Naparstek et al, 1986; Prabhakar et al, 1984). EBV infection, however, can lead to autoantibody production by other mechanisms as well. EBV induces the nuclear expression of 6 virally encoded proteins, the EBNAs. Antibodies to the EBNA 2 antigen appear shortly after infection and disappear after 1-2 months. Antibodies to the EBNA 1 protein appear later and persist in EBV-infected subjects (Henle et al, 1987). The EBNA 1 has an unusual structureconsisting in its central portion of a number of Gly-Ala repeats (Hennessy et al, 1983) which represent the dominant B epitope in the anti-EBNA 1 immune response (Dillner et al, 1984; Rhodes et al, 1985). A peptide corresponding to a sequence contained in this region has been used to affinity-purify antibodies from rheumatoid arthritis sera. The anti-peptide antibodies react with the viral protein as well as with collagen and keratin: Gly-Ala sequences are, in fact, highly represented in these proteins. In the anti-EBNA 1 immune response, therefore, anti-collagen and anti-keratin autoantibodies can be produced by a mechanism of molecular mimicry (Baboonian et al, 1991; Birkenfeld et al, 1990). The EBNA 1 contains other sequences of great interest. The 35-58 sequence bears a striking homology to the C-terminal SmD region 95-119: 11 consecutive amino acids and 4 non-consecutive are identical and 4 show conservative substitutions over a total lenght of 24 amino acids (Rokeach et al, 1988).

Anti-EBV antibodies in patients with connective tissue disorders

We measured IgG antibodies specific for the EBNA 35-58 sequence in the sera of patients with connective tissue disorders. Nine out of 36 rheumatoid arthritis sera, 15/40 SLE, 4/12 Sjogren'syndrome and 6/20 normal subjects contain antibodies to this peptide. Anti-EBNA 35-58 antibodies were affinity-purified on a peptide column from lupus, rheumatoid arthritis and normal sera and their specificity was evaluated by a liquid-phase competition assay. The binding of anti-peptide antibodies purified from normal and rheumatoid arthritis sera to EBNA

35-58 is inhibited by the EBNA peptide, but not by the homologous SmD . On the contrary, anti-EBNA 35-58 antibodies affinity-purified from SLE sera have a similar affinity for the viral and the SmD C-terminal peptides. The results indicate that, although antibody production to EBNA 35-58 is a common finding, only SLE sera contain anti-EBNA I antibodies cross-reactive with an autoantigen.

To demostrate a possible role of EBV infection in the induction of anti-SmD antibodies, we have to prove that i) the anti-SmD immune response in lupus is mainly focused on the epitope homologous to the viral protein, ii) the anti-SmD antibodies also bind the viral protein.

To map the epitopes recognized by SLE autoantibodies on SmD, we synthesized seven 20 amino acid-long overlapping peptides corresponding to the entire SmD protein. The peptides were used as antigens on the solid-phase in an ELISA assay to measure IgG antibodies in lupus sera (Sabbatini et al, in press). As shown in Table 1, the highest percentage of positivity was obtained with the C-terminal peptide. Moreover, we detected high levels of anti-peptide antibodies (3-4 times the amount detectable in normal sera) only with the C-terminal peptide.

Table 1: Percentage of SLE sera containing antibodies against seven synthetic peptides corresponding to the whole sequence of SmD.

Peptides	1-19	15-34	30-49	45-64	60-79	75-94	95-119
Positive sera	16%	12%	5%	12%	12%	17%	25%

Sera diluted 1/500 were tested by ELISA on peptide-coated plates and bound antibodies were detected by alkaline phosphatase-labeled anti-human IgG. The sera were considered positive when their optical density was higher than the mean plus two standard deviations of normal sera.

Anti-SmD antibodies bind the EBNA antigen

Therefore, we can conclude that a dominant epitope in the anti-SmD immune response is indeed located in the C-terminal region of the molecule homologous to the viral protein EBNA 1. The antibodies specific for this sequence were purified by affinity-chromatography on a peptide-Sepharose column. The eluted antibodies bound the SmD peptide as well as the EBNA 1, in direct and competitive assays. Their ability to recognize the intact SmD and EBNA 1 proteins was evaluated by immunoblotting. As shown in Figure 1, the anti-peptide antibodies detect on the lysate of EBNA 1-expressing lines a band of 76 kD, corresponding to the molecular weight of EBNA 1 and identified by reference anti-EBNA 1 antibodies. On the same blot, anti-peptide antibodies react also with a 13 kD band corresponding to SmD. The results show that antibodies reacting with the C-terminal region of SmD can bind the EBNA 1 protein as well (Sabbatini et al, 1993).

Figure 1: Binding of anti-SmD 95-119 antibodies to EBNA I.
Total cell lysate from EBNA I-negative and EBNA I-expressing lines were separated on a 12,5% acrylamide gel under reducing conditions and transferred to nitrocellulose. Nitrocellulose strips were probed with reference anti-EBNA I antibodies **(1)** or affinity-purified anti-SmD 95-119 antibodies **(2, 3, 4)**. The cell lines used were: **A)** DG75; **B)** DG75 FA; **C)** DG75 EBNA I. Anti-SmD 95-119 antibodies detect a 76kD band present only in EBNA I-expressing lines (lane C); they react also with a 13 kD band expressed in all the lines and corresponding to SmD.

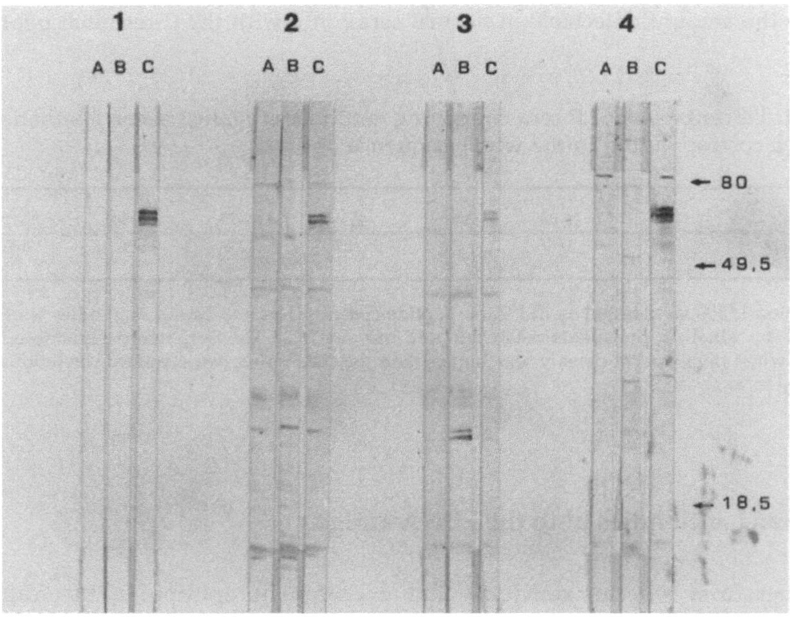

Immunization with EBNA induces anti-SmD antibodies

We then immunized BALB/c mice with the EBNA 1 peptide coupled to KLH. All mice produced antibodies reacting with EBNA 1 and SmD peptides and by immunoblotting sera from immunized mice reacted with recombinant SmD as well. Monoclonal antibodies were obtained from these animals (5 IgM and 1 IgG): they all bound both peptides and reacted with recombinant SmD by immunoblotting (Figure 2). Immunizing mice with a sequence from a viral protein, we induced the production of autoantibodies. Taken together, these results show that EBV infection has the potential to generate anti-SmD antibodies by molecular mimicry.

Figure 2: Binding of immunization-induced anti-EBNA I monoclonal antibodies to recombinant SmD.

Monoclonal antibodies were derived from mice immunized with peptide EBNA 1 35-58 coupled to KLH; they were selected on the basis of their peptide-binding ability. The antibodies were concentrated from culture fluid by ammonium sulfate precipitation (10x) and diluted 1:2 in TBS 5% dry non-fat milk for immunoblotting. Lysates from bacteria transfected with the SmD gene were run on a 15% SDS gel under reducing conditions and transferred onto nitrocellulose. The filters were probed with a control monoclonal antibody (**lane 1**); with the IgM (**lanes 2 to 5**) and with the IgG (**lane 6**) monoclonal anti-EBNA antibodies. The upper band corresponds to the molecular weight of the recombinant antigen.

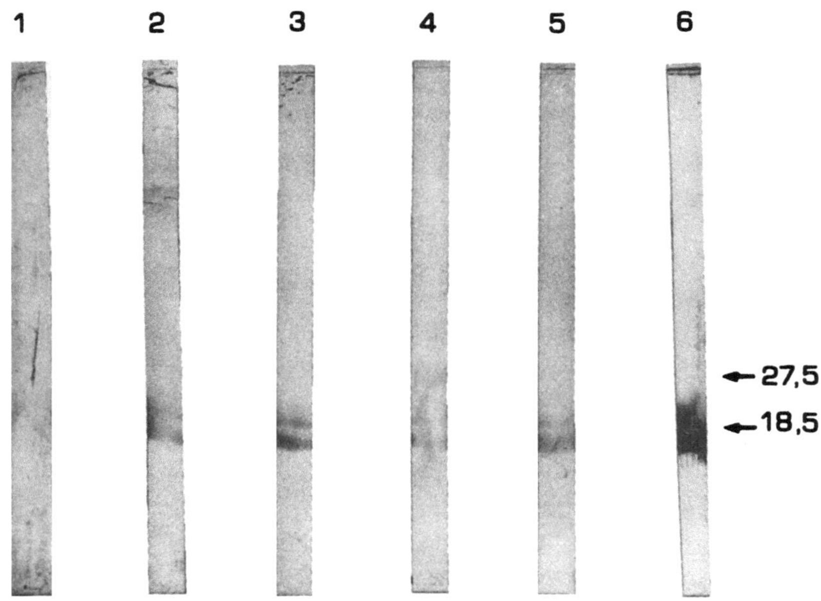

Conclusion

The cellular circuits involved in the production of anti-SmD antibodies are still unclear. Preliminary evidence indicates the presence on SmD of multiple T cell epitopes that can associate with different MHC haplotypes. These T epitopes are located in the central portion of SmD and do not overlap with the C terminal dominant B cell epitope. We can now try to connect all these findings in a unifying hypothesis for anti-SmD antibody production. In some individuals, viral infection might expand B cells specific for a sequence of EBNA I which is crossreactive with SmD. The EBNA 1 specific B cells might use their surface immunoglobulins to uptake SmD protein normally released through cell turnover, process and present it to SmD specific T cells. These T cells might help and induce expansion of B cells specific for SmD. Anti-SmD antibody production, initially elicited by virus-specific "crossreactive" B cells, helped by virus-specific T cells, would then be maintained by autoantigen-specific T cells. After EBV clearance, the autoantigen which is released in higher amounts in sites of inflammation would sustain autoantibody production.

References

Baboonian C, Venables PJW, Williams DG, Williams RO, Maini RN (1991) Cross reaction of antibodies to a glycine/alanine repeat sequence of Epstein-Barr virus nuclear antigen-1 with collagen, cytokeratine, and actine. Ann Rheum Dis 50: 772-775

Birkenfeld P, Haratz N, Klein G, Sulitzeanu D (1990) Cross-reactivity between the EBNA I p107 peptide, collagen, and keratin: implication for the pathogenesis of rheumatoid arthritis. Clin Immunol Immunopathol 54: 14-25

Dighiero G, Lymberi P, Holmberg D, Lundquist T, Coutinho A, Avrameas S (1985) High frequency of natural autoantibodies in normal newborn mice. J Immunol 134: 765-771

Dillner J, Sternas L, Kallin B, Alexander H, Ehlin-Henriksonn B, Jornvall H, Klein G, Lerner L (1984) Antibodies against a synthetic peptide identify the Epstein-Barr virus-determined nuclear antigen. Proc.Natl.Acad.Sci.USA. 81: 4652-4656

Guldner HH, Netter HJ, Szostecki C, Jaeger E, Will H (1990) Human anti-p68 autoantibodies recognize a common epitope of U1 RNA containing small nuclear ribonucleoprotein and influenza B virus. J Exp Med 171: 819-829.

Henle WG, Andersson J, Ernberg I, Klein G, Horwitz CA, Marklund G, Rymo L, Wellinder C, Straus SE (1987) Antibody responses to Epstein-Barr virus-determined nuclear antigen (EBNA)-I and EBNA-2 in acute and chronic Epstein-Barr virus infection. Proc Natl Acad Sci USA 84: 570-574

Hennessy K, Kieff E (1983) One of two Epstein-Barr virus nuclear antigens contains a glycine-alanine copolymer domain. Proc Natl Acad Sci USA 40: 5665-5669

Klein G (1989) Viral latency and transformation: the strategy of Epstein-Barr virus. Cell 58: 5-8

Naparstek Y, André-Schwartz J, Manser T., Wysocki L.J., Breitman L., Stollar B.D., Gefter M., Schwartz RS (1986) A single germline VH gene segment of normal A/J mice encodes autoantibodies characteristic of systemic lupus erythematosus. J Exp Med 164: 614-626.

Prabhakar BS, Segusa J, Onodera T, Notkins AL (1984) Lymphocytes capable of making monoclonal autoantibodies that react with multiple organs are a common feature of the normal B cell repertoire. J Immunol 133: 2815-2820.

Puccetti A, Koizumi T, Migliorini P, André-Schwartz J, Barrett KJ, Schwartz RS (1990) An immunoglobulin light chain from a lupus prone mouse induces autoantibodies in normal mice. J Exp Med 171: 1919-1930

Rhodes G, Carson DA, Valbracht J, Houghten R, Vaughan JH (1985) Human immune responses to synthetic peptides from the Epstein-Barr nuclear antigen. J Immunol 134: 211-216

Rhodes G, Carson DA, Valbracht J, Houghten R, Vaughan JH (1985) Human immune responses to synthetic peptides from the Epstein-Barr nuclear antigen. J Immunol 134: 211-216

Rokeach LA, Haselby J A, Hoch SO (1988) Molecular cloning of a cDNA encoding the human SmD autoantigen. Proc.Natl.Acad.Sci.USA. 85: 4832-4836

Sabbatini A, Bombardieri S, Migliorini P (1993) Autoantibodies from patients with systemic lupus bind a shared sequence of SmD and Epstein-Barr-virus encoded nuclear antigen EBNA I. Eur J Immunol 23: 1146-1152

Sabbatini A, Dolcher MP, Marchini B, Bombardieri S, Migliorini P (1993) Mapping of epitopes on the SmD molecule: the use of multiple antigen peptides to measure autoantibodies in systemic lupus.
J Rheumatol, in press.

The Expression of Heat Shock Proteins in Systemic Lupus Erythematosus

Veena B. Dhillon[1,2], David S. Latchman[1] and David A. Isenberg[2]
Division of Molecular Pathology and [2]Department of Rheumatology Research, University College and Middlesex School of Medicine, London [1]W1P 6DB and [2]W1P 6PG, U.K..

INTRODUCTION

Heat shock proteins

The heat shock response (HSR) was discovered in *Drosophila buksii* over 30 years ago [Ritossa 1962], and subsequently characterized by Tissières et al [1974] who showed that it consisted of the rapid synthesis of a small number of new proteins - the heat shock proteins (hsps) - against a background of repressed general protein synthesis, upon exposure of organisms to a variety of insults. These include hypoxia, glucose deprivation or supplementation, amino acid analogues, oxygen free radicals, cytokines, heavy metals, calcium ionophores, ethanol and many infectious pathogens [Linquist 1986]. The common denominator appears to be the accumulation of unfolded or malfolded proteins in the cell [Gething and Sambrook 1992]. The HSR has been observed in all cell systems, tissues and organisms studied. There is a high degree of homology of hsps phylogenetically. They thus appear to have an important role to play in cell survival. This is borne out by the fact that both the protein-coding and the regulatory sequences of some heat shock genes have been very highly conserved throughout evolution. However, while the HSR is ubiquitous, it is also finely tuned and versatile.

Some hsps are expressed to high levels constitutively, when they have roles to play in normal cellular functioning, growth and development. With the delineation of these roles, it has become possible to explain the initial view that

NATO ASI Series, Vol. H 80
Autoimmunity: Experimental Aspects
Edited by Moncef Zouali

hsps help to renature proteins that have been denatured by heat [Pelham 1988]. There is considerable evidence to show that constitutively expressed hsps are involved in resistance to heat shock [Bansal et al 1991, Angelidis et al 1991]. Hsps are classified into families based upon their molecular weight (MW). In mammals,. the important families are the approximately 90kD hsps (hsp90), the approximately 70kD hsps (hsp70), the approximately 60kD hsps (hsp60), the 22 to 32kD hsps (small hsps), and ubiquitin (MW 7 to 8kD). Many hsps share structural and biochemical properties, and it is possible that all hsps evolved from a few genes [Welch 1987]. Table 1 provides a summary of the main hsp families and their principle functions.

Family	Members	Function/localization
Hsp90	Hsp90a	Most abundant constitutive hsp.
	Hsp90ß	Hsp90 more inducible than hsp90ß,Cytoplasmic anchor for target proteins. Maintains steroid receptors, Ah receptor, pp60-src and other kinases inactive until appropriate. May have other functions in cell cycle regulation, development, immune function. Binds ATP, no known ATPase activity.
	GRP94	Similar functions to hsp90, within Golgi and at plasma membranes.
	Hsp110	Association with rRNA upon heat shock within nucleolus.
Hsp70	Hsp72	Hsp72 largely inducible, hsp73 largely constitutive.
	Hsp73	Both members located in cytoplasm/nucleus. Protein binding/transport/folding/assembly/turnover. Cell-cycle regulation. Binds ATP, strong ATPase activity.
	GRP78	ER and lysosomes. High basal levels in secretory cells. Protein import/glycosylation/assembly.
	GRP75	Mitochondria. Protein import/folding/assembly.
Hsp60		Mitochondria. Present as large oligomers. Assembly of macromolecular protein structures. Cellular respiration. Binds ATP, weak ATPase activity.
Small hsps	Hsp28	Large oligomer in peri-nuclear cytoplasm and Golgi. Embryogenesis, development, proliferation, cell transformation.
	Ubiquitin	Cytoplasm/nucleus. Facilitates protein degradation by ATP-dependent conjugation. Association with histone H2A - possible role in gene regulation.

Table 1: The major mammalian hsps

The functions of individual hsp families are interlinked. There is the sequential action of the hsp70 family - facilitating transport of polypeptides into mitochondria - followed by assembly by hsp60 of protein complexes [Manning-Krieg et al 1991]. Similarly, there may be a synergistic role for the involvement of hsp90, the hsp70 family and a 56kD hsp (hsp56) in the function of the glucocorticoid receptor (GR) and pp60-src [Sorger 1991, Whitelaw et al 1991]. Hsp90 may also be involved in the response to immunomodulatory drugs - as suggested by the association of the 59kD immunophilin with the inactive hsp90-GR complex [Tai et al 1992].

The HSR is an extremely complex phenomenon [Lindquist 1986]. Study of the HSR has led to the concept of "thermotolerance" (TT), whereby cells exposed to moderately elevated temperatures are more likely to survive when exposed to further, higher temperatures than those which have not been pre-exposed. This tolerance extends to other forms of stress and cross-tolerance to different stresses may develop [Jenkins et al 1988]. Hsps are intimately involved with other mechanisms which enable cells to cope with stress, including DNA repair and anti-oxidant systems, and can be regarded as the common theme in the cellular defence network. Thus, hsps might also be expected to have an important role in cellular growth and differentiation, and evidence for this comes from studies of embryonic and tumour cell lines [Gruppi et al 1991, Kohda et al 1991]. Control of the expression of hsps may occur at the transcriptional, post-transcriptional or translational levels [Schlesinger 1990]. There is a common regulatory mechanism for the transcription of all (constitutive and inducible) hsp genes in eukaryotes [Lindquist 1986]. Hsp gene sequence data show the presence of specific nucleotide sequences upstream of the transcription start sites which are essential for induction. These heat shock elements (HSEs) bind the heat shock transcription factors (HSFs) which activate hsp gene transcription [Burdon 1987].

Hsps and immune function

In view of what is known and has been postulated regarding their constitutive functions, it can be envisaged that hsps may play a vital role in antigen processing and presentation [Pierce et al 1991]. There is already some evidence for this [Flynn et al 1989, Vanbuskirk et al 1989]. Further evidence for the involvement of hsps in immune recognition comes from the mapping of hsp genes to regions within the major histocompatibility complex (MHC) in the human, mouse, goat and rat [Günther 1991]. Despite these findings, it is not yet clear that linkage of genes with the MHC implies that their products are functionally related. Perhaps the MHC region could be envisaged as the meeting place for genes encoding the members of many different cell defence systems for example, specific immunity, complement, inflammation and the HSR.

There is considerable evidence that hsps are major antigenic determinants of pathogenic bacteria and parasites. They play an important role in both the humoral and cell-mediated limbs of the immune system in this respect [Young and Mehlert 1990, Hoffman et al 1990]. An important subset of T cells which are stimulated by mycobacterial hsps are those bearing T cell receptors (TCR) composed of the disulphide linked g/d chain heterodimer (g/dTCR), and which are CD4/CD8-negative. It has been postulated that these cells, by virtue of their cytolytic properties, exert an important surveillance function both phylogenetically and ontogenetically [Janeway et al 1988, Born et al 1990a]. Since hsps are similarly highly conserved, and since they are induced by a wide variety of insults including infection, it would seem reasonable to suggest that they are the antigens recognized by these g/d T cells, and there is evidence to support this [Born et al 1990b]. There is also evidence that amongst these g/d T cells are a significant number with restricted g and d chain usage which are directed against shared epitopes of mammalian, mycobacterial and *E. coli* hsps [Koga et al 1989, Rajesekar et al 1990, Fisch et al 1990]. Such reactivity may resemble the massive proliferative responses produced by superantigens [Janeway 1991].

Research into hsp expression in the host as a result of intracellular parasitic infection has been neglected, although it seems reasonable to envisage that the variety of stress stimuli imposed upon the host at such a time would result in induction of host hsps. Since viruses do not themselves carry genes encoding hsps, much more information on host (self)-hsp expression is available from studies of viral infection. Thus, heat shock has a role in the reactivation of some latent viral infections [Zerbini et al 1985, Geelen et al 1988]. Cellular heat shock gene induction and protein synthesis is highly specific with respect to both the genes induced and the infecting virus [LaThangue and Latchman 1988, Phillips et al 1991]. These studies also demonstrate that cell surface expression can occur as a consequence of over-expression of self-hsps. Such findings have important implications for the role of recognition of self-hsps in immunity and autoimmunity.

It would also seem reasonable to assume that hsps in the context of fever play a key role, but most studies of the heat shock response have involved subjecting cells and tissues to temperatures higher, and for shorter periods, than those often seen during febrile episodes. Thus this question has only recently been addressed in *in vivo* studies [Polla and Kantengwa 1991, Ciavarra and

Simeone 1990]. There is considerably more evidence to support the notion that hsps play a pivotal role in enhancing the inflammatory response through their interactions with inflammatory mediators [Chang et al 1991, Hansen et al 1991]. In addition, Hansen et al [1991] showed that there is very fine tuning of hsp expression which occurs at the level of hsp gene transcription in unstressed cells. Indeed, the kinetics of mRNA expression in this study suggested that hsp73 and hsp90 gene expression in human T cells are driven by an interleukin-2 (IL-2)/IL-2 receptor-dependent pathway. In summary, perturbation of hsp expression, by whatever mechanism, might be expected to have a profound, and possibly deletrious, effect upon the immune system.

Hsps and autoimmune disease

There are a number of connected ways in which hsps may be involved in the generation of autoimmune disease. The simplest mechanism involves the concept of molecular mimicry. Thus, since g/d T cells reactive with microbial hsps are positively selected for in the neonatal thymus, and since hsps are immunodominant antigens of microbes, infection might act as a trigger to the development of autoimmune disease by stimulating the over-expression of homologous host and foreign hsps, with resultant reactivity to self-hsps. Self-hsps are induced by a variety of stresses, any one of which could trigger the onset of autoreactivity. An extension of this hypothesis suggests that hsps may act as superantigens in diseases such as SLE [Friedman et al 1991]. The occurrence of autoimmune diseases in only some individuals implicates the involvement of other factors, for example HLA haplotype. Once the cascade of autoimmune tissue damage is in motion, it may be fuelled by continued overexpression of hsps. However, the ubiquity of hsps does not explain the organ specificity of some autoimmune diseases, unless there is differential overexpression of one or more hsp in different organs. Alternatively, autoimmune disease could arise as a result of immune dysregulation, perhaps at the level of hsp gene expression. If such alterations in hsp gene expression were tissue-specific, this might help to explain the organ specificity of some autoimmune diseases. Genetic upregulation of hsp expression could also give rise to autoimmune disease by stimulating autoreactive T cell clones.

A role for hsps in the pathogenesis of autoimmune diseases such as insulin-dependent diabetes and reactive arthritis has been suggested [Elias et al 1990, Keat et al 1989]. Many *in vitro* and *in vivo* studies exploring the role of hsps in rheumatoid arthritis (RA) have also been reported - in particular demonstrating T cell proliferative responses to the mycobacterial hsp60 homologue, and transfer of arthritis or protection against disease by individual T cell clones, in animal models of RA including adjuvant arthritis [van Eden et al 1988, 1989], pristane-induced arthritis [Thompson et al 1990] and streptococcal cell wall-induced arthritis [van den et al Broek 1989]. The T cell-reactive mycobacterial peptide has been loalized to amino acid position 180-188, and was shown to confer protection against all forms of experimentally-induced arthritis mentioned above [van Eden et al 1988, Yang et al 1990]. Moreover, a related peptide containing conserved sequences of mycobacterial hsp60 (M. hsp60) (amino acid position 180-196) stimulated murine neonatal thymus derived g/d T cell hybridomas [O'Brien et al 1989].

The relevance of these findings to human rheumatoid disease is not clear. However, g/d T cells isolated from human RA synovium have increased reactivity to M. hsp60, in particular to the 180-188 nonapeptide, in an MHC unrestricted fashion [Holoshitz et al 1989]. Cytoplasmic staining of autologous hsp60 in the cartilage/pannus junction and in rheumatoid nodules of patients with RA and rats with experimental arthritis has been demonstrated [Karlsson-Parra et al 1990]. Such staining, though, is not in excess of that expected from the disorganized cellular architecture associated with disease [Sharif et al 1992]. Bright surface staining of macrophages from RA joint fluid of human hsp70 and hsp90 has also been demonstrated [Kaufmann et al 1990]. IgG antibody titres to M. hsp60 are significantly elevated in RA [Tsoulfa et al 1989a], with hsp- and species-specificity [Tsoulfa et al 1989b]. However, it is difficult to isolate T cell clones from synovial fluid or tissue of RA patients which responds to M. hsp60. When such reactivity does occur, it appears to be an atypical response in unusual patients, and in these cases it is often difficult to map the antigenic peptide [Dr H. Gaston - personal communication]. In other cases, there is often considerable background reactivity to the *E. coli* hsp60 homologue [Gaston 1991]. This may be related to the problem of the purity of preparations of human hsp60 which have been expressed in *E. coli*, and then tested for their ability to stimulate T cells. Furthermore, the frequency of hsp60-reactive T cell clones is at least as common in reactive arthritis, suggesting that it is not a disease-specific response [Life et al 1991]. Thus, although there is considerable *in vitro* evidence for the role of hsp60 in adjuvant arthritis, it is difficult to implicate hsp60 directly in the pathogenesis of RA.

Hsps and systemic lupus erythematosus

Deguchi et al [1987] demonstrated increased synthesis (occurring at the level of gene transcription) of two proteins of MW 70kD and 85kD in pooled peripheral blood mononuclear cells (PBMC) from 20 patients with systemic lupus erythematosus (SLE) without significant fever compared with those from healthy age and sex matched volunteers. Deguchi and Kishimoto [1990] used nuclear run-on assays to demonstrate that PBMC from five patients with SLE had spontaneously higher rates of transcription of the hsp70 gene, compared with asthmatics on high dose oral prednisolone and healthy volunteers. Levels of hsp90 in PBMC of 15% of SLE patients (n=57) were found to be raised compared with normal controls, in studies performed in our laboratories [Norton et al 1989]. A disease control group of patients with RA was included, and none of these 12 patients had elevated hsp90. However, the age range of the RA group was higher (52-66 years) than that of the healthy control or SLE groups (20-67 and 21-50 years respectively). Norton et al [1989] found the highest hsp90 levels in SLE patients with active disease, although not all individuals with active disease had high hsp90 levels. Neither the presence of fever nor corticosteroid therapy correlated with elevated hsp90 in this study. Deguchi [1991] found a ten-times greater transcription rate of an hsp70 gene in kidney lymphoid cells from MRL/*lpr-lpr* mice compared with those from control (MRL/+-+) mice. This elevation was enhanced during the growing period, and appeared to correlate with a deterioration in renal function, and with the development of renal lesions in the lupus-prone mice.

Minota et al [1988a] found that 38% of SLE patients had IgM and 15% had IgG anti-hsp73 autoantibodies. They were also detected in 13% of patients with childhood SLE, and 10% of patients with infectious mononucleosis; they were not found in patients with RA, ankylosing spondylitis (AS), acute hepatitis A, or in healthy volunteers. They also cross-reacted with the peptide PBP72/74 described by Vanbuskirk et al [1989] and [Winfield and Jarjour (1991). The same group also demonstrated the presence of autoantibodies to hsp90 in 50% of SLE patients, and 33% of polymyositis patients, but not in patients with RA or scleroderma nor in healthy volunteers [Minota et al 1988b]. These workers subsequently reported rather different results using a similar non-quantitative method of immunoblotting: they described a much lower frequency of these autoantibodies [Jarjour et al 1991]. The presence of antibodies to hsp60 and hsp70 in SLE has also been investigated in 18 patients, albeit as a disease control, in studies by Tsoulfa et al [1989a, 1989b]. 78% of SLE patients had raised IgA antibody levels to recombinant M. hsp70, and 72% had raised IgA levels to recombinant human hsp70. Levels of IgG antibodies to recombinant M. hsp70 and *E. coli* hsp70, and IgA antibodies to recombinant M. hsp60 and *E. coli* hsp60 were also raised in the SLE group, but to a lesser extent. Muller et al [1988] demonstrated that nearly 80%

of lupus patients had anti-ubiquitin antibodies compared with 3% of normals, whereas only 55% of them possessed anti-dsDNA antibodies. Seventy-four percent of these patients had antibodies to a synthetic ubiquitin peptide fragment, compared with 5% of normals and 16% or less of various disease controls. Results from serial bleeds in individual patients showed that high levels of anti-ubiquitin antibodies often correlated inversely with active SLE and anti-dsDNA antibody levels in individual patients [Muller et al 1990]. Furthermore, histone and ubiquitin have been detected in immune complex deposits in human and murine lupus nephritis biopsies [Batsford et al 1992]. The nuclear autoantigen Ki has amino acid homology with the 90-100kD tumour antigen which is expressed in the nuclei of SV40 transformed cells, and which is related to the viral SV40 T antigen [Anderson et al 1977]. This tumour antigen may also be structurally related to hsp90 [Srivastava and Maki 1991, Takasaki et al 1992]. Autoantibodies to Ki in SLE patients have been detected, and these have been shown to be to an epitope which is highly homologous to part of the SV40 T antigen [Takasaki et al 1992]. If the presence of autoantibodies to hsps in SLE can be correlated with overexpression of hsps, a mechanism by which the former occurs might be the expression of these proteins on the cell surface. Minota and Winfield [1987] showed that a 90kD protein, subsequently identified as hsp90, is a lymphocyte surface membrane target of anti-lymphocyte antibodies found in SLE. Immunofluorescence has shown that hsp90 remains intracytoplasmic, with staining along the periphery of the cell adjacent to the plasma membrane [Koyasu et al 1986].

METHODS

Pattern of hsp expression in SLE

We wished to establish that hsp90 is elevated in SLE blood samples compared with normal controls - and to determine whether this reflects general induction of hsps. This was achieved by quantitating levels of the other hsps in a representative selection of the same SLE and control samples in which hsp90 had been quantitated. Levels of hsps90, 72, 73 and 60 in PBMC isolated from heparinized venous blood were quantitated by a combination of sodium dodecyl sulphate polyacrylamide gel electrophoresis and Western blotting, followed by scanning densitometry, as described in detail elsewhere [Norton et al 1989, Dhillon et al 1993a and 1993b].One hundred and two SLE patients were studied, median age 39 (range 17-70), 97 female and five male, diagnosed according to the American Rheumatism Association (ARA) 1982 revised criteria for the classification of SLE [Tan et al 1982]. Samples of venous blood were taken during morning outpatient attendances at the Bloomsbury Rheumatology Unit from 1987 to 1992. Fifty-nine normal control subjects were studied, median age 29 (range 16-65), 47 female and 12 male, who were healthy hospital and laboratory staff volunteers. Blood samples were taken in the morning. Patient and control groups had roughly equal ethnicity: 67% of each group were White Caucasians, the remainder were Non-White - Indian, Afro-Caribbean, Chinese, Philipino or of mixed ethnic origin. Hsp90 was quantitated in 222 samples from the SLE patients and in 102-samples from the controls.

Clinical relevance of hsp overexpression in SLE

Several clinical and laboratory parameters were recorded at the time of sampling of patients' blood for this study. Such data was available for 95% of the SLE samples. Disease activity was initially analyzed using the overall UCH/Middlesex scoring system [Morrow et al 1982, 1983]. However, SLE is a multi-system disease with extremely variable expression both within individual patients over a period of time, as well as within populations of patients. We therefore also used the British Isles Lupus Assessment Group's (BILAG) system for analyzing disease activity, which has been validated internationally against other methods of scoring SLE activity, including the UCH/Middlesex scoring system [Gladman et al 1992, Symmons et al 1988]. BILAG defines activity within each of eight disease categories or organ systems: Score A or B denote active disease; C and D denote barely noticeable and no/never any disease respectively (subsequently E has been added to distinguish no previous disease). We compared active (A or B) with inactive (C or D) disease in each of the eight disease categories, for example, active mucocutaneous SLE was compared with inactive mucocutaneous SLE. Thus, we were able to examine the relationship of hsp levels to each of eight individual disease categories in which SLE could be active or inactive in a patient at the time of sampling.

Hsp expression in relation to other variables in SLE

The influence of glucocorticoids upon hsp expression was studied. The relationship between hsp overexpression and the following serological features was also examined: ANA and anti-dsDNA antibody titre, anti-Sm, anti-La, anti-Ro and anti-RNP antibodies - (Department of Immunology, University College and Middlesex School of Medicine (UCMSM)); erythrocyte sedimentation rate (ESR), thrombocytopaenia (platelet count of less than $150 \times 10^9/l$), lupus anticoagulant (LAC) - (Department of Haematology, University College Hospital);

anti-cardiolipin antibody (ACL) - (Department of Rhematology Research, UCMSM); HLA allo- and haplotype - (Professor R. Batchelor, Hammersmith Hospital, London).

Disease specificity of hsp overexpression

Patient details are given in Table 2. Except where stated otherwise, patients were attending the Bloomsbury Rheumatology Unit. As with the SLE and normal control groups, hsp90 was quantitated in all subjects; the other hsps were quantitated in a selection of the same samples, the limiting factor being the amount of protein sample available after hsp90 had been quantitated.

Group	n	Age	Median Range	Sex F	M
Dermato/polymyositis (D/PM) *[Bohan and Peter 1975]*	12	32	25-67	6	6
Primary Sjögren's (SJ) *[Isenberg et al 1984]*	15	51	17-71	15	0
Systemic sclerosis (SS) *[Masi et al 1981]*	11	35	22-71	9	2
Behçet's syndrome (BS) *[Lehner and Barnes 1979]*	10	45	39-63	7	3
Rheumatoid arthritis (RA) *[Arnett et al 1988]*	17	62	28-77	13	4
Ankylosing spondylitis (AS)	11	45	30-66	2	9
Osteoarthritis (OA)	10	62	39-86	8	2
Multiple sclerosis (MS)	9	41	20-54	7	2
Febrile infection (FI)	11	31	23-89	4	7
Hospital controls (HC)	12	71	21-83	7	5

Table 2 : Details of disease groups and controls

All patients with febrile infection (FI) were bed-bound with a fever of 37.5∞C or greater at the time of sampling, and all were diagnosed as having a definite infectious cause for their pyrexia (septicaemia, bacterial pneumonia, gastroenteritis, dengue, typhoid and tick typhus). The non-febrile hospital controls (HC) largely bed-bound, had no concurrent infection, and had a variety of medical problems (peptic ulceration, coronary artery disease, pulmonary embolism, uncontrolled diabetes mellitus, epilepsy, autonomic neuropathy).

Statistical analysis

The data was not normally distributed. Therefore, either Student's two-tailed t tests for unpaired samples were performed upon $\log_{10}$-normalized data, or two-tailed Mann-Whitney tests were applied to raw data; often both parametric and non-parametric tests were done. For comparisons of groups which contained small numbers, the results of Mann-Whitney testing were supplemented with 95% confidence intervals (CI) for the difference between medians [Campbell and Gardner 1989]. Where appropriate, the chi-squared test, Fisher's exact test and Spearman's rank correlation were also employed. We performed multiple sampling from individuals on separate occasions, in the SLE patients and in normal controls. All samples were included in comparisons of SLE patients with normal controls, and in correlations of hsp levels with variables which were measured on multiple occasions, such as anti-dsDNA antibody titre and ESR. Analysis of data from patients with other diseases was performed by taking the first sample only of each SLE patient or normal control subject in which the maximum number of hsps had been quantitated, to render these groups comparable with the other disease groups, in which only one sample per patient had been taken. This approach was also used in analyses of the SLE samples involving laboratory data which were invariant with time.

RESULTS

Pattern of hsp expression in SLE

Figure 1 shows that hsp90 and hsp72 were significantly elevated in samples from SLE patients compared with those from normal controls. Thirty-two percent of lupus patients studied had elevated hsp90 in at least one sample at some time during the study. There was no correlation between elevated hsp90 and hsp72 in SLE or control samples, nor indeed between levels of any of the hsps [Dhillon et al 1993a].

Clinical relevance of hsp overexpression in SLE

Hsp90 was significantly elevated in samples from SLE patients with severely active compared with inactive or mildly active disease using the UCH/Middlesex scoring system ($P<0.02$, $P<0.04$ respectively - Student's t test on $\log_{10}$-data). There was no difference in mean hsp72 or hsp60 levels across the four UCH/Middlesex activity grades, but mean hsp73 was higher (not significantly so) in inactive and mild lupus compared with more severe disease (data not shown). Figure 2 shows that hsp90 was significantly higher in samples specifically from patients with BILAG active neuro-psychiatric (NP) and cardio-respiratory (CR) lupus compared with inactive disease in these categories. There were no differences in hsp72 or hsp60 levels in active compared with inactive BILAG disease categories, but hsp73 was higher in inactive compared with active categories, significantly so in NP and renal BILAG categories (P<0.02, P<0.025 respectively - Student's t test on $\log_{10}$-data) [Dhillon et al 1993a].

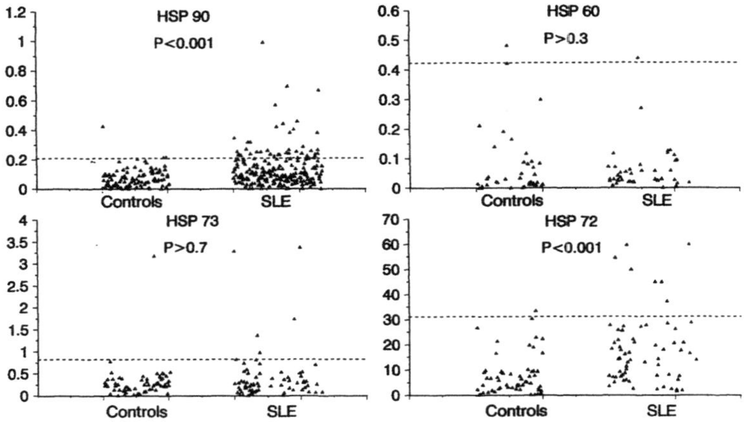

Figure 1: Hsp levels in SLE and normal control samples P values: two-tailed Student's t tests on $\log_{10}$-data. Dotted line: 97.5th centile of controls. Hsp values were calculated relative to a positive control (Jurkat cell line), used as the standard , and equalized for protein loading by scanning the actin band of the corresponding destained gel. Hsp values are expressed in arbitrary units and levels of one hsp are not directly comparable with those of another [Dhillon et al 1993a].

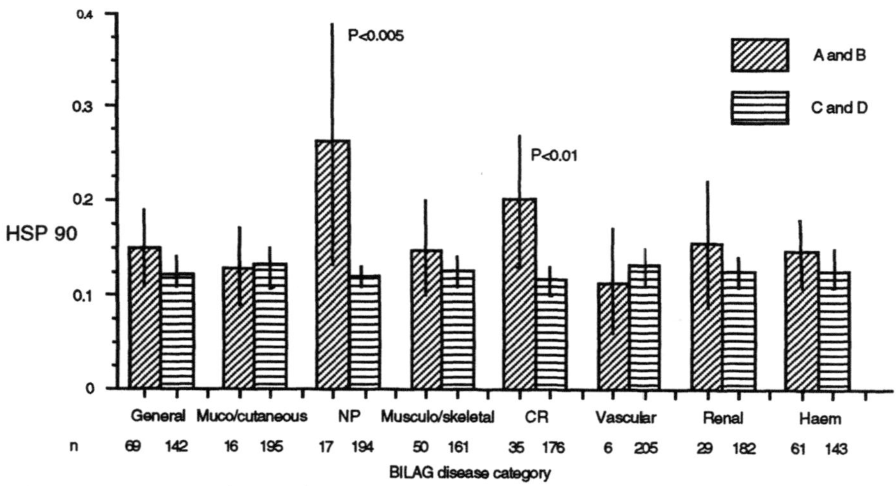

Figure 2: Mean (with 95% CI mean) hsp90 levels in active (A and B) compared with inactive (C and D) BILAG disease categories (P values: results of two-tailed Student's t test on $\log_{10}$-data).

Serial studies of the relationship between changes in hsp90 levels and changes in BILAG activity showed that there were sensitivities of 69% for NP, 64% for haematological (haem), 54% for CR and 56% for renal BILAG categories - where an increase in activity was accompanied by a rise in hsp90 level. In general, there were high negative predictive values, low specificities and low negative predictive values - emphasizing the influence of other, individual factors upon hsp90 levels. There were high sensitivities of 80% for NP, 100% for haem and 67% for renal categories - where a rise in activity was accompanied by a fall in hsp73 levels, although numbers here were rather small.

Hsp expression in relation to other variables in SLE

There were no differences in levels of any hsps in patients on no, low doses or high doses of oral prednisolone. There was no significant correlation between hsp levels and anti-dsDNA antibody titre, or ESR. There were no significant differences in hsp levels in patients who possessed ANA, anti-Sm, anti-Ro, anti-La or anti-RNP antibodies. The level of hsp90, but not the other hsps, was significantly higher in patients who had LAC, ACL or thrombocytopaenia compared with patients who did not have these laboratory features of anti-phospholipid syndrome (APS) (P=0.03, P=0.03, P=0.008 respectively - two-tailed Mann-Whitney U test) [Krilis and Hunt 1992, Alarçon-Segovia 1992]. Figure 3 summarizes these latter findings.

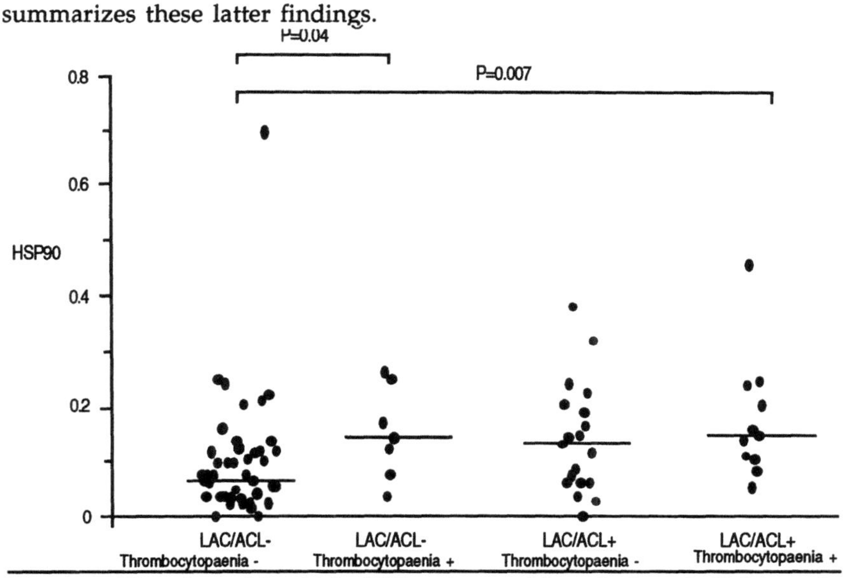

Figure 3: Elevation of hsp90 in SLE patients with anti-phospholipid syndrome (P values: results of two-tailed Mann-Whitney U significance testing).

Figure 4 shows that hsp90 levels, but not levels of the other hsps (data not shown), were influenced by patients' HLA type. Hsp90 levels were higher in patients who did not possess HLA types which have been shown, in this cohort, to be associated with SLE [Worrall et al 1990]. These findings were significant in patients who were negative for the HLA allele DR3 and haplotype A1/B8/DR3, and were independent of patients' ethnic group.

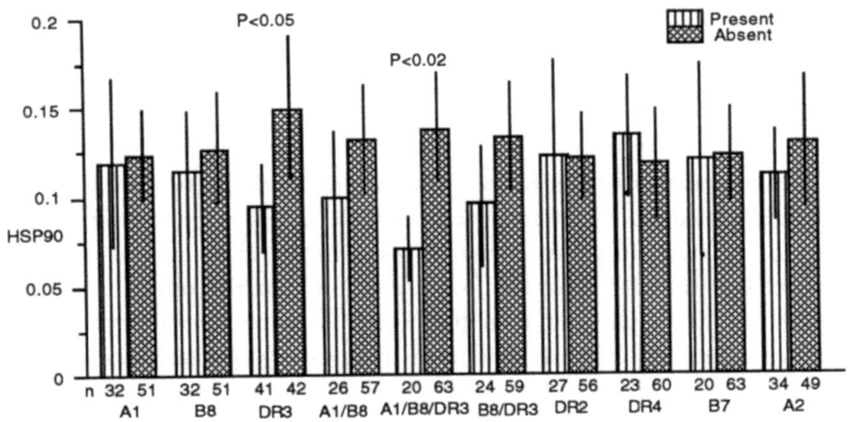

Figure 4 Mean (with 95%CI mean) hsp90 levels with respect to SLE patients' HLA haplotype (P values: two-tailed Students' t test on $\log_{10}$-data).

Disease specificity of hsp overexpression

Figure 5 shows that the overall pattern of overexpression of hsp90 and hsp72 seen in SLE was not seen in the other disease groups, with the possible exception of D/PM. In the whole study population, there were no samples where elevation of one hsp was matched by elevation of another [Dhillon et al 1993b]. We expected all hsps to be overexpressed in the FI patients, but Figure 5 shows that this was not the case, and hsp73 was even underexpressed. Nor did the severity of fever correlate with the degree of hsp elevation (data not shown). Three patients were sampled again after their temperature had been normal for three or more days. Hsp90 did not alter significantly, but in all cases, hsp73 levels rose (data not shown). In several samples from the whole study population, there appeared to be a reciprocal relationship between hsp72 and hsp73 levels, although there was no statistically significant inverse correlation overall between these two hsps (Figure 6).

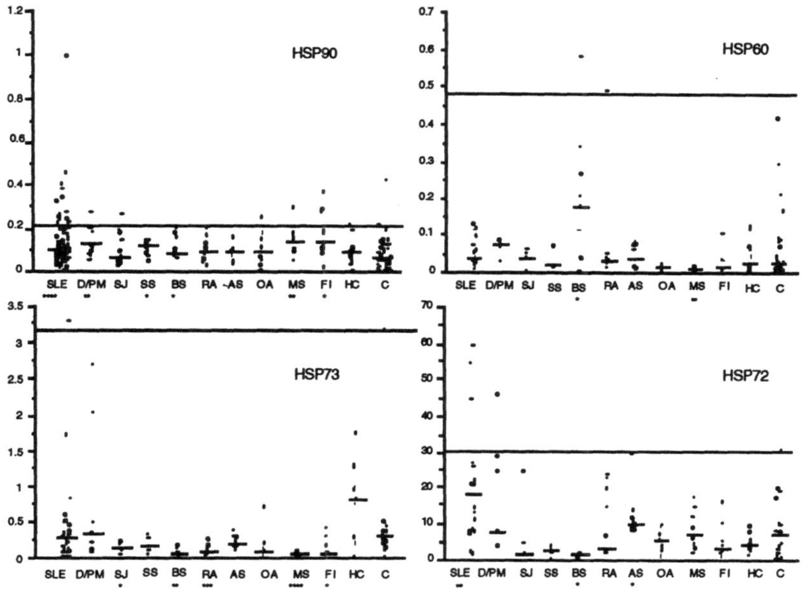

Figure 5 : Scatter plots of hsp levels for all disease groups
(P values - **** P<0.001, *** P<0.005, ** P<0.01, * P<0.05: results of two-tailed Mann-Whitney U tests for differences between disease groups and controls. Dotted line: 97.5th centile of controls. Solid line: median hsp levels.

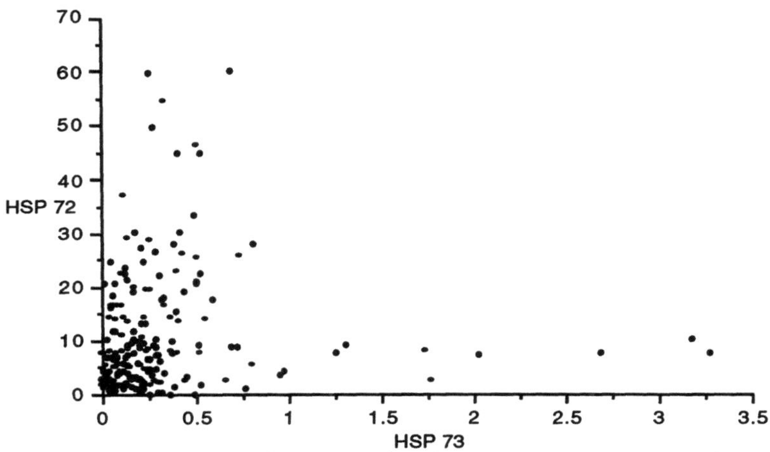

Figure 6: Scatter plot of hsp72 against hsp73 levels in individual study samples

DISCUSSION

Mechanisms of hsp overexpression in SLE

We have demonstrated that hsp90 and hsp72 are differentially overexpressed in SLE, and that all cell-types in individual SLE patients' PBMC are capable of hsp90 overexpression (latter data not shown here) [Dhillon et al 1993a]. It is possible that this overexpression may be occurring at the level of gene transcription. Indeed, we have shown that transcription rates of the hsp90ß gene but not those of the hsp90a gene are significantly elevated in several of the same SLE PBMC samples compared with normal control samples in which hsp90 was quantitated in the study described here [Twomey et al 1993]. Moreover, hsp90ß gene transcription, but not that of hsp90a, correlates significantly with elevated hsp90 protein levels in these samples (r=0.64, P=0.003). Elevation of transcription of the hsp70 gene did occur in some patients, but not in the same patients with elevated hsp90ß transcription, and there was no correlation between hsp90ß transcription and that of the hsp70 gene. Nor was there a correlation between hsp70 transcription and hsp72 protein levels, implying that post-transcriptional mechansims may be responsible for the elevation of hsp72 seen in SLE. Further evidence for the differential overexpression of hsps in SLE comes from our report that transcription rates of genes which encode ubiquitin are not elevated in SLE patients compared with normal controls [Twomey et al 1992]. Our study described in the previous paragraph also provided data confirming this finding, and showed that there was no correlation between ubiquitin gene transcription and that of the hsp90 genes or of the hsp70 gene. Hansen et al [1991] showed that hsp90 gene expression in human T cells is driven by an IL-2/IL-2 receptor-dependent pathway, and Mouzaki et al [1992] showed that the IL-2 gene "transcriptional silencer" is inactive in lupus T cells. Interestingly, although hsp90a and hsp90ß genes are both upregulated by heat shock and mitogen stimulation, hsp90a is induced to a greater degree by heat shock, and hsp90ß is more inducible by mitogen stimulation [Hansen et al 1991]. Thus, the upregulation of hsp90ß that we have observed in SLE is likely to be mediated via cytokine stimulation, or it may be due to a primary genetic aberration.

Further evidence in support of our findings, and for a primary genetic aberration of hsp expression comes from studies in MRL/*lpr-lpr* mice which have been performed in our laboratories [Faulds et al 1993]. Hsp90 and hsp72, but not hsp73 or hsp60, have been shown to be elevated in the spleens of these mice, compared to MRL/+-+ mice (which develop no or very mild disease when much older than six months) or BALB/c mice. This elevation in hsp levels in the

MRL/*lpr-lpr* mice occurred prior to the onset of overt clinical or serological disease (as judged by elevated anti-DNA autoantibody titres), at two months of age. There appeared to be different expression patterns for hsp90 and hsp72. Thus, the hsp90 elevation in spleen was shown to occur in a tissue-specific manner, in contrast to that of hsp72 which occurred in several different tissues including heart, brain, kidney and liver. The primary genetic defect in MRL/*lpr-lpr* mice is the inactivation of the gene encoding the Apo-1/Fas antigen [Watanabe-Fukanaga et al 1992]. This results in a decreased rate of apoptosis, which is mediated by this antigen in lymphoid cells leading to the observed lymphoproliferation in these mice. A report from our laboratory has recently shown that hsp overexpression can protect cells from apoptosis [Mailhos et al 1993]. Thus, it is possible that hsp90 elevation in SLE may play a role in the pathogenesis of the disease by enhancing the survival of lymphoid cells in this manner, and that the inactivation of the Apo-1 antigen may partially mediate its inhibitory effect upon apoptosis via hsp90 overexpression.

Surface expression of hsps in SLE

One event linking overexpression of cellular hsps to the breakdown of tolerance would be the expression of hsps on the cell surface. We have shown that surface expression of hsp90 does indeed occur on lymphocytes and monocytes of patients with SLE, compared with normal controls ($P<0.006$), using MoAb AC88 and fluorescein activated cell sorting (FACS) analysis [Erkeller-Yüksel et al 1992]. Monocytes tended to have a higher level of surface expression of hsp90 (shsp90) than lymphocytes. There was an heterogenous pattern of shsp90 expression in different cell types - occurring on monocytes, B cells, CD3+ T cells and CD4+, but not on CD8+ T cells. This parallels our reported findings of a heterogenous pattern of hsp90 expression in cell-types fractionated from PBMC of individual lupus patients [Dhillon et al 1993a]. In addition, there was no surface expression of hsp60 in SLE patients, supporting our findings regarding hsp60 levels in SLE. There also appeared to be two groups of SLE patients: those in whom lymphocyte shsp90 expression did not occur, and those in whom it did occur. In the latter group there appered to be a correlation with hsp90 levels ($r=0.5$, with regression analysis).

There are several reasons why hsp90 might occur at the cell surface. Hsp90 may be involved directly in antigen presentation, or indirectly by chaperoning other proteins to the cell surface. Jardetsky et al [1991] have shown that peptide products of both hsp90a and hsp90ß genes are among several endogenous proteins whose peptides are bound to HLA-B27 in a lymphoblastoid cell line.

However, the surface expression of the whole hsp90 molecule might need to be demonstrated, if functional hsp90 is to be implicated in antigen presentation. It is possible that hsp90 peptide products are good candidates for being bound to HLA molecules simply because hsp90 is an abundant cellular protein. Alternatively, hsp90 released from damaged cells binds to viable cells, and it is the viable cells that are detected by FACS analysis. Experiments are underway to distinguish these possibilities using internally-labelled hsp90 and immunoprecipitation techniques. Another means by which breakdown of tolerance might occur as a result of overexpression and/or surface expression of hsps might be the development of anti-hsp autoantibodies. There is already some evidence for this (see Introduction), and it is currently the object of further investigation in our laboratories, with studies being performed upon serum from the same blood samples in which hsps were quantitated in this study, and in the shsp90 study.

Our findings of differential overexpression of hsps in SLE are reflected in *in vitro* studies performed by Ferrarini et al [1992], who compared several human tumour cell lines with normal cells, and found that most tumour cell lines constitutively express 2-10 times higher levels of hsp90. In the absence of stress, two out of six lines also overexpressed hsp72. This hsp72 was predominantly nuclear and, in contrast to normal cells, it was not cell-cycle regulated. Hsp90 and, to a lesser extent, hsp72 were localized at the cell surface. In immunoprecipitation experiments, it appeared possible to co-precipitate hsp90 with MHC I molecules, suggesting that cell surface expression of hsp90 may indeed be occurring.

In summary, a criticism of efforts to test the hypothesis that hsp overexpression may play a fundamental role in autoimmunity is that one might expect all hsps to be overexpressed in concert merely as a result of the ongoing disease process - a "stressful" situation. The evidence presented here suggests that this is not the case.

Clinical relevance of hsp overexpression in SLE

We have shown that hsp90 is significantly elevated in severely active SLE, in particular active NP and CR lupus [Norton et al 1989, Dhillon et al 1993a]. Our study of shsp90 also showed significantly higher lymphocyte shsp90 in patients with active disease (BILAG global score >6, P<0.009) [Erkeller-Yüksel et al 1992]. Hsp73 is higher in inactive and mild SLE, significantly so in inactive NP and renal lupus. Underexpression of hsp73 in active disease, as well as hsp73 overexpression in inactive disease may be equally important (data not shown). In contrast, there is no relationship between hsp72 levels and overall disease activity or BILAG disease activity. Serial studies showed that the sensitivities of shifts in hsp90 and hsp73 levels for changes in NP and haem SLE activity, and NP and renal SLE activity, respectively, are comparable to/better than results obtained from many other studies which attempted to correlate laboratory parameters with lupus activity [for example: Morrow et al 1982, Gordon et al 1992, Aderka et al 1992]. The generally poor specificities from our serial studies reflects the influence of other factors upon hsp levels. One of these, as discussed, might be genetic. The lack of correlation between hsp levels and markers of general disease activity such as ESR and anti-dsDNA antibody titres is paralleled by findings from our FACS studies of shsp90 [Erkeller-Yüksel et al 1992]. There was no correlation between shsp90 expression and ESR, anti-dsDNA titres, C3 levels, or indeed with markers of T cell activation such as HLA-DR and IL-2.

The possibility that hsp72 and hsp73 overexpression occurs in general and in quiescent disease states, respectively, has already been alluded to. A correlation between these hsps and the above variables might therefore have been expected. There was an insignificant negative correlation between hsp73 levels and ESR ($r=-0.2$, $P=0.17$). The lack of correlation again might suggest that there are other factors at play which determine hsp expression. Conversely, ESR and anti-dsDNA titres may, in fact, be rather poor indicators of general disease activity. Indeed, the correlation between ESR and anti-dsDNA titres was only modest ($r=0.25$). Interestingly, Faulds et al [1993] found that hsp72 levels in MRL/*lpr-lpr* mice rose with age (which was a reflection of general disease as confirmed by elevated anti-DNA autoantibody titres), and that, in contrast, the elevation of hsp90 was not age-dependent.

There was significant elevation of hsp90 in patients with anti-phospholipid antibodies (assayed by coagulation studies or by ELISA - LAC and ACL, respectively [Krilis and Hunt 1992]). There was also significant elevation of hsp90 in patients with thrombocytopaenia. There was a high frequency of anti-phospholipid antibodies (APA) in patients with active NP SLE, compared with the frequency of thrombocytopaenia in these patients, and compared with the frequency of APA or thrombocytopaenia in patients with active CR SLE (data not shown). Conversely, being of Non-White ethnic origin was significantly associated with elevation of hsp90 in active CR SLE (data not shown). Taken together, our findings suggest that hsp90 is particularly elevated in patients with the anti-phospholipid syndrome. The role for some APA, via their reactivity with the cofactor β_2 glycoprotein-I (β_2GPI), in generating disease by promoting thrombosis is established [Schousboe 1983, Nimpf et al 1986, Cuesta et al 1992]. However, thrombosis in association with ACL in our lupus cohort was rare (7%) [Worrall et all 1990]. Furthermore, there is a wide range of reactivity of APA in

SLE, and binding with ß2GPI is not an absolute requirement [Loizou et al 1990, Pierangeli et al 1992, Domenéch et al 1992]. Thus, if hsp90 is related to the pathogenesis of APS in SLE, it is likely that its role involves factors other than, or additional to, thrombosis. Indeed, Goldstein et al [1992] found that ACL/LAC/positive VDRL were associated with active NP SLE whether or not this was thrombotic.

One mechanism for disease in APS may occur by virtue of a shared antigenicity of the phosphodiester groups which form the core of nucleic acid molecules and phospholipids (in particular, cardiolipin) and with which anti-dsDNA antibodies may react [Morrow and Isenberg 1987]. There is considerable evidence for commonality of anti-endothelial cell antibodies (AECA) and ACL, both in the genetic background driving their production, and in their reactivities [Galeazzi et al 1992, Mackworth-Young et al 1992, Lindsey et al 1993]. Thus, there is increasing evidence to suggest that APA may also contribute to disease by a mechanism other than thrombosis. Hsp90 may be involved in this mechanism, by perhaps sharing an epitope with antigens with which ACL and/or AECA react, in patients who have a genetic aberration which results in the overexpression of this hsp. It would be interesting to study patients with primary antiphospholipid syndrome (PAPS), who do not have SLE, in order to determine the strength of the relationship between hsp90 overexpression and APS in the context of SLE. However, Ehrenstein et al [1993] have recently reported that 53% of patients with PAPS have anti-DNA antibodies, supporting the view that PAPS and SLE are probably overlapping syndromes. With the advent of an anti-hsp90 autoantibody ELISA assay in our laboratories, it would be intriguing to discover whether anti-hsp90 auoantibodies also had APA or AECA activity.

The results of our analysis of hsp90 levels in relation to HLA haplotype are interesting but difficult to explain. Since Batchelor et al [1987] demonstrated that a significant number of DR3-negative SLE patients possess C4 null alleles, and that the link between DR3 and SLE is secondary to the association with complement null alleles, it is possible that hsp90 is overexpressed in SLE patients who suffer from complement deficiency. Alternatively, A1/B8/DR3-negativity and/or DR3-negativity may be linked with the genetic upregulation of hsp90 in subsets of SLE. It is possible that negativity for these allo-/haplotypes may be in linkage disequilibrium with some other, as yet unknown, genetic factor which would predispose to either overexpression of hsp90 or to hsp90 autoreactivity.

Differential expression of hsps in disease

The large range of hsp levels seen in all groups again suggests that variables are also operating which influence the expression of hsps in an individual with a particular disease. This is illustrated by the finding that no two samples have the same pattern of hsp expression. Hsp90 was overexpressed in febrile patients with infections, although there was no direct correlation with the degree of fever. In keeping with *in vitro* evidence, it may be that hsp90 is down-regulated at very high, as opposed to moderately high, temperatures [Yost et al 1990]. There are several reports of *in vitro* studies which demonstrate the specific expression of endogenous hsps in different viral infections or tumour cell lines [La Thangue et al 1988, Phillips et al 1991, Ferrarini et al 1992]. Differential hsp overexpression has also been shown to occur in conditions such as acute lymphoblastic leukaemia and Graves' disease [Strahler et al 1990, Heufelder et al 1991]. Our findings are in concordance with these studies, suggesting that hsps are differentially expressed in different diseases, and not supporting the notion that all hsps are overexpressed merely a result of being ill [Dhillon et al 1993b].

The overexpression of hsp90 in SLE, as well as the proportion of patients with elevated hsp90, was also seen in D/PM (primarily polymyositis). Autoantibodies to hsp90 have also been demonstrated in polymyositis as they have in SLE, but not in systemic sclerosis (SS). In our study there was an overall elevation of the median hsp90 level in SS, but there were no SS samples in which hsp90 was actually elevated beyond the 97.5th centile of the control group [Minota et al 1988b].

That hsp90 is overexpressed in multiple sclerosis (MS) is intriguing, in view of our similar findings in active NP SLE. All of the MS patients studied had "active" disease, but we did not find particular elevation in those who were actually experiencing a relapse at the time of sampling. Perhaps overexpression of hsp90 is related to the pathology of autoimmue NP disease. It has been shown that one of the targets of anti-lymphocytotoxic antibodies (LCA) in SLE is hsp90, bound tightly to the inner surface of the plasmalemma [Minota and Winfield 1987, Koyasu et al 1986]. LCA are sometimes considered to be a marker for NP SLE [Temesvari et al 1983, Bresnihan et al 1979]. One possibility is that there may be cross-reactivity between epitopes of hsp90 and myelin basic protein. Myelin basic protein is recognized by specific T cells in experimental allergic encephalomyelitis, a laboratory model for MS [Vandenbark et al 1989, Cross et al 1990].

That hsp90 overexpression was not seen in the PBMC of rheumatoid arthritis (RA) patients supports the findings of Norton et al [1989]. We also found no overexpression of hsp90 in synovium taken from joints involved in RA, but this did occur in osteoarthritis (OA) (data not shown). Kubo et al [1985] showed that OA chondrocytes overexpress a member of the hsp70 family, although hsp90 expression was not studied in that report. Although 16% of patients in the SLE cohort had (secondary) Sjögren's syndrome, there was no overall overexpression of hsp90 in primary Sjögren's (SJ). In addition, there was no difference in hsp90 expression between SLE patients with or without Sjögren's, compared with normal controls. Thus, SLE patients both with and without Sjögren's had significant hsp90 elevation compared with control subjects ($P<0.01$ Mann-Whitney U test, data not shown). Furthermore, there was no elevation of hsp90 in a small number of Sjögren's histology-positive lip biopsy specimens studied, compared with those which showed chronic inflammatory changes upon histological examination (data not shown). Thus, overexpression of hsp90 does not seem to play an important role in RA or in Sjögren's syndrome. Franz and Grominca-Ihle [1992] reported that although 34% of patients with Sjögren's syndrome have ACL, these antibodies are not associated with APS manifestations. This may be seen to provide further evidence in support of the finding from our study that hsp90 may be elevated in patients with ACL only in the context of SLE, and that this elevation may have a direct bearing upon the disease as a result of the presence of ACL. However, this argument is countered by Franz and Grominca-Ihle's [1992] simultaneous finding that 19% of RA patients also have ACL, and these antibodies are associated with "thrombophilia" (a propensity to have thromboses), sicca symptoms and other extra-articular problems in these patients.

No overall overexpression of hsp72 was seen in those diseases where hsp90 overexpression occurred (with the possible exception of D/PM), in contrast to the findings in SLE. The overexpression of hsp72 seen in SLE has been shown to be unrelated to disease activity, in whatever manner this was analyzed. As with the hsp90 data, there was no difference in hsp72 expression in SLE patients with or without Sjögren's compared with normal controls. Hsp72 was significantly elevated in both cases compared with controls ($P<0.01$ Mann-Whitney U test, data not shown). The underexpression of hsp73 in several disease groups is interesting. It is notable that the only patient in this study who had an elevated hsp73 had SLE, and she had inactive disease in all eight disease categories as analyzed by the BILAG system. It is possible that the stress of ongoing disease sometimes results in the down-regulation of the almost wholly constitutive hsp73, and up-regulation of the almost wholly inducible hsp72. Thus, in general, relative underexpression of hsp73 and overexpression of hsp72 may be non-specific reflections of ongoing disease states. This is supported by the finding of a reciprocal relationship between hsp72 and hsp73 in several individuals. There is further evidence for this inverse relationship from studies in embryonic carcinoma cells which show a reciprocal induction of hsp72 and hsp73 [Mezger et al 1989].

The finding that the Behçet's patients have significant overexpression of hsp60 is an interesting one in view of a report that elevated circulating antibody levels to the streptococcal hsp60 homologue occur in Behçet's syndrome [Lehner et al 1991]. It is notable that changes in hsp60 expression in PBMC in RA in particular did not occur, in view of the implication of hsp60 and the mycobacterial homologue in adjuvant arthritis and RA. Our findings perhaps support the more recent doubts which have been cast upon the role of hsp60 in RA (see Introduction).

Several reports suggest that hsp expression may be impaired in senescent drosophila and rats, as well as in normal human PBMC and upon stimulation of normal human T cells from aged donors [Deguchi et al 1988, Faassen et al 1989, Niedzweski et al 1991, Blake et al 1991]. Our studies suggest that this does not apply in disease states. There was no correlation of hsp levels with age (data not shown), nor was there any overall significant difference in hsp expression in hospital control patients (HC - median age 71) compared with healthy control subjects (C - median age 29). In addition, the highest hsp90 level in any disease group in the study, with the exception of SLE, occurred in an 89 year old man with FI whose hsp90 level, when febrile, was 0.37. However, the upward trend of hsp73 values which occurred in HC, taken together with the impression of down-regulation of hsp73 in several diseases, might be taken to imply that there is some loss in control of the postulated reciprocality of hsp73 and hsp72 with increasing age. Another example of the influence of physiological variables upon hsp expression lies in our finding that hsp90 levels vary with the menstrual cycle in normal women: we have found that there is mid-late follicular dip in hsp90 levels, usually followed by a luteal phase rise (unpublished observations).

CONCLUSION

Figure 7 is a schematic representation of the mechanisms by which hsps may be involved in autoimmune tissue damage. We consider the hypothesis of molecular mimicry likely to be only a small part of the whole picture. It seems much more likely that, as physiologically important molecules, the genetically determined differential expression of hsps (for example, in the case of SLE, the overexpression of hsp90) effects changes in immune function. Alternatively, by virtue of their high degree of conservation and their immunodominant nature, hsps may themselves effect the breakdown of tolerance. This may occur either by the formation of autoantibodies, or by the stimulation of cell-mediated immunity. It is unlikely that the elucidation of the role of hsps in a syndrome as clearly multi-factorial as SLE will have a direct effect upon the development of a more specific treatment for lupus. However, the issues discussed here may help to draw together many of the factors which interact to generate this intriguing disease. In this way, perhaps more rational therapies may be devised.

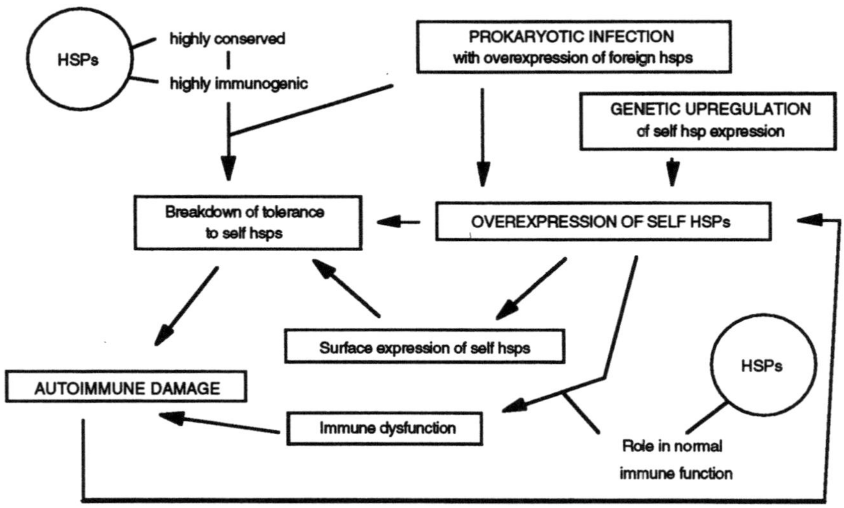

Figure 7: A model for the link between hsps and autoimmunity

ACKNOWLEDGEMENTS

We wish to thank Dr Breda Twomey, Stuart McCallum, Dr David Wilks, Dr Alan Ebringer, Professor Tom Lehner, Dr Jia Newcombe and Dr David Toft for their assistance and collaboration. VBD was a Wellcome Trust Research Fellow.

REFERENCES

Anderson JL, Martin RG, Chang C, Mora PT, Livingston DM. (1977). Nuclear preparations of SV40 transformed cells contain tumor-specific transplantation antigen activity. Virology, 76:420-5

Angelidis E, Lazardis I, Pagoulatos GN. (1991). Constitutive expression of hsp70 in mammalian cells confers thermoresistance. J Biochem FEBS, 199:35-9

Arnett FC, Edworthy SM, Bloch DA. (1988). The American Rheumatism Association 1987 revised criteria for the classification of rheumatoid arthritis. Arthritis Rheum, 31:315-24

Bansal GS, Norton PM, Latchman DS. (1991). The 90-kDa heat shock protein protects mammalian cells from thermal stress but not from viral infection. Exp Cell Res, 195(2):303-6

Batchelor JR, Fielder AHL, Walport MJ. (1987). Family study of the major histocompatibility complex in HLA DR3 negative patients with systemic lupus erythematosus. Clin Exp Immunol, 70:364-71

Blake MJ, Fargnoli J, Gershon D, Holbrook NJ.(1991). Concomitant decline inhyperthermia and hsp70 mRNA expression in aged rats. Am J Physiol, 260:663-7

Bohan A, Peter JB. (1975). Polymyositis and dermatomyositis. New Engl J Med, 292:344-7

Born W, Happ MP, Dallas A, et al. (1990a). Recognition of heat shock proteins and gamma delta cell function. Immunol Today, 11(2):40-3

Born W, Hall L, Dallas A, et al. (1990b). Recognition of a peptide antigen by heat shock-reactive gamma/delta T lymphocytes. Science, 249:67-9

Bresnihan B, Oliver M, Williams B, Hughes GRV. (1979). An anti-neuronal antibody cross-reacting with erythrocytes and lymphocytes in systemic lupus erythematosus. Arthritis Rheum, 22(4):313-20

Burdon RH. (1987). Thermotolerance and the heat shock proteins. Symp Soc Exp Biol, 41:269-83

Campbell MJ, Gardner MJ. (1989). Calculating confidence intervals for some non-parametric analyses. In: Gardner MJ, Altman DG, eds. Statistics with Confidence. Br Med J, London. Pp 71-9.

Chang CC, Konno S, Wu JM. (1991). Enhanced expression of heat shock protein and mRNA synthesis by type I interferon in human HL-60 leukemic cells. Biochem Int, 24(2):369-77

Ciavarra RP, Simeone A. (1990). T lymphocyte stress response. I. Induction of heat shock protein synthesis at febrile temperatures is correlated with enhanced resistance to hyperthermic stress but not to heavy metal toxicity or dexamethasone-induced immunosuppression. Cell Immunol, 129(2):363-76

Cross AH, Dolich S, Raine CS. (1990). Antigen processing of myelin basic protein is required prior to recognition by T cells inducing EAE. Cell Immunol, 129:22-31

Deguchi Y. (1991). Enhanced expression of hsp gene in kidney lymphoid cells of lupus prone mice during growing process. Autoimmunity, 10:1-5

Deguchi Y, Kishimoto S. (1990). Enhanced expression of hsp gene in peripheral mononuclear cells of patients with active SLE. Ann Rheum Dis, 49:893-5

Deguchi Y, Negoro S, Kishimoto S. (1988). Age related changes of hsp gene transcription in human peripheral blood mononuclear cells. Biochem Biophys Res Comm, 157(2):580-4

Deguchi Y, Negoro S, Kishimoto S. (1987). Heat shock protein synthesis by human peripheral mononuclear cells from SLE patients. Biochem Biophys Res Comm, 148(3):1063-8

Dhillon VB, McCallum S, Norton P, et al. (1993a). Differential heat shock protein overexpression and its clinical relevance in systemic lupus erythematosus. Ann Rheum Dis, 52:436-442

Dhillon VB, McCallum S, Wilks D, Twomey B, Latchman D, Isenberg D. (1993b). The differential expression of heat shock proteins in rheumatic disease . Brit J Rheum (In press)

Eglin RP, Lehner T, Subak-Sharpe JH. (1982). Detection of RNA complementary to HSV in mononuclear cells from patients with Behcet's syndrome and recurrent oral ulceration. Lancet, ii:1356-61

Ehrenstein MR, Swana M, Keeling D, Asherson R, Hughes GRV, Isenberg DA. (1993). Anti-DNA antibodies in the primary anti-phospholipid syndrome (PAPS). Br J Rheumatol, 32:362-5

Elias D, Markovits D, Reshef T, van der Zee R, Cohen IR. (1990). Induction and therapy of autoimmune diabetes in the non-obese diabetic (NOD/Lt) mouse by a 65kDa heat shock protein. Proc Natl Acad Sci USA, 87:1576-80

Erkeller-Yüksel F, Isenberg DA, Dhillon VB, Latchman DS, Lydyard PM. (1992). Surface expression of heat shock protein 90 by blood mononuclear cells from patients with systemic lupus erythematosus. J Autoimmunity, 5:803-14

Faassen AE, O'Leary JJ, Rodysill KJ, Bergh N, Hallgren HM. (1989). Diminished heat-shock protein synthesis following mitogen stimulation of lymphocytes from aged donors. Exp Cell Res, 183(2):326-34

Faulds GB, Isenberg DA, Latchman DS. (1993). The tissue-specific elevation in synthesis of the 90kD heat shock protein preceeds the onset of disease in lupus-prone MRL/lpr mice. J Rheumatol. (In press).

Ferrarini M, Heltai S, Zocchi MF, Rugarli C. (1992). Unusual expression and localization of heat-shock proteins in human tumor cells. Int J Cancer, 51:613-19

Fisch P, Malkovsky M, Kovats S, et al. (1990). Recognition by human Vgamma9/Vdelta2 T cells of a GroEL homologue on Daudi Burkitt's lymphoma cells. Science, 250(250):1269-73

Flynn GC, Chappell TG, Rothman JE. (1989). Peptide binding and release by proteins implicated as catalysts of protein assembly. Science, 245:385-8

Friedman SM, Posnett DN, Tumang JR, Cole BC, Crow MK. (1991). A potential role for microbial superantigens in the pathogenesis of systemic autoimmune disease. Arthritis Rheum, 34(4):468-80

Gaston JS. (1991). Heat shock proteins and autoimmunity. Semin Immunol, 3(1):35-42

Geelen JLMC, Minnaar RP, Boom R, van der Noordaa J, Goudsmit J. (1988). Heat shock induction of the human immunodeficiency virus long terminal repeat. J Gen Virol, 69:2913-7

Gething M-J, Sambrook J. (1992). Protein folding in the cell. Nature, 355:33-45

Gladman DD, Goldsmith CH, Urowitz MB. (1992c). Cross-cultural validation and reliability of three disease activity indices in SLE. J Rheumatol, 19:608-11

Gruppi CM, Zakeri ZF, Wolgemuth DJ. (1991). Stage and lineage-regulated expression of two hsp90 transcripts during mouse germ cell differentiation and embryogenesis. Mol Reprod Dev, 28(3):209-17

Günther E. (1991). Heat shock protein genes and the major histocompatibility complex. Curr Top Microbiol Immunol, 167:57-68

Hansen LK, Houchins JP, O'Leary JJ. (1991). Differential regulation of HSC70 HSP70 HSP90 alpha and HSP90 beta mRNA expression by mitogen activation and heat shock in human lymphocytes. Exp Cell Res, 192(2):587-96

Heufelder AE, Wenzel BE, Gorman CA, Bahn RS. (1991). Detection cellular localization and modulation of heat shock proteins in cultured fibroblasts from patients with extrathyroidal manifestations of Graves' disease. J Clin Endocrinol Metabol, 73(4):739-45

Hoffman P, Houston L, Butler CA. (1990). Legionella pneumophilia htpAB heat shock operon: nucleotide sequence and expression of the 60 kDa antigen in L. pneumophilia-infected HeLa cells. Infect Immun, 58(10):3380-7

Holoshitz J, Koning F, Coligan JE, De Bruyn J, Strober S. (1989). Isolation of CD4-CD8- mycobacteria-reactive T lymphocyte clones from rheumatoid arthritis synovial fluid. Nature, 339:226-9

Isenberg DA, Rowe A, Tookman A. (1984). An immunohistological study of secondary Sjögren's syndrome. Ann Rheum Dis, 43:470-476

Janeway C. (1991). Mls: makes a little sense. Nature, 349:459-61

Janeway CA Jnr, Jones B, Hayday A. (1988). Specificity and function of T cells bearing gamma/delta receptors. Immunol Today, 9:73-6

Jardetzky TS, Lane WS, Robinson RA, Madden DR, Wiley DC. (1991). Identification of self peptides bound to purified HLA-B27. Nature, 353:326-9

Jarjour WN, Jeffries BD, Davis JS 4th, Welch WJ, Mimura T, Winfield JB. (1991). Autoantibodies to human stress proteins. A survey of various rheumatic and other inflammatory diseases. Arthritis Rheum, 34(9):1133-8

Jenkins DE, Schultz JE, Matin A. (1988). Starvation-induced cross protection against heat or hydrogen peroxide challenge in Escherichia coli. J Bacteriol, 170:3910-4

Karlsson-Parra A, Söderström K, Ferm M, Ivanyi J, Kiessling R, Klareskog L. (1990). Presence of human 65kD heat shock protein (hsp) in inflamed joints and subcutaneous nodules of RA patients. Scand J Immunol, 31:283-8

Kaufmann SH, Schoel B, Wand Wurttenberger A, Steinhoff U, Munk ME, Koga T. (1990). T-cells, stress proteins, and pathogenesis of mycobacterial infections. Curr Top Microbiol Immunol, 155:125-41

Keat A, Thomas B, Hughes R, Taylor Robinson D. (1989). Chlamydia trachomatis in reactive arthritis. Rheumatol Int, 9:197-200

Koga T, Wand-Württenberger A, Debruyn J, Munk M, Schoel B, Kaufmann SHE. (1989). T cells against a bacterial heat shock protein recognize stressed macrophages. Science, 245:1112-5

Kohda T, Kondo K, Oishi M. (1991). Cellular HSP90 (HSP86) mRNA level and in vitro differentiation of mouse embryonal carcinoma (F9) cells. FEBS Lett, 290(1-2):107-10

Koyasu S, Nishida E, Kadowaki T, et al. (1986). Two mammalian heat shock proteins HSP90 and HSP100 are actin-binding proteins. Proc Nat Acad Sci USA, 83(21):8054-8

Kubo T, Towle CA, Mankin HJ, Treadwell BV. (1985). Stress induced proteins in chondrocytes from patients with osteoarthritis. Arthritis Rheum, 28(10):1140-5

LaThangue NB, Latchman DS. (1988). A cellular protein related to heat-shock protein 90 accumulates during herpes simplex virus infection and is overexpressed in transformed cells. Exp Cell Res, 178(1):169-79

Lehner T, Barnes CG. (1979). Criteria for the diagnosis and classification of Behçet's syndrome. In: Lehner T, Barnes CG, eds. Behçet's Syndrome: Clinical and Immunological Features. Academic Press, London.

Lehner T, Lavery E, Smith R, van der Zee R, Mizushima Y, Shinnock T. (1991). Association of the 65-kD heat shock protein of Streptococcus sanguis and the corresponding antibodies in Behçet's syndrome. Infect Immunol, 59:1434-41.

Life PF, Bassey EOE, Gaston JSH. (1991). T cell recognition of bacterial hsps in inflammatory arthritis. Immunol Rev, 121:113-35

Lindquist S. (1986). The heat shock response. Ann Rev Biochem, 55:1151-91

Lindsey NJ, Dawson RA, Henderson FI, Greaves M, Hughes P. (1993). Stimulation of von Willebrand Factor antigen release by immunoglobulin from thrombosis prone patients with systemic lupus erythematosus and the anti-phospholipid syndrome. Br J Rheumatol, 32:123-6

Loizou S, Mackworth-Young CG, Cofiner C, Walport MJ. (1990). Heterogeneity of binding reactivity to different phospholipids of antibodies from patients with systemic lupus erythematosus (SLE) and with syphilis. Clin Exp Immunol, 80:171-6

Mailhos C, Howard MK, Latchman DS. (1993). Heat shock protects neuronal cells from programmed cell death by apoptosis. Neurosciences. (In press).

Manning-Krieg UC, Scherer PE, Schatz G. (1991). Sequential action of protein chaperones in protein import into the matrix. EMBO J, 10(11):3273-80

Masi AT, Rodnan GP, Medsger TA. (1981). Preliminary criteria for the classification of systemic sclerosis (scleroderma). Bull Rheum Dis, 31:1-6

Mezger V, Bensaude O, Morange M. (1989). Unusual levels of heat shock element-binding activity in embryonal carcinoma cells. Mol Cell Biol, 9:3888-96

Minota S, Cameron B, Welch WJ, Winfield JB. (1988a). Autoantibodies to the constitutive 73 kD member of the hsp70 family of hsps in systemic lupus erythematosus. J Exp Med, 168:1475-80

Minota S, Koyasu S, Yahara I, Winfield J. (1988b). Autoantibodies to the heat-shock protein hsp90 in systemic lupus erythematosus. J Clin Invest, 81(1):106-9

Minota S, Winfield JB. (1987). IgG anti-lymphocyte antibodies in systemic lupus erythematosus react with surface molecules shared by peripheral T cells and a primitive T cell line. J Immuol, 138(6):1750-6

Morrow J, Isenberg D. (1987). Systemic lupus erythematosus. In: Morrow WJW, Isenberg DA, eds. Autoimmune Rheumatic Diseases. Blackwell, Oxford. Pp 48-147.

Morrow WJW, Isenberg DA, Todd-Pokropeck A, Parry HF, Snaith ML. (1982). Useful laboratory measures in the management of systemic lupus erythematosus. Quart J Med, 202:125-38

Morrow WJW, Williams DJP, Ferec C, et al. (1983). The use of C3d as a means of monitering clinical activity in systemic lupus erythematosus and rheumatoid arthritis. Ann Rheum Dis, 42:668-71

Muller S, Briand J-P, van Regenmortel MHV. (1988). Presence of antibodies to ubiquitin during the autoimmune response associated with systemic lupus erythematosus. Proc Natl Acad Sci USA, 85:8176-80

Muller S, Barakat S, Watts R, Joubaud P, Isenberg DA. (1990). Longitudinal analysis of antibodies to histones, Sm-D peptides and ubiquitin in serum of patients with systemic lupus erythematosus, rheumatoid arthritis and tuberculosis. Clin Exp Rheumatol, 8:445-53

Niedzweski A, Konpachith AM, Fleming JE. (1991). Aging affects expression of 70 kD heat shock proteins in Drosophila. J Biol Chem, 266(4):9332-8

Nimpf J, Bevers EM, Bomans PHH. (1986). Prothrombinase activity of human platelets is inhibited by beta-2 glycoprotein I. Biochim Biophys Acta, 884:142-9

Norton PM, Isenberg DA, Latchman DS. (1989). Elevated levels of the glucocorticoid-associated 90 kDa heat shock protein in a proportion of SLE patients with active disease. J Autoimmunity, 2:187-95

O'Brien RL, Happ MP, Dallas A, Palmer E, Kubo R, Born WK. (1989). Stimulation of a major subset of lymphocytes expressing T cell receptor gamma/delta by an antigen derived from mycobacterium tuberculosis. Cell, 57:667-74

Pelham HR. (1988). Heat shock proteins: coming in from the cold. Nature, 332:776-7

Phillips B, Abravaya K, Morimoto RI. (1991). Analysis of the specificity and mechanism of transcriptional activation of the human hsp70 gene during infection by DNA viruses. J Virol, 65(11):5680-92

Pierce SK, DeNagel DC, VanBuskirk AM. (1991). A role for heat shock proteins in antigen processing and presentation. Curr Top Microbiol Immunol, 167:83-92

Polla BS, Kantengwa S. (1991). Heat shock proteins and inflammation. Curr Top Microbiol Immunol, 167:93-105

Rajasekar R, Sim GK, Augustin A. (1990). Self heat shock and gamma/delta T cell reactivity. Proc Natl Acad Sci USA, 87:1767-71

Ritossa F. (1962). A new puffing pattern induced by temperature shock and DNP in Drosophila. Experentia, 18:571-3

Schlesinger MJ. (1990). Heat shock proteins. J Biol Chem, 265(21):12111-4

Schousboe I. (1983). Effect of beta2-glycoprotein I on the activity of adenylate cyclase in platelet membranes. Thromb Res, 32:291-9

Sharif M, Worrall JG, Singh B, et al. (1992). The development of monoclonal antibodies to the human mitochondrial 60-kd heat-shock protein, and their use in studying the expression of the protein in rheumatoid arthritis. Arthritis Rheum, 35:1427-33

Sorger PK. (1991). Heat shock factor and the heat shock response. Cell, 65(3):363-6

Srivastava PK, Maki RG. (1991). Stress-induced proteins in immune response to cancer. Curr Top Microbiol Immunol, 167:109-23

Strahler JR, Kuick R, Eckerskorn C, et al. (1990). Identification of two related markers for common acute lymphoblastic leukaemia as heat shock proteins. J Clin Invest, 85:200-7

Symmons DPM, Coppock JS, Bacon PA, et al. (1988). Development and assessment of a computerized index of clinical disease activity in systemic lupus erythematosus. Quart J Med, 69(258):927-37

Tai P-KK, Albers MW, Chang H, Faber LE, , Schreiber SL. (1992). Association of a 59-kilodalton immunophilin with the glucocorticoid receptor complex. Science, 256:1315-8

Tan EM, Cohen AS, Fries JF, et al. (1982a). The 1982 revised criteria for the classification of systemic lupus erythematosus. Arthritis Rheum, 25(11):1271-7

Temesvari P, Denburg J, Denbuerg S, Carbotte R, Bensen W, Singal D. (1983). Serum lymphocytotoxic antibodies in neuropsychiatric lupus: a serial study. Clin Immunol Immunopathol, 28:243-51

Thompson SJ, Bedwell D, Hooper DC, et al. (1990). Pristane-induced arthritis: a possible role for the mycobacterial 65kD heat shock protein. In: Rice-Evans, ed. Stress Proteins in Inflammation. Richlieu, London.

Tissières A, Mitchell HK, Tracy UM. (1974). Protein synthesis in salivary glands of Drosophila melanogaster: relation to chromosome puffs. J Mol Biol, 84:389-98

Tsoulfa G, Rook GAW, van Embden JDA, et al. (1989a). Raised serum IgG and IgA antibodies to mycobacterial antigens in rheumatoid arthritis. Ann Rheum Dis, 48:118-23

Tsoulfa G, Rook GAW, Bahr GM, et al. (1989b). Elevated IgG antibody levels to the mycobacterial 65 kD heat shock protein are a characteristic of patients with rheumatoid arthritis. Scand J Immunol, 30:519-29

Twomey B, McCallum S, Dhillon V, Isenberg D, Latchman D. (1992). Transcription of the genes encoding the small heat shock protein ubiquitin is unchanged in patients with systemic lupus erythematosus. Autoimmunity, 13:197-200

Van den Broek MF, Hogervorst EJ, Van Bruggen MC, Van Eden W, van der Zee R, van den Berg WB. (1989). Protection against streptococcal cell wall-induced arthritis by pretreatment with the 65-kD mycobacterial heat shock protein. J Exp Med, 170:449-66

Van Eden W, Thole JER, van der Zee R, et al. (1988). Cloning of the mycobacterial epitope recognized by T lymphocytes in adjuvant arthritis. Nature, 331:171-3

Van Eden W, Hogervorst EJ, Hensen EJ, van der Zee R, van Embden JD, Cohen IR. (1989). A cartilage-mimicking T-cell epitope on a 65K mycobacterial heat-shock protein: adjuvant arthritis as a model for human rheumatoid arthritis. Curr Top Microbiol Immunol, 145:27-43

Vanbuskirk A, Crump BL, Margoliash E, Pierce SK. (1989). A peptide binding protein having a role in antigen presentation is a member of the hsp70 heat shock family. J Exp Med, 170:1799-809

Vandenbark AA, Hashim G, Offner H. (1989). Immunization with a synthetic T-cell receptor V-region peptide protects against experimental autoimmune encephalomyelitis. Nature, 341:541-4

Watanabe-Fukunaga R, Brannan CI, Copeland NG, Jenkins NA, Nagata S. (1992). Lymphoproliferation disorder in mice explained by defects in Fas antigen that mediates apoptosis. Nature, 356:314-7

Welch WJ. (1987). The mammalian heat shock (or stress) response: a cellular defense mechanism. Adv Exp Med Biol, 225:287-304

Whitelaw ML, Hutchison K, Perdew GH. (1991). A 50-kDa cytosolic protein complexed with the 90-kDa heat shock protein (hsp90) is the same protein complexed with pp60v-src hsp90 in cells transformed by the Rous sarcoma virus. J Biol Chem, 266(25):16436-40

Winfield JB, Jarjour WN. (1991). Stress proteins, autoimmunity, and autoimmune disease. Curr Top Microbiol Immunol, 167:161-89

Worrall JG, Snaith ML, Batchelor JR, Isenberg DA. (1990). SLE: a rheumatological view Analysis of clinical features, serology and immunogenetics of 100 SLE patients during longterm followup. Quart J Med, 74(275):319-30

Yang X-D, Gasser J, Feige U. (1990). Prevention of adjuvant arthritis in rats by a nonapeptide from the 65 kD mycobacterial heat shock protein. Clin Exp Immunol, 81:189-94

Yost HJ, Petersen RB, Lindquist S. (1990). RNA metabolism: strategies for regulation in the heat shock response. Trends Genet, 6(7):223-7

Young D, and Mehlert A. (1990). Stress proteins and infectious diseases. In: Morimoto RI, Tissières A, Georgopoulos C, eds. Stress Proteins in Biology and Medicine. Cold Spring Harbour Laboratory Press, USA. Pp 131-65.

Zerbini M, Musiani M, La Placa M. (1985). Effect of heat-shock on Epstein-Barr virus and cytomegalovirus expression. J Gen Virol, 66:633-6

The heavy chain variable region genes of human lupus autoantibodies

Christophe Demaison, Patricia Chastagner, Céline Bensimon and Moncef Zouali
Laboratoire d'Immunogénétique Cellulaire
Institut Pasteur
25-28 rue du Dr Roux
75015 Paris
France

INTRODUCTION

An important role of the immune system is to distinguish "self" from "non-self" and to act as a first-line of defense against external agents. To cope with the diversity of the potentially antigenic molecules of the environment and those of self-components, the immune system requires a powerful discriminatory ability. In normal conditions, mature B cells respond to antigen stimulation, yet must avoid high-affinity recognition of self-antigens that may lead so aggressive autoimmunity. No mechanism fully accounts for this phenomenon.

Because molecular characterization of the immunoglobulin variable regions expressed by a B cell clone can provide information on the mechanisms of its selection and activation *in vivo* (Rajewsky et al., 1987), studies of the Ig receptors may help understanding the processes that lead to autoimmune diseases.

Here, we review data on the molecular characterization of pathogenic autoantibodies found in human systemic lupus erythematosus (SLE), a prototype of systemic autoimmune disease of unknown etiology.

NATO ASI Series, Vol. H 80
Autoimmunity: Experimental Aspects
Edited by Moncef Zouali

The primary antibody repertoire

The human IgH cluster of approximately 2500 Kb maps to chromosome 14, band 14q32.33, and contains at least 130 V_H gene segments (Matsuda et al., 1993), over 30 D_H segments (Chastagner et al., 1992), 6 functional J_H segments and a constant region complex of 9 functional genes (μ, δ, $\gamma 3$, $\gamma 1$, $\alpha 1$, $\gamma 2$, $\gamma 4$, ε and $\alpha 2$). The organization of the V_H locus is extremely complex (Pascual and Capra, 1991) with a potential germline repertoire of functional V_H genes that can be categorized into 7 families based on at least 80% nucleotide sequence homology. Upstream of the most 3' V_H gene segment (V_H6), members of the other families are extensively intermingled and some of them have not been mapped. It is of note that while V_H genes are highly polymorphic, the D_H locus exhibits little heterogeneity (Figure 1).

At the pre-B cell stage, the recombination machinery brings together a D_H segment with one of the 6 functional J_H elements. This D_H-J_H complex subsequently joins to a V_H gene segment to generate a complete heavy chain variable region (Tonegawa, 1983). Recombination events that give rise to the third complementarity determining region (CDR3), provide additional combinatorial diversity by generating unique amino acid residues at the V_H-D_H and D_H-J_H junctions during the process of gene rearrangement (Alt et al., 1992; reviewed by Schatz et al., 1992).

The antibody response to an immunogen is dictated, in part, by the repertoire expressed by B cells at the time of stimulation. All possible combinations of germline V gene segments, including the additional diversity generated during CDR3 formation, constitue the potential naive repertoire. However, the number of individual functional V genes and the presence of transcriptional regulatory sequences can influence this potential repertoire which is also dependent on positive and negative selection mechanisms that operate on newly formed B cells in primary lymphoid organs and continue to act during the life of the B lymphocyte (Berek and Milstein, 1988). Specific interaction with an exogenous antigen may result in selection, proliferation and terminal differentiation of B cells. Subsequently, a number of the selected cells return to the available B cell pool as memory cells. In response to an antigenic stimulus, positive selection of B cells operates on both the expressed, naive and activated cell repertoires, including those of the memory pool. The V_H-D_H-J_H rearrangements expressed by memory cells differ from those of naive B cells by the presence of mutations introduced during antigen-driven selection.

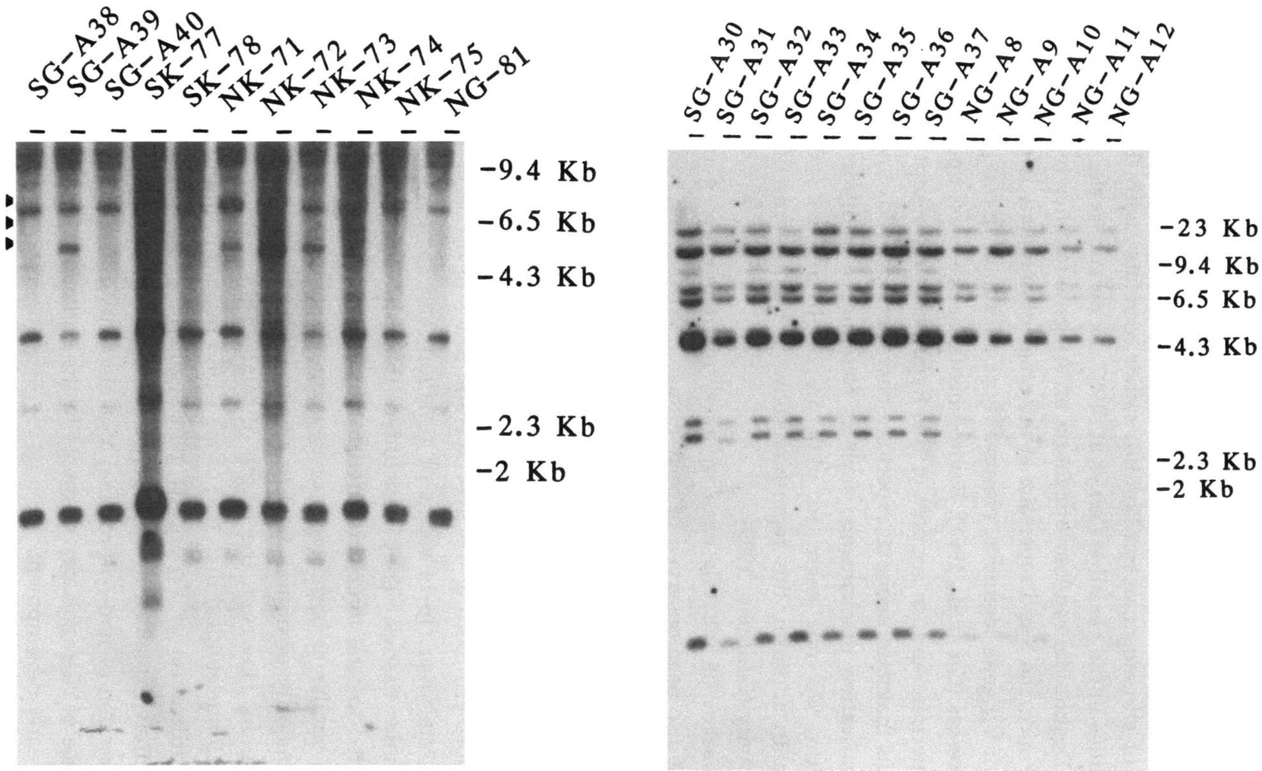

Figure 1: Southern blot hybridization of human genomic DNA from unrelated individuals digested with the restriction enzymes *Bam HI* (left) or *Bgl II* (right) and hybridized to a human V_H1 gene-family probe (left) or to a D_H2 cDNA probe (right). Polymorphic restriction bands are marked by arrowheads.

In the murine system, the fetal and the neonatal repertoires are skewed towards expression of the D_H-proximal V_H gene-families. This bias changes at about 4-7 days after birth and, by day 7-14, the V_H gene repertoire becomes independent of chromosomal position but rather correlates with the size of the V_H gene-families in the germline (Alt et al., 1987; Jeong and Teale, 1988; Malynn et al., 1990). Despite the random utilization of V_H gene families in peripheral B cells of adult mice, the B-cell repertoire at this stage of ontogeny is dominated by a limited set of V_H gene segments, presumably as a result of Ig receptor-mediated selection procèsses (Gu et al., 1991). Environmental antigens act upon this newly generated repertoire and, in support of this hypothesis, mice kept under germ-free conditions maintain neonatal repertoire characteristics (Bos and Meeuwsen, 1989). Diversification mechanisms, such as N-sequence addition and somatic hypermutation, do not seem to play an important role during early development. In addition, this repertoire shows a high degree of connectivity and self-reactivity (Coutinho, 1989).

Reminiscent of what is seen in mice, restricted V_H gene utilization patterns have also been observed in early human ontogeny and interconnectivity is also readily detectable. Analyzing human cDNA libraries derived from fetal liver at 104 and 130 days of gestation, Schroeder et al. (1987, 1990) provided evidence that a small set of V_H3 gene segments, together with the most J_H-proximal D_H element (DQ52) and two functional J_H elements (J_H3 and J_H4), contribute disproportionally to the B cell fetal repertoire. Berman and Alt (1990) studied by Northern blotting the hybridization ratios of different V_H probes to mRNA from fetal liver (16-24 weeks of gestation) and compared it to different adult lymphoid tissues. They concluded that the V_H6 gene segment is overrepresented in fetal liver compared to adult tissues whereas V_H3 and V_H4 expression is similar at both stages of development. In contrast to the murine system, the majority of human fetal Ig rearrangements contain N sequences in the CDR3 region. However, the length of fetal CDR3 regions is significantly shorter (6 to 15 nucleotides) than that of the CDR3 regions found in adult B cells. Preferential utilization of V_H and J_H gene segments at this stage of ontogeny cannot be entirely explained by their genomic location. Chen (1990) hypothesized that overexpression of non-J_H-proximal V_H3 elements in fetal liver could be due to the presence of *cis*-regulatory elements in the 5' flanking region of V_H genes expressed in early development.

To analyze the diversity of the primary antibody repertoire, Mortari et al. (1992, 1993) cloned IgM, IgG and IgA transcripts from a cord-blood cDNA library. They found that the IgM repertoire differs significantly from that expressed in the second-trimester fetal liver. Of thirteen cord-blood sequences, only one V_H gene had been previously described in fetal liver. The neonatal IgM transcripts were enriched in J_H4, J_H5 and J_H6 elements and, in contrast to the overexpression of the DQ52 gene element in fetal rearrangements, they used a number of different D_H gene segments. As a result of enhanced N-region addition and use of longer D_H and J_H elements, the cord-blood CDR3 regions contain three additional amino acids as compared with their fetal counterparts. This mechanism leads to an increase of the size of the theoretical cord-blood repertoire. However, the molecular characteristics of the V_H IgG neonatal transcripts are more similar to those of the IgM fetal liver repertoire. The IgG and IgA neonatal V_H gene repertoires appear to represent an intermediate stage of diversification between the fetal and the neonatal IgM repertoires.

Maturation of the human antibody repertoire

In adults, our knowledge of the molecular mechanisms involved in the expression of the repertoire were initially derived from studies of B cell malignancies where B cell differentiation is frozen by malignant transformation. In this regard, acute lymphoblastic leukemia and chronic lymphocytic leukemia, which are the malignant correlates of pre-B and intermediate B cell stages, respectively, are restricted in their V_H family usage and frequently rearrange genes from the fetal V_H gene repertoire (Chen et al., 1990). Other studies focused on expression of V_H families in EBV-transformed B cell lines and B cell hybridomas derived from peripheral blood B cells. The results obtained with B cell clones tended to support the hypothesis of a random utilization of V_H gene families in the expressed repertoire of normal adults (Stewart et al., 1992; Young et al., 1990). In these studies, the possibility that cell fusion or EBV transformation may have introduced a bias in clonal analysis cannot be formally excluded. Using *in situ* hybridization, Zouali et al. (1991) found that V_H gene expression in the peripheral blood repertoire is dominated by the V_H3 family.

To investigate the immunoglobulin V-gene repertoire of normal adults, Huang et al. (1992) developed a two-step PCR approach to analyze cDNA libraries from unmanipulated peripheral blood B cells. The expressed IgM V_H gene repertoire of two normal individuals was studied and marked differences in the pattern of V_H gene expression between the two subjects were seen. In the cDNA library of one subject, a non-random representation of V_H gene segments was noted, with disproportionate representation of V_H5 and V_H3 gene family members. In contrast to this repertoire, which resembles in its overall composition the fetal IgM libraries analyzed by Schroeder et al. (1987, 1990), the V_H usage from the second donor reflects more closely the genomic complexity of V_H gene-families. It seems therefore that these libraries provide a "snapshot" of the expressed repertoire at a single time-point and, with this technique, the V_H gene repertoire was demonstrated to vary with time. Another noteworthy aspect of this study is that the V region genes showed evidence of somatic mutations. Molecular evidence for clonal selection in circulating B cells has also been reported by an other group (Logtenberg et al., 1992). It is of note that this procedure may introduce a bias towards isolation of transcripts expressed by plasma B cells that may contain up to a 1000-fold more Ig transcripts than resting B cells. In addition to this apparently biased utilization of V_H gene-families, evidence favoring a preferential expression of individual D_H and J_H genes was also noted (Sanz, 1991; Yamada et al., 1991). In these studies, analysis of CDR3 sequences from peripheral B cells of normal adults revealed a high representation of J_H4 and J_H6 segments and overexpression of the DXP and DLR gene families.

THE ANTIBODY REPERTOIRE IN SYSTEMIC LUPUS ERYTHEMATOSUS

Systemic lupus erythematosus (SLE) is an autoimmune disease characterized by the presence of a wide variety of autoantibodies, including antibodies to DNA, RNA, ribonucleoproteins and histones (Tan, 1989). Antibodies to double-stranded DNA (dsDNA) are generally regarded as a marker of SLE and of disease activity. These antibodies appear in the circulation of lupus patients and differ from anti-DNA antibodies that can be found in non-autoimmune individuals. They bind dsDNA with high-affinity, are often IgG and cationic in charge and do not cross-react with a wide spectrum of unrelated antigens. A direct role for IgG anti-dsDNA autoantibodies in the pathogenesis of SLE has been established by finding a correlation of serum antibody level with disease activity and deposition of DNA/anti-DNA immune complexes at the sites of tissue damage (Madaio et al., 1987; Susuki et al., 1993). However, it is not unusual to detect high levels of serum anti-dsDNA autoantibodies in lupus patients with inactive disease. Other studies showed that IgG anti-DNA antibodies during active and inactive stages of the disease are clonally and qualitatively different (Suenaga et al., 1991).

To understand the genetic origin of pathogenic anti-dsDNA antibodies and the structural basis for DNA binding, we undertook a molecular analysis of anti-DNA antibodies expressing a disease-related idiotypic marker. Idiotypic determinants, also referred to as idiotopes (Id), are fingerprints of the variable region of the antibody molecule. Initially, idiotypic studies revealed that anti-dsDNA autoantibodies express a restricted set of idiotopes (Zouali et al., 1984) and it was hypothesized that this apparent idiotypic dominance probably reflects the use of a limited repertoire of V_H gene segments and/or three-dimensional motifs in the pathogenic anti-DNA response. Schematically, the various anti-DNA Id can be divided into two representative families (Table 1). The first set, defined by IgM anti-DNA autoantibodies with low-affinity, is encoded by germline Ig-variable gene elements and can be found in a number of normal and pathological conditions. The second set of Id, more restricted to lupus disease, is associated with mature IgG anti-dsDNA. These autoantibodies probably result from preferential antigen- or idiotype-driven clonal expansion of a selected number of clonotypes. The relationship between these two autoantibody populations remains to be determined and whether autoantibodies of the first kind represent the precursors of pathogenic anti-dsDNA antibodies remains open (Zouali and Diamond, 1990).

	Ab of the preimmune repertoire	Aggressive Ab
Specificity	multireactive	mono/oligo reactive
Affinity	low	high
pHi	?	cationic
Isotype	IgM	IgM/IgG
Idiotype	ubiquitous - 16/6 (Shoenfeld et al., 1983)	highly restricted - F4 (Davidson et al., 1989) - TOF (Zouali et al., 1984) - O-81/NE-1 (Muryoi et al., 1990)
V genes	germinal	?

Table 1: Distinctive properties of antibodies of the primary repertoire and those that play a pathogenic role.

To approach this question, we generated a large panel of Epstein-Barr virus-transformed B cell clones from five SLE patients. Twelve B-cell clones with dsDNA binding specificity and expressing the O-81 idiotype were selected for further study. Because elevated titers of the O-81 idiotype marker are present in SLE patients with active disease, and since this idiotype is recurrently expressed in serum immune complexes and in renal lesions of nephritic patients, the twelve anti-DNA antibodies analyzed may be considered representative of a subset of aggressive antibodies involved in the active phase of the disease (Muryoi et al., 1990). Their Ig variable region genes were amplified by PCR, cloned and the complete sequences were determined (Chastagner et al., 1991; Zouali et al., 1992; Demaison et al., 1993). The V_H gene segments encoding the heavy chains of the $0\text{-}81^+$ anti-DNA antibodies are derived from the V_H3, V_H4 and V_H2 gene-families (Table 2). The V_H genes used by five antibodies are derivative of the V_H56P1 gene, which is expressed during fetal life.

When compared with germline genes, our sequence data suggest that the heavy chains are somatic mutants of V_H genes (Figure 2). The differences observed between the expressed V_H genes and the corresponding germline V_H genes may reflect true mutations and/or errors introduced by experimental techniques (EBV immortalization, reverse transcription, *Taq* polymerase activity, cloning and/or sequencing). Comparative PCR amplification of genomic rearranged DNA from B cell clones and of germline DNA from non-lymphoid cells was used to demonstrate that these expressed genes are the somatically mutated versions of the corresponding germline genes. One primer was from the 5′ leader sequence and the other was from a conserved sequence of the third framework region (FR3). This combination of primers enabled amplification of a V_H gene product from genomic DNA (lymphoid and non-lymphoid) that was detected by a V_H-family-specific cDNA probe. By contrast, the somatic CDR2 probe corresponding to the sequence of the V_H gene expressed by the B cell clone hybridized only to the lymphoid genomic DNA, suggesting that these V_H gene segments were acquired by somatic mutations of germline variable region genes (Demaison et al., 1993).

Somatic mutation in Ig V-regions is a well-known phenomenon, but its mechanism is not yet clearly understood (French et al., 1989). Analysis of Ig heavy chain transgenes during an immune response (Sohn et al., 1993) and characterization of monoclonal antibodies derived from animals immunized with exogenous antigens (Kocks and Rajewsky, 1989; Rada et al., 1991) revealed that somatic mutation: i) occurs at a rate of 10^{-4}/base pair/generation, which is

Clonotype (Isotype)	V_H Gene-family	Closest germline relative (% homology)	Amino acid homology	R/S ratio		D_H gene segment	Germline J_H (% homology)
				FR	CDR		
T20-11 (γ)	VH3	56P1 (94%)	93%	1	5	Q52/13P1*	JH6 (98%)
T20-24 (μ)	VH4	58P2 (85%)	90%	1,1	3,7	?	JH4 (98%)
T21-9 (μ)	VH3	56P1 (90%)	90%	1	2,3	2P1	JH3 (95%
T24-3 (μ)	VH3	60P2 (88%)	82%	1,25	3	15P1	JH3 (98%)
T24-9 (μ)	VH3	56P1 (89%)	86%	1,5	2	DN4	JH4 (95%)
T33-1 (μ)	VH4	V71-4 (98%)	96%	0,25	> 4	DN1	JH6 (96%)
T33-2 (μ)	VH2	M60 (94%)	86%	1,8	7	DN1	JH4 (91%)
T33-4 (γ)	VH4	V71-2 (91%)	84%	1,25	3	DK4	JH5 (93%)
T33-16 (μ)	VH3	56P1 (89%)	87%	0,8	3,3	20P1*	JH4 (95%)
T33-17 (μ)	VH3	56P1 (88%)	86%	1	3,3	20P1*	JH4 (95%)
T33-20 (μ)	VH4	V71-4 (99%)	97%	0,25	> 4	DN1	JH6 (95%)
T34-1 (μ)	VH3	VH26 (95%)	93%	1	6	?	JH4 (86%)

Table 2: Characterization of the V_H gene segments of the O-81$^+$ anti-DNA autoantibodies

(*): Indicates a D_H gene segment used in inverted orientation

equivalent to one base change in the V region during each cell division; ii) is a relatively random process that is restricted to the V region and its flanking sequences; iii) requires T-cell factors and iiii) begins with B cell activation and continues after class-switching, but ends before the plasma cell stage.

The contribution of somatic mutations occurring during an immune response can be analyzed by comparing the ratio of replacement (R) to silent (S) mutations in the CDRs and in the framework regions (FRs). Assuming that mutations occur randomly, Shlomchik et al. (1987) calculated that there would be 2.9 times more R than S mutations. However, a compilation of R/S ratios in the FRs reveals a ratio of 1.5, presumably reflecting negative selection against amino acid substitutions to prevent destabilization of the Ig molecule structure. Therefore, a ratio of 3 or higher is presumed to reflect positive selection for mutations in regions of the antibody molecule that are in direct contact with the antigen. In this study, all the expressed V_H genes, except T24-9, have a R/S ratio < 2 in the framework regions and > 2.9 in the CDRs (Table 2).

The J_H gene segments utilized by the antibody panel are derived from J_H4, J_H6, J_H3 and J_H5 genes (Table 2). These expressed J_H gene segments exhibit diversity at the 5' end, with evidence for both somatic mutations and nucleotide trimming. There is also heterogeneity at the level of both the length and the sequences of the D_H genes used. Some of the D_H elements are utilized in an inverted orientation and one antibody, T20-11 clone, expresses a D_H-D_H fusion, a previously reported feature of human antibodies (Figure 3). The D_H region sequences of T20-24 and T34-1 clones could not be assigned to known D_H genes. It is possible that these D_H segments are highly diversified through somatic mechanisms. Alternatively, they could be derived from germline D_H gene that have not been yet described.

The predicted amino acid sequences indicated that some of the V_H replacement mutations yielded an arginine, asparagine or lysine residue, whereas additional basic amino acids were introduced during the V_H-D_H-J_H rearrangement in the CDR3. Because these amino acids may allow hydrogen bonding with the phosphate oxygen on the DNA backbone to stabilize the DNA-antibody specific interactions (Marion et al., 1992), they have been shown to play an important role in determining specificity of anti-DNA antibodies from SLE-prone mice (Shlomchik et al., 1990).

									-10									-1	*
	Glu	Phe	Gly	Leu	Ser	Trp	Val	Phe	Leu	Val	Ala	Ile	Leu	Glu	Gly	Val	Gln	Cys	Glu
T21-9	GAG	TTT	GGG	CTG	AGC	TGG	GTT	TTC	CTT	GTT	GCT	ATT	TTA	GAA	GGT	GTC	CAG	TGT	GAG
56P1	...	...	...	...	...	...	...	...	...	...	...	...	...	...	...	...	...	...	C..

									10	*					*				20
	Val	Gln	Leu	Val	Glu	Ser	Gly	Gly	Gly	Leu	Val	Gln	Pro	Gly	Gly	Ser	Leu	Arg	Leu
T21-9	GTG	CAG	CTG	GTG	GAG	TCT	GGG	GGA	GGC	TTG	GTC	CAG	CCT	GGG	GGG	TCC	CTG	AGA	CTC
56P1	...	...	...	...	...	...	...	...	...	G..	...	...	...	...	.A.	...	...	...	...

											<----	CDR1	-----	-----	---->				
										30			*		*				
	Ser	Cys	Ala	Ala	Ser	Gly	Phe	Thr	Phe	Ser	Ser	Tyr	Trp	Met	Ser	Trp	Val	Arg	Gln
T21-9	TCC	TGT	GCA	GCC	TCT	GGA	TTC	ACC	TTT	AGT	AGC	TAT	TGG	ATG	AGC	TGG	GTC	CGC	CAG
56P1	...	...	...	...	...	...	...	...	..C	...	...	...	GCT	...	CAG	...	...	...	...

											<----	-----	-----	-----	-----	-----	CDR2	-----	-----
	*	41									*		*	*				*	
	Leu	Pro	Gly	Lys	Gly	Leu	Glu	Trp	Val	Ala	Asn	Ile	Lys	Gln	Asp	Gly	Ser	Glu	Lys
T21-9	CTC	CCA	GGG	AAG	GGG	CTG	GAG	TGG	GTG	GCC	AAC	ATA	AAG	CAA	GAT	GGA	AGT	GAG	AAA
56P1	GCT	...	..C	...	...	...	...	...	...	..A	GTT	...	TCA	TAT	...	...	..C	AAT	...

	-----	-----	-----	-----	-----	-----	-----	---->											
			*										70						
	Tyr	Tyr	Val	Asp	Ser	Val	Lys	Gly	Arg	Phe	Thr	Ile	Ser	Arg	Asp	Asn	Ala	Lys	Asn
T21-9	TAC	TAT	GTG	GAC	TCT	GTG	AAG	GGC	CGA	TTC	ACC	ATC	TCC	AGA	GAC	AAC	GCC	AAG	AAC
56P1	...	TAC	GCA	...	..C	...	...	...	...	...	...	...	...	...	...	..T	...	...	...

	*			80													90		
	Ser	Leu	Tyr	Leu	Gln	Met	Asn	Ser	Leu	Arg	Ala	Glu	Asp	Thr	Ala	Val	Tyr	Tyr	Cys
T21-9	TCA	CTG	TAT	CTG	CAA	ATG	AAC	AGC	CTG	AGA	GCC	GAG	GAC	ACG	GCT	GTG	TAT	TAC	TGT
56P1	ACG	...	...	...	...	...	...	...	...	...	..T	...	...	...	...	...	...	...	...

	Ala	Arg
T21-9	GCG	AGA
56P1	...	...

								+1									10		*
	Ile	Ile	Lys	Gly	Val	Gln	Cys	Gln	Val	Gln	Leu	Val	Glu	Ser	Gly	Gly	Gly	Leu	Val
T24-3	ATT	ATA	AAA	GGT	GTC	CAG	TGT	CAG	GTG	CAG	CTG	GTG	GAG	TCT	GGG	GGA	GGC	TTG	GTC
60P2	...	...	...	...	...	...	...	...	...	...	...	...	...	...	..A	...	...	...	A..

																			<----
	*	*						20									*	30	
	Lys	Ser	Gly	Gly	Ser	Leu	Arg	Leu	Ser	Cys	Ala	Ala	Ser	Gly	Phe	Thr	Leu	Ser	Ser
T24-3	AAG	TCT	GGA	GGG	TCC	CTG	AGA	CTC	TCC	TGT	GCA	GCC	TCT	GGA	TTC	ACG	TTG	AGT	AGC
60P2	C..	C..	..G	...	...	...	...	...	...	...	...	...	...	..G	...	..C	G..	...	...

	-----	CDR1	-----	---->															<----
	*		*			*			40									49	*
	Tyr	Tyr	Val	Ser	Trp	Ile	Arg	Gln	Ala	Pro	Gly	Lys	Gly	Leu	Glu	Trp	Val	Ser	Tyr
T24-3	TAC	TAC	GTG	AGC	TGG	ATC	CGC	CAG	GCT	CCA	GGG	AAG	GGG	CTG	GAG	TGG	GTC	TCA	TAC
60P2	A..	...	A..	...	...	G..	...	...	...	...	...	...	...	...	...	...	...	...	GTT

	----	-----	-----	-----	-----	CDR2	-----	-----	-----	-----	-----	-----	-----	-----	-----	---->			
	*	*		*		*	*	*											
	Met	Ser	Ser	Ser	Gly	Arg	Thr	Met	Tyr	Tyr	Ala	Asp	Ser	Val	Lys	Gly	Arg	Phe	Thr
T24-3	ATG	AGC	AGT	AGT	GGG	AGA	ACC	ATG	TAC	TAC	GCA	GAC	TCT	GTG	AAG	GGC	CGA	TTC	ACC
60P2	..T	TAT	..C	G..	..T		.G.	.CA	...	...	...	...	..C	...	...	...	...	...	...

		70				*			*		*	80							*
	Ile	Ser	Arg	Asp	Asn	Ala	Lys	Asn	Ser	Leu	His	Leu	Gln	Met	Asn	Ser	Leu	Arg	Val
T24-3	ATC	TCC	AGG	GAC	AAC	GCC	AAG	AAC	TCA	CTG	CAT	CTG	CAA	ATG	AAC	AGC	CTG	AGA	GTC
60P2	...	...	..A	...	..T	T..	...	...	A.G	...	T..	..T	...	...	...	...	...	...	.C.

						90	*			
	Glu	Asp	Thr	Ala	Val	Tyr	Phe	Cys	Ala	Arg
T24-3	GAG	GAC	ACG	GCC	GTA	TAT	TTC	TGT	GCG	AGA
60P2	...	...	...	...	..G	...	.A.	...	...	...

Figure 2: Nucleotide and deduced amino acid sequences of the VH genes expressed by the SLE-derived T21-9 and T24-3 nephritogenic antibodies. Asterisks denote replacement mutations. The first and second complementarity-determining regions (CDR) are overlined. Numbering is according to Kabat et al., 1991. The sequence of VH56P1 and VH60P2 genes are from Schroeder et al., 1987.

CONCLUSIONS

We found that lupus pathogenic high-affinity autoantibody secreting B cells arise as a result of antigen-driven B cell activation, presumably through T helper cell-dependent mechanisms. A hallmark of these processes is the positive selection for replacement mutations in the V regions CDRs of antigen-activated B cells, leading to higher affinity for the antigen.

Because in normal conditions, B cells expressing autoreactive Ig receptors are thought to be negatively down-regulated through inactivation and/or clonal deletion (Gay et al., 1993; Tsao et al., 1993) and since the majority of lupus autoantibodies with various specificities (dsDNA, cardiolipin or Sm ribonucleoprotein) are somatically mutated (van Es et al., 1992; Manheimer-Lory et al., 1991; van Es et al., 1991; Winkler et al., 1992; Zouali, 1992, Demaison et al., 1993), we favor the view that there might be a break or a defect of tolerance in this autoimmune disease. In this regard , it is notable that in a lupus-prone mouse model, there is evidence that a similar mechanism is involved. In the MRL/Mp-*lpr/lpr* strain which develops a lymphoproliferative syndrome and a clinical profile similar to human lupus erythematosus, Watanabe-Fukunaga et al. (1992) demonstrated that the *lpr* mutant has undergone a genetic rearrangement within the *Fas* gene that prevents expression of the Fas protein. It is of further interest that Russell et al.,(1993) found that T cells from MRL/Mp-*lpr/lpr* mice exhibit a normal activation pattern, but once activated, they have a defect in antigen-stimulated suicide. These studies suggest that antigen-stimulated death of mature cells may be important in establishing peripheral tolerance and in preventing autoimmunity during normal immune responses. In this model, B cells producing antibodies to foreign antigens may, as a result of somatic mutation, generate antibodies that react with self and, then, proliferate to participate in an autoimmune response.

It is clearly important to understand what triggers the response to the multiplicity of autoantigens in SLE, why peripheral tolerance is broken and what continues to sustain the response after it has been initiated. How SLE autoantigens with apparently unrelated structures, such as dsDNA, cardiolipin, chromatin and RNA-associated proteins, but not other endogenous proteins, become antigenic remains unclear. Brendel et al. (1991) found that most of these autoantigens display a high number of charges, either in a form of very long stretches of charged residues or in the form of charged three-dimensional motifs.

	3' end of VH gene	N + DH + N sequence	JH gene sequence
T21-9	AGA	AAAAGGGCTGGGAAAGCG	GCTTTTAATATCTGGGGCCAAGGGACAATGGTCACCGTCTCTTCA
2P1		AgaGGCTGGGgtA	
Germline JH3			GCTTTTgATgTCTGGGGCCAAGGGACAATGGTCACCGTCTCTTCA
T24-3	AGA	GAGAATTACTATGACAACCCCACCATGAAA	GCTTTTGATATCTGGGGCCAAGGGACAATGGTCACCGTCTCTTCA
15P1		AGAAaTcagAgGggAACCCgggCggGAAcca	
Germline JH3			GCTTTTGATgTCTGGGGCCAAGGGACAATGGTCACCGTCTCTTCA
T34-1	AGA	CGAACAATATACGGTGGCTCGACTCGT	CGCGTGTTCGCCTGGGGCCAGGGAAGCCTGGTCACCGTCTCCTCA
DHFL16		tgacTACGGTGGtaatcCcaccggtt	
Inverted DN1		cctGCTCGACgatatgag	
Germline JH4			taCtTtgaCTaCTGGGGCCAaGGAAcCCTGGTCACCGTCTCCTCA
T20-24	AGA	CATCTTTTCCCCGGAAACTACGCAGCA	CCTTTTGTCTACTGGGGCCGGGAAACCTGGTCACCGTCTCCTCA
Inverted 38P1		ggagccaaagTCTTTTCgagGA	
Inverted DIR		ggacTCgagaCCCCGGAggaccCaCccCcg	
Germline JH4			tacTTTGaCTACTGGGGCCaGGAAcCCTGGTCACCGTCTCCTCA
T20-11	AAA	GATGGGGAGGGGTGGGGGCTA	TACTACTACTACGGTATGGACGTCTGGGGCCAAGGGACCACGGTCACCGTCTCCTCA
DHQ52		ctaacTGGGGA	
Inverted 13P1		cGGCGGGGGCTctt	
Germline JH6			tacTACTACTACTACGGTATGGACGTCTGGGGgCAAGGGACCACGGTCACCGTCTCCTCA

Figure 3: VH-DH-JH junctionnal sequences of 5 rearranged IgH chain genes expressed in lupus antibodies. Shown are the 3' end of the VH genes, the DH elements plus the N region nucleotides, and the JH gene segments. Under each diversity element expressed by a B cell clones, the germline DH or the part of the DIR segment that could account for this sequence is depicted. Possible DH-DH fusions are indicated by showing the homologous DH in direct or inverted positions. Upper case : nucleotide that match; lower case: mismatches.

The importance of charge-charge interactions can explain the immunological cross-reactivities between a long charged exoantigen and several self charge-rich structures, such as nucleic acids, negatively charged phospholipids, proteoglycans or heat shock proteins. Viral or bacterial charged molecules could initially break the peripheral tolerance towards charged self-antigens; the biochemical structure of these self-antigens renders the target host-proteins strongly immunogenic. In this model, the high-affinity autoantibody profile seen in lupus patients might reflect the chronic exposure to intracellular molecules, in conjunction with cross-reactive viral or bacterial antigens.

ACKNOWLEDGMENTS

We would like to thank G. Fournié at the "Centre Hospitalier Universitaire Purpan" (Toulouse, France) for providing us with the blood samples of the SLE patients, T. Sasaki (Tohoku, Japan) for gift of the anti-idiotypic antibodies, F.W. Alt (Boston, USA), T. Honjo (Kyoto, Japan), P. Leder (Boston,USA) and M.-P. Lefranc (Montpellier, France) for the cDNA probes. C.D. is a pre-doctoral fellow of the Ligue Nationale de Recherche contre le Cancer (Paris).

REFERENCES

Alt F.W., T.K. Blackwell and G.D. Yancopoulos (1987). Developpement of the primary antibody repertoire. *Science* **238**, 1079-1081.

Alt W.F., E.M. Oltz, F. Young, J. Gorman, G. Taccioli and J. Chen. (1992). VDJ recombination. *Immunol. Today* **13**, 306-314.

Berek C. and C. Milstein (1988). The dynamic nature of the antibody repertoire. *Immunol. Rev.* **105**, 5-26.

Berman J.E. and F.W. Alt (1990). Human heavy chain variable region gene diversity, organization and expression. *Int. Rev. Immunol.* **5**, 203-214.

Bos N.A. and C.G. Meeuwsen (1989). B cell repertoire in adult antigen-free and conventional neonatal BALB/c mice. Preferential utilization of the CH-proximal VH gene family PC7183. *Eur. J. Immunol.* **19**, 1811-1815.

Brendel V., J. Dohman, B.E. Blaisell and S. Karlin (1991). Very long charge runs in systemic lupus erythematosus-associated antigen. *Proc. Natl. Acad. Sci. USA.* **88**, 1536-1540.

Chastagner P., J. Thèze and M. Zouali (1991). Molecular cloning of a lupus-associated human autoantibody VK region using the polymerase chain reaction and degenerate primers. *Gene.***101**, 305-306.

Chastagner P., J. Thèze and M. Zouali (1991). Monomorphic organization of human diversity (DH) heavy chain variable genes. *Eur. J. Immunogenetics* **19**, 303-310.

Chen P.P. (1990). Structural analyses of human developmentally regulated VH3 genes. *Scand. J. Immunol.* **31**, 257-267.

Chen P.P., N.J. Olsen, P.M. Yang, R.W. Soto-Gil, T. Olee, K.A. Siminovitch and D.A. Carlson (1990). From human autoantibodies to the fetal antibody repertoire to B cell malignancy: It's a small word after all. *Int. Rev. Immunol.* **5**, 239-251.

Coutinho A. (1989). Beyond clonal selection and network. *Immunol. Rev.* **110**, 63-88.

Davidson A., A. Smith, J. Katz, J.L. Preud'homme, A. Solomon, B. Diamond (1989). A cross-reactive idiotype on anti-DNA antibodies defines an H chain determinant present almost exclusively on IgG antibodies. *J. Immunol.* **143**, 174-180.

Demaison C., P. Chastagner, J. These and M. Zouali (1993). Somatic diversification in the heavy chain variable genes expressed by human autoantibodies bearing a lupus associated nephritogenic anti-DNA idiotype. *Proc. Natl. Acad. Sci. USA.* in press.

French D.L., R. Laskov and M.D. Scharff (1989). The role of somatic hypermutation in trhe generation of antibody diversity. *Science* **244**, 1152-1157.

Gay D., T. Saunders, S. Camper and M. Weigert (1993). Receptor editing: An approach by autoreactive B cells to escape tolerance. *J. Exp; Med.* **177**, 999-1008.

Gu H., D. Kitamura and K. Rajewsky (1991). B cell development regulated by gene rearrangement: Arrest of maturation by membrane-bound Dμ protein and selection of DH element reading frames. *Cell* **65**, 47-54.

Huang C., A.K. Stewart, R.S. Schwartz and B.D. Stollar (1992). Immunoglobulin heavy chain gene expression in peripheral blood B lymphocytes. *J. Clin. Invest* **89**, 1331-1343.

Jeong H.D. and J.M. Teale (1988). Comparison of the fetaland adult functional B cell repertoires by analysis of V_H gene family expression. *J. Exp. Med.* **168**, 589-603.

Kabat E.A., T.T. Wu, H.M. Perry, K.S. Gottesman and C. Foeller (1991). *Sequences of Proteins of Immunological Interest.* US Department of Health and Human Services, Bethesda, 5th edition

Kocks C. and K. Rajewsky (1989). Stable expression and somatic hypermutation of antibody V regions in B-cell developmental pathways. *Annu. Rev. Immunol.* **96**, 537-559.

Logtenberg T., M.E. Schutte, S.B. Ebeling, F.H.J. Gmelig-Meyling and J.H. Van Es (1992). Molecular approaches to the study of human B-cell and (auto) antibody repertoire generation and selection. *Immunol. Rev.* **128**, 23-47.

Madaio M.P., J. Carlson, J. Cataldo, A. Ucci, P. Migliorini and O. Pankewyez (1987). Murine monoclonal anti-DNA antibodies bind directly to glomerular antigens and form imune deposits. *J. Immunol.* **138**, 2883-2889.

Malynn B.A., G.D. Yancopoulos, J.E. Barth, C.A. Bona and F.W. Alt (1990). Biased expression of JH-proximal V_H genes occurs in the newly generated repertoire of neonatal and adult mice. *J. Exp. Med.* **171**, 843.

Manheimer-Lory A., J.B.Katz, M. Pillinger, C. Ghossein, A. Smith and B. Diamond (1991). Molecular characteristics of antibodies bearing an anti-DNA-associated idiotype. *J. Exp. Med.* **174**, 1639-1652.

Marion T.N., D.M. Tillman, N.T. Jou and R.J. Hill (1992). Selection of immunoglobulin variable regions in autoimmunity to DNA. *Immunol. Rev.* **128**, 123-147.

Matsuda F., E.K. Shin, H. Nagaoka, R. Matsumura, M. Haino, Y. Fukita, S. Takaishi, T. Imai, J.H. Riley, R. Anand, E. Soeda and T. Honjo (1993). Structure and physical map of the 64 variable segments in the 3' 0.8-megabase region of the human immunoglobulin heavy chain locus. *Nature Genetics* **3**, 88-94.

Mortari F., J.A. Newton, J.Y. Wang and H.W. Schoeder (1992). The human cord blood antibody repertoire. Frequent usage of the VH7 gene family. *Eur. J. Immunol.* **22**, 241-245.

Mortari F., J.Y. Wang and H.W. Schoeder (1993). Human cord blood antibody repertoire. *J. Immunol.* **150**, 1348-1357.

Muryoi T., T. Sasaki, A. Hatakeyama, S. Shibata, M. Susuki, J. Seino and K. Yoshinaga (1990). Clonotypes of anti-DNA antibodies expressing specific idiotypes in immune complexes of patients with active lupus nephritis. *J. Immunol.* **144**, 3856-3861.

Pascual V. and J.D. Capra (1991). Human immunoglobulin heavy-chain variable region genes: organization, polymorphism, and expression. *Adv. Immunol.* **49**, 1-74.

Rada C., S.K. Gupta, E. Gherardi and C. Miltein (1991). Mutation and selection during the secondary response to 2-phenyl oxazalone. *Proc. Natl. Acad. Sci. USA* **88,** 5508-5512.

Rajewsky J., I. Forster and A. Cumano (1987). Evolutionary and somatic selection of the antibody repertoire in the mouse. *Science (Wash. DC)* **238**, 1088.

Russell J.H., B. Rush, C. Weaver and R. Wang (1993). Mature t cells of autoimmune *lpr/lpr* mice have a defect in antigen-stimulated suicide. *Proc. Natl. Acad. Sci. USA.* **90**, 4409-4413.

Sanz I. (1991). Multiple mechanisms participate in the generation of diversity of human H chain CDR3 regions. *J. Immunol.* **147**, 1720-1729.

Schatz D.G., M.A. Oettinger and M.S. Schlissel (1992). V(D)J recombinaison: molecular biology and regulation. *Annu. Rev. Immunol.* **10**, 359-383.

Schroeder H.W., L. Hillson and R.M. Perlmutter (1987). Early restriction of the human antibody repertoire. *Science* **238**, 791-793.

Schroeder H.W. and J.Y. Wang (1990). Preferential utilization of conserved immunoglobulin heavy chain variable gene segments during human fetal life. *Proc. Natl. Acad. Sci. USA* **87**, 6146-6150.

Shlomchik M., M. Mascelli, H. Shan, M.Z. Radic, D. Pisetsky, A. Marshak-Rothstein and M. Weigert (1990). Anti-DNA antibodies from autoimmune mice arise by clonal expansion and somatic mutation. *J. Exp. Med.* **171**, 265-292.

Shlomchik M.S., A. Marshak-Rosthstein, C.B. Wolfowicz, T.L. Rothstein and M. Weigert (1987). The role of clonal selection and somatic mutation in autoimmunity. *Nature* **328**, 805-811.

Shoenfeld Y., D. Isenberg, A. Rauch, M. Madaio, B.D. Stollar, R.S. Schwartz (1983). Idiotypic cross reactions of monoclonal human lupus autoantibodies. *J. Exp. Med.* **158**, 718-730.

Sohn J., R.M. Gerstein, C.-L. Hsieh, M. Lemer and E. Selsing (1993). Somatic hypermutation of an immunoglobulin μ heavy chain transgene. *J. Exp. Med.* **177**, 493-504.

Stewart A.K., C. Huang, A.A. Long, B.D. Stollar and R.S. Schwartz (1992). VH-gene representation in autoantibodies reflect the normal human B cell repertoire. *Immunol. Rev.* **128**, 101-122.

Suenaga R., M. Evans and N.I. Abdou (1991). Idiotypic and immunochemical differences of anti-DNA antibodies of a lupus patient during active and inactive disease. *Clin. Immunol. Immunopathol.* **61**, 320-331.

Susuki N., T. Harada, Y. Mizushima and T. Sakane (1993). Possible pathogenic role of cationic anti-DNA autoantibodies in the development of nephritis in patients with systemic lupus erythematosus. *J. Immunol.* **151**, 1128-1136.

Tan E.M. (1989). Antinuclear antibodies: diagnostic markers for autoimmune diseases and probes for cell biology. *Adv.Immunol.* **44**, 93-151.

Tonegawa S. (1983). Somatic generation of antibody diversity. *Nature* **302**, 575-581.

Tsao B.P., A. Chow, H. Cheroutre, Y.W. Song, M.E. McGrath and M. Kronenberg (1993). B cells are anergic in transgenic mice that express IgM anti-DNA antibodies. *Eur. J. Immunol.* **23**, 2332-2339.

van Es J.H., F.H.J.G. Meyling, W.R.M. van de Akker, H. Aanstoot, R.H.W.M. Derksen and T. Logtenberg (1991). Somatic mutatons in the variable regions of a human IgG anti-double-stranded DNA autoantibody suggest a role for antigen in the induction of systemic lupus erythematosus. *J. Exp. Med.* **173**, 461-470.

van Es J.H., H. Aanstoot, F.H.J. Gmelig-Meyling, R.H.W.M. Derksen and T. Logtenberg (1992). A human systemic lupus erythematosus-related anti-cardiolipin/single stranded DNA autoantibody is encoded by a somatically mutated variant of the developmentally restricted 51P1 VH gene. *J. Immunol.* **149**, 2234-2240.

Watanabe-Fukunaga R., C.I. Brannan, N.G. Copeland, N.A. Jenkins and S. Nagata (1992). Lymphoproliferation disorder in mice explained by defects in Fas antigen that mediates apoptosis. *Science* **356**, 314-317.

Winkler T.H., H. Fehr and J.R. Kalden (1992). Analysis of immunoglobulin variable region genes from human IgG anti-DNA hybridomas. *Eur. J. Immunol.* **22**, 1719-1728.

Yamada M., R. Wasserman, B.Z. Reichard, S. Shane, A.J. Caton and R. Giovanni (1991). Preferential utilization of specific immunoglobulin heavy chain diversity and joining segments in adult human peripheral blood B lymphocytes. *J. Exp. Med.* **173**, 395-407.

Young F.M., L. Tucker, D.B. Rubinstein, T. Guillaume, J.André-Schwartz, K.J. Barrett, R.S. Schwartz and T. Logtenberg (1990). Molecular analysis of a germ line-encoded idiotypic marker of pathogenic human lupus autoantibodies. *J. Immunol.* **145**, 2545-2553.

Zouali M. and A. Eyquem (1983). Idiotypic-anti-idiotypic interactions in systemic lupus erythematosus. Demonstration of oscillatory levels of anti-DNA auto-antibodies and reciprocal anti-idiotypic activity in a single patient. *Ann. Immunol. (Institut Pasteur)* **134C**, 377-391.

Zouali M., J.M. Fine and A. Eyquem (1984). Anti-DNA autoantibody activity and idiotypic relationships of human monoclonal proteins. *Eur. J. Immunol.* **14**, 1085-1089.

Zouali M. and B. Diamond (1990). Idiotype-mediated intervention in systemic lupus erythematosus. *J. Autoimmunity* **3**, 381-388.

Zouali M. and J. Thèze (1991). Probing VH gene family utilization in human peripheral B cells by *in situ* hybridization. *J. Immunol.* **146**, 2855-2864.

Zouali M. (1992). Development of human antibody variable genes in systemic autoimmunity. *Immunol. Rev.* 73-79.

Molecular analysis of anti-dsDNA autoantibodies from patients with systemic lupus erythematosus

Holger K. Fehr, Thomas H. Winkler and Joachim R. Kalden
Institute for Clinical Immunology and Rheumatology,
Dept. Medicine III, University of Erlangen-Nuremberg,
Krankenhausstrasse. 12, D-91054 Erlangen, Germany

Introduction

Autoantibodies against double-stranded DNA (dsDNA) are a characteristic marker of human systemic lupus erythematosus (SLE) and its murine models. These antibodies play a major role in the pathogenic mechanisms, since their titers correlate with disease activity and deposits of DNA/anti-DNA immune complexes are found in diseased organs (Tan, 1989). Furthermore, nephritis has been induced in SLE-prone mice by injection of monoclonal anti-dsDNA antibodies (Vlahakos et al, 1992). The typical SLE anti-dsDNA antibody is of high-affinity and of the IgG isotype. This contrasts with the IgM anti-DNA antibodies of the so-called "natural autoantibody repertoire" which can be isolated not only from SLE patients but also from normal donors. These antibodies are of lower affinity, polyreactive and prefer single-stranded DNA (ssDNA) to dsDNA as the target antigen. We have generated human monoclonal anti-dsDNA antibodies of the IgG isotype by direct fusion of peripheral blood lymphocytes isolated from patients with SLE with the cell line CB-F7. The the variable region genes of the antibodies were sequenced and compared to the most homologous germline genes.

Molecular analysis of anti-DNA antibodies

Cryoconserved mononuclear cells isolated from the blood or leukapheresis material of three patients with a highly active state of disease were fused directly with the cell line CB-F7 (Grunow et al, 1989). 17 of a few hundred

NATO ASI Series, Vol. H 80
Autoimmunity: Experimental Aspects
Edited by Moncef Zouali

When tested for binding to a variety of antigens, clone 33.F12 was found to be crossreactive with cardiolipin, while clone 33.C9 showed weak reactivity to poly-ribose phosphate coupled to diphtheria toxoid. For the remaining four clones, no crossreactive antigen was found. This clearly contrasts with the binding characteristics of IgM anti-DNA clones isolated by us and by others, which where typically polyreactive. The dissociation constants of the monoclonal antibodies to dsDNA and ssDNA were measured by the method of Friguet et al. (Friguet et al, 1985). For the calculation of the K_D-values an average molecular weight of 660 (dsDNA) or 330 (ssDNA) per monomeric unit was used. This assumption was made because the size of the epitopes as well as the number of antibodies able to bind to a single DNA molecule are unknown. Comparable results were obtained using heterogeneous calf thymus DNA or plasmid DNA of defined length (data not shown). The K_D-values for anti-ssDNA of the IgG clones were approximately 2 to 3 magnitudes lower than the K_D values measured for monoclonal IgM anti-DNA antibodies. No affinity to dsDNA was detectable for the IgM clones (data not shown). A summary of the binding characteristics of the six IgG clones is given in Table 1. Four of the established clones orginated from one patient (S.S.). With the exception of one clone belonging to the VH4 family all others could be defined as members the VH3 family.

Table 1: Characteristics of the six human monoclonal IgG anti-dsDNA antibodies

Clone	Patient	Isotype	ELISA	Crithidia	Farr	dsDNA	ssDNA	Cardiolipin	PRP-D	K_D dsDNA	K_D ssDNA
35.21	H.E.	IgG2/λ	+	+	+	++	+	-	-	1.9×10^{-7}	2.2×10^{-7}
19.E7	M.H.	IgG1/κ	+	+	+	++	+	-	-	3.9×10^{-6}	7.9×10^{-6}
32.B9	S.S.	IgG3/λ3 *	+	+	+	++	++	-	-	9.6×10^{-9}	1.8×10^{-8}
33.F12	S.S.	IgG1/κ	+	+	-	++	++	++	-	1.8×10^{-6}	7.4×10^{-7}
33.C9	S.S.	IgG2/κ	+	+	-	++	++	-	+	9.1×10^{-9}	1.6×10^{-8}
33.H11	S.S.	IgG1/λ	+	-	-	++	++	-	-	6.6×10^{-7}	2.7×10^{-7}

Given are the source and the isotype of the antibodies as well as their reaction in three anti-dsDNA antibody assays of different stringency. Additonally the antibodies were tested in ELISA systems for their reactivity to dsDNA/ssDNA and various other antigens. Two of them crossreacted with different antigens. The K_D-values of the antibodies with ds- and ss-DNA as antigens are listed. No reactivity was found with chondrotin sulphate, tetanus toxoid, myosin, actin, phosphatidylcholine and vimentin. PRP-D is an acronym for poly-ribose phosphate coupled to diphtheria toxoid. * as verified by sequencing the 5′-region of intron B.

To analyze the composition of the antibodies, their variable region genes were cloned after amplification of first strand cDNA with adequate pairs of PCR primers (Larrick et al, 1989). In addition, genomic DNA of the hybridomas was amplified with primers homologous to the 5′-UTRs of the leader regions and primers homologous to the JH-regions or 3′-UTRs, respectively, and the leader exon and intron A were sequenced.

Figure 1: Comparison of the deduced amino acid sequences of five hybridomas VH genes with germline VH26 of patient S.S.

	FR1	CDR 1	FR2
VH26(S.S.)	EVQLVESGGGLVQPGGSLRLSCAASGFTFS	SYAMS	WVRQAPGKGLEWVS
32.B9	-----------------**Q**----**V**------**N**	**N**-**V**--	--------------
35.21	--**R**---------------------------	--W-H	----------**M**---
19.E7	Q---------V----R----------**I**•--	--G-**D**	-------------A
33.H11	---------------------------**I**•-	**NF**W-**T**	-------------A
33.F12	----K-------------------	D-Y--	-I------------

	CDR2	FR3
VH26(S.S.)	AISGSGGSTYYADSVKG	RFTISRDNSKNTLYLQMNSLRAEDTAVYYCAK
32.B9	**V**-----•----------	--------------**F**-----------•-----
35.21	R-N**A**D-S**NRR**--**G**----	---**V**----A------------**P**---------R
19.E7	V-WYD-**R**NK--------	-------------------------------R
33.H11	N-**N**Q**G**-S**KRS**-V-----	-------**V**--S--------------------R
33.F12	Y-**T**S-S**T**Y**R**N-------	-------A--S----------**G**-•-------R

The deduced amino acid sequence of the germline gene VH26, as cloned from the patient S.S., is compared to the deduced amino acid sequences of the clones 32.B9, 35.21, 19.E7, 33.H11 and 33.F12, all belonging to the VH3 family. Identity is marked with a dash and differences due to somatic mutations are marked in bold. Silent somatic mutations are marked by dots or underlined letters.

To compare the sequences obtained with their germline counterparts, libraries of the VH3 and VH4 gene families were generated. Genomic DNA of patient S.S. was amplified with appropriate degenerate PCR primers for the leader regions and the heptamer-nonamer motif. The PCR products were subsequently cloned into a plasmid vector. After transformation of *E. coli,* the bacteria were plated and the libraries were screened for clones containing hybridomas VH sequences (Winkler et al, 1992). Most of the replacement mutations reside in the CDR regions. Statistical analysis of the ratio of silent to replacement mutations in the CDR and framework regions showed a low to very low probability indicative of a random distribution and suggesting that at least four of the antibodies have underwent an antigen-driven process of affinity-maturation (Table 2). By comparing the genes isolated from the libraries and sequences obtained from the literature with the variable regions of the anti-DNA antibodies we were able to identify somatic mutations during the maturation of the antibodies. In Figure 1 the amino acid sequence of the VH26 gene as isolated from patient S.S. is compared to those of the VH3 regions of five clones.

Table 2: Silent to replacement mutations (S/R) ratios in the CDR and framework (FR) regions

Clone	**S/R ratio in FRs**	**S/R ratios in CDRs**	**p-values***
32.B9	2/5	2/7	0.047
33.F12	1/1	0/4	0.005
33.H11	1/2	0/11	< 0.001
35.21	4/4	1/6	0.081
33.C9	9/9	3/5	> 0.2
19.E7	1/1	0/2	-
total	18/22	6/35	< 0.001

* probability to find replacement mutations in the CDRs at random

Analysis of the nature of the somatic mutations showed an overrepesentation of replacement mutations to particularly positive charged basic amino acids in the CDR regions. These amino acids are known to be involved in protein-DNA interactions in various models (Kostrewa et al, 1991) as well as in anti-DNA antibodies (Radic et al, 1993).

Table 3: Changes in amino acid composition in the CDR and FR regions.

Amino acid	Framework		CDR	
changes	silent	replacement	silent	replacement
total	14	18	4	23
conserved		11		12
charged		2		7
Arg/Lys/Asn		2		10

Amino acid changes due to somatic mutations in the six anti-dsDNA hybridomas were grouped by changes of conservative nature and changes introducing charged or basic residues.

Characterization of antibody 32.B9

Hybridoma 32.B9 was analysed to a larger extent. The comparison of the variable region of the heavy chain with published data gave the strongest homologies to the sequence of VH26 gene (Chen et al, 1988). By screening our VH3 gene library with oligonucleotides specific for an area in FR1 which differs in 32.B9 heavy chain (32.B9H) and VH26 by a single nucleotide, we were able to show that VH26 gene is most probably the corresponding germline gene. In patient S.S. this gene shows an allelic polymorphism (T->G at nucleotide position 13). This polymorphism was seen in 32.B9 clone as well. Sequence analysis of a part of the 5′-UTR, the leader region and intron A of the 32.B9 heavy chain showed identity to the published sequence of VH26. A genomic library of hybridoma 32.B9 in phage l Fix II (Stratagene, Heidelberg, FRG) was generated and clones containing rearranged light and heavy chain variable regions were isolated. Restriction mapping of approximately 3 kb 5′-UTR of 32B9H showed identity to the restriction pattern of the corresponding region of VH26 (data not shown).

Figure 2: Comparison of the VH of 32.B9 antibody with germline VH26 gene

```
                                                                Gln
32B9 GAG GTG CAG CTG GTG GAG TCT GGG GGA GGC TTG GTA CAG CCT GGG GGG TCC CAG AGA CTC
VH26 GAG GTG CAG CTG GTG GAG TCT GGG GGA GGC TTG GTA CAG CCT GGG GGG TCC CTG AGA CTC
                                                                Leu

             Val                         Asn ┌Asn     Val CDR1  ┐
32B9 TCC TGT GTA GCC TCT GGA TTC ACC TTT AAC │ AAC TAT GTC ATG AGC │ TGG GTC CGC CAG
VH26 TCC TGT GCA GCC TCT GGA TTC ACC TTT AGC │ AGC TAT GCC ATG AGC │ TGG GTC CGC CAG
             Ala                         Ser └Ser     Ala           ┘

                                             ┌Val             Gly   CDR2
32B9 GCT CCA GGG AAG GGG CTG GAG TGG GTC TCA │ GTT ATT AGT GGT AGT GGC GGT AGC ACA TAC
VH26 GCT CCA GGG AAG GGG CTG GAG TGG GTC TCA │ GCT ATT AGT GGT AGT GGT GGT AGC ACA TAC
                                             └Ala             Gly

                                  ┐
32B9 TAC GCA GAC TCC GTG AAG GGC │ CGG TTC ACC ATC TCC AGA GAC AAT TCC AAG AAC ACG CTG
VH26 TAC GCA GAC TCC GTG AAG GGC │ CGG TTC ACC ATC TCC AGA GAC AAT TCC AAG AAC ACG CTG
                                  ┘

     Phe                                             Val                     ┌ CDR3
32B9 TTT CTG CAA ATG AAC AGC CTG AGA GCC GAG GAC ACG GCC GTT TAT TAC TGT GCG AAA │
VH26 TAT CTG CAA ATG AAC AGC CTG AGA GCC GAG GAC ACG GCC GTA TAT TAC TGT GCG AAA │
     Tyr                                             Val                     └ H/N
```

Comparison of the VH gene of antibody 32.B9 and its corresponding germline gene VH26 as isolated from patient S.S.. In both sequences an allelic polymorphism is found at nucleotide position 15. Somatic mutations are marked in bold and the resulting amino acids are indicated.

The sequence of the light chain of 32.B9 antibody had only weak homologies to published sequences. We therefore isolated the corresponding germline gene from a human genomic library in phage λ. Southern blot analysis showed that this gene is the only member of a new Vλ gene family which we have designated Vλ8. When we isolated and sequenced this gene from the genomic DNA of the patient S.S. as well as from a number of normal donors, we did not find any differences to the sequence of the λ clone. For the above reasons, we feel sure about the germline gene used by the hybridoma 32.B9. A comparison of the sequences of the light and heavy chain and the corresponding germline genes is given in Figures 2 and 3.

Figure 3: Comparison of the light chain of 32.B9 with the germline gene Vλ8.

```
32B9 TCT CAG ACT GTG GTG ACC CAG GAG CCA TCG TTC TCA GTG TCC CCT GGA GGG ACA GTC ACA
VL8  TCT CAG ACT GTG GTG ACC CAG GAG CCA TCG TTC TCA GTG TCC CCT GGA GGG ACA GTC ACA

                                       CDR1        Ser
32B9 CTC ACT TGT GGC TTG AGC TCT [ GGC TCA GTC TCT ACT AGT TCT TAC CCC AGC ] TGG TAC
VL8  CTC ACT TGT GGC TTG AGC TCT [ GGC TCA GTC TCT ACT AGT TAC TAC CCC AGC ] TGG TAC
                                                        Tyr

                             Gly                 Ile      Asn          CDR2
32B9 CAG CAG ACC CCA GGC CAG GGT CCA CGC ACG CTC ATT TAC [ AAC ACA AAC ACT CGC TCT TCT ]
VL8  CAG CAG ACC CCA GGC CAG GCT CCA CGC ACG CTC ATC TAC [ AGC ACA AAC ACT CGC TCT TCT ]
                             Ala                 Ile      Ser

32B9 GGG GTC CCT GAT CGC TTC TCT GGC TCC ATC CTT GGG AAC AAA GCT GCC CTC ACC ATC ACG
VL8  GGG GTC CCT GAT CGC TTC TCT GGC TCC ATC CTT GGG AAC AAA GCT GCC CTC ACC ATC ACG

                                                    Leu          Val   CDR3
32B9 GGG GCC CAG GCA GAT GAT GAA TCT GAT TAT TAC TGT [ TTG CTG TAT GTG GGT AGT GGC
VL8  GGG GCC CAG GCA GAT GAT GAA TCT GAT TAT TAC TGT [ GTG CTG TAT ATG GGT AGT GGC
                                                    Val          Met
```

The sequence of the light chain of 32.B9 antibody is compared to the sequence of the corresponding germline gene Vλ8, the only member of a new Vλ family. Somatic mutations are marked in bold and the corresponding amino acids are indicated.

If one assumes that anti-dsDNA antibodies are antigen-selected and antigen-driven, the question immediately arises: what triggers the antibody response, since as was shown repeatedly, natural double-stranded cellular DNA *per se* is not immunogenic (Kalden et al, 1991). The presence of circulating free DNA has been demonstrated repeatedly in SLE sera, ranging from 2 mg/ml to more than 300 mg/ml. Sano and Morimoto (1982) discussed the possibility that the enrichment of guanine/cytosine blocks in RNA isolates from DNA-anti-DNA complexes from SLE patients could turn non-immunogenic B-DNA into an immunogenic form. More recently, Li and Steinmann (1989) provided evidence that SLE plasma contains a non-random selection of human genomic base sequences, which makes it highly unlikely that plasma DNA in patients originates from foreign DNA, and is responsible for triggering anti-dsDNA antibody responses by molecular mimicry. However, the authors could not exclude the possibility that the isolated plasma DNA might consist of rather small fractions of non-human genomic sequences, which could account for the induction of anti-dsDNA antibodies. To address further the problem of the unknown antigen-inducing anti-dsDNA antibodies in SLE, our group isolated and analyzed plasma nucleic acids obtained from circulating anti-dsDNA immune complexes from SLE patients. Biochemical analysis of the plasma nucleic acids revealed a DNA component of about 15-20 Kb with a riboguanosine/cytosine ratio of about 15 %, which is significantly increased as compared to normal random human cellular DNA. In addition to the DNA component, 60-200 bases long RNA molecules were also identified. Furthermore, it could be demonstrated that immunization of mice and rabbits with the DNA/RNA containing plasma nucleic acids resulted in the induction of antibodies reactive with dsDNA (Krapf et al, 1989).

Discussion

In sera of patients with SLE a large variety of autoantibodies against numerous antigens, especially against nuclear antigens, are found. Only a few of them, including the high-affinity anti-dsDNA antibodies, are specific for SLE and are thought to be involved in pathogenic mechanisms. The titer of anti-DNA antibodies correlates with disease activity and DNA/anti-DNA immune complexes were demonstrated in skin lesions as well as in the glomeruli of patients with lupus nephritis (Tan, 1989). In the murine system, lupus nephritis was induceable by the injection of monoclonal anti-dsDNA antibodies (Vlahakos et al, 1992). Since anti-DNA antibodies play an important role in SLE, the analysis of monoclonal anti-dsDNA antibodies could give new insights into important pathogenic events. We therefore generated and analyzed human monoclonal anti-dsDNA antibodies of the IgG isotype by immortalization of B-cells from SLE patients.

In the present experiments fusion rates increased significantly when the patients were in an activated state of disease (Winkler et al, 1991), which might be due to the increased number of dsDNA-secreting B cells in the peripheral blood. Similarly it was shown in the mouse model that fusion rates using activated B-cell were approximately 100 times higher than using inactive B-cells (Andersson and Melchers, 1978). Since the B-cells used were obtained from SLE patients in an highly active stage of disease, we presume that our hybridomas represent a subfraction of the pathogenic autoantibody secreting B-cells. The pattern of the somatic mutations found suggests an antigen driven process of affinity-maturation. This was confirmed by the statistical analysis of the ratios of silent to replacement mutations in the CDR and framework regions as well as by the fact that there was a high rate of amino acid exchanges relevant for DNA-protein interaction (Kostrewa et al, 1991; Radic et al, 1993) in the CDRs. The statistical analysis indicated a high probability for a non-random pattern of mutation for clones 33.H11, 32.B9, 33.F12 and 35.21, which is in favour of a positive, antigen-driven selection of these clones. Also the heavy chain of clone 19.E7 contained only few mutations which fit very well into the pattern of a typical affinity-selection. Similar results were described by van Es et al. (1991) for an anti-dsDNA antibody secreting B-cell, immortalized from the blood of a patient with SLE by transformation with EBV. Their antibody, T14, as well as our antibody 33.F12 used the Vk kv325 gene. Interestingly this gene was also found to be used by rheumatoid factors (Chen et al, 1990).

The analysis of antibody 32.B9 demonstrated that the variable region genes and their germline correlates were comparable, as shown by cloning, sequencing and restriction mapping of the variable region genes and the germline genes. Antibody 32.B9 used gene VH26, which was previously shown to be used in the restricted foetal repertoire (Chen et al, 1990), the natural autoantibody repertoire and by normal adults (Stewart et al, 1993). Antibodies encoded by VH26 and a number of highly homologous genes bear the idiotype 16/6, which was found to be elevated in sera of patients with SLE (Silvestris et al, 1989). However, in contrast to this result Stewart et al. (1993) found no difference of the expression of this gene in normal donors and patients with SLE. The λ-light chain of clone 33.H11 was encoded by a mutated Vλ2 gene known to correlate with the 8.12 idiotype. This idiotype has previously reported in anti-DNA antibodies of patients with SLE (Paul et al, 1992). The somatic mutations found in our anti-DNA antibodies were homologous to the somatic mutations found in murine monoclonal antibodies isolated from mice suffering from SLE like syndrome. Radic et al. (1993), analyzing the influence of the somatic mutations, were able to show that a single Gly to Arg mutation was able to enhance the affinity of an anti-

DNA antibody several fold. By introducing of arginine residues into the VH-gene Radic et al. were additionally able to enhance the affinity of the antibody to DNA, showing that basic residues like arginine could play a major role in antibody-protein interactions. As already mentioned, the trigger mechanism leading to anti-dsDNA antibodies is still not understood. From the available data it seems that DNA complexed to carrier proteins or RNA, as shown in our experiments, or in the form of nucleosomes (Rumoe and Steinmann, 1990) could induce anti-dsDNA antibodies. That nucleosomes might induce an anti-dsDNA antibody response has recently been demonstrated by Mohan et al. (1993). Furthermore, the possibility has been demonstrated that anti-dsDNA antibodies might have a direct cytopathic effect on cells, via a specific DNA membrane receptor (Bennett et al, 1987).

In summary our data show that anti-dsDNA antibodies from SLE-patients are antigen-selected. That this process of affinity maturation is assisted by T-cells seems likely but has still to be proven. Although the trigger mechanisms of anti-dsDNA antibody are not yet completely understood, increasing evidence suggests that complexes containing ds-B-DNA) such as nucleosomes might play an important role. It is of interest that murine T-cells can give help to anti-DNA secreting B-cells of the same origin (Mohan et al, 1993; Naiki et al, 1992; Rajagopalan et al, 1990). Whether there are similar "anti-DNA" T-cells in patients with SLE remains to be investigated. Also other autoantibodies, such as human IgG rheumatoid factors, were found to be somatically mutated and probably affinity-maturated (Randen et al, 1993). Experiments to analyze the functional relevance of the somatic mutations found in antibodies 32.B9 and 33.F12 similar to those of Radic et al. (1993) are now in progress.

References

Andersson J. and Melchers F. (1978). The antibody repertiore of hybrid cell lines obtained by fusion of X63-AG8 myeloma cells with mitogen activated B-cell blasts. Curr. Top. Microbiol. Immunol. 81, 130-139.

Bennett R.M., Kotzin B.L. and Merrit M.J. (1987) . DNA receptor dysfunction in systemic lupus erythematosus and kindred disorders. Introduction by anti-DNA antibodies, antihistone antibodies, and anti-receptor antibodies. J. Exp. Med.166, 850-858.

Chen P.P., Liu M.F., Sinha S. and Carson D.A. (1988). A 16/6 idiotype positive anti-DNA antibody is encoded by a conserved VH gene with no somatic mutation. Arthitis Rheum. 31, 1429-1431.

Chen P.P., Olsen N.J., Yang P.M., Soto-Gil W., Olee T., Siminovitch K. A. and Carson D.A. (1990). From human autoantibodies to the fetal antibody repertoire to B cell maignancy: It´s a small world after all. Intern. Rev. Immunol. 5, 239-251.

Friguet B., Chaffotte A.F., Djavadi-Ohaniance L. and Goldberg M.E. (1985). Measurements of the true affinity constant in solution of antigen-antibody complexes by enzyme linked immunosorbent assay. J. Immunol. Meth. 77, 305-319.

Grunow R., Jahn S., Porstman S., Kiessig S.S., Steinkellner H., Steindl F., Mattanovich D., Gurtler L., Deinhardt F., Kattinger H. and von Baer R. (1988). The high efficency, human B cell immortalizing heteromyeloma CB-F7. J. Imm. Meth. 106, 257-265

Kalden J.R., Winkler T.H., Herrmann M. and Krapf F. (1991). Pathogenesis of SLE: immunopathology in man. Rheumatol. Int. 11, 95-100.

Kostrewa D, Granzin J., Koch C., Choe H.-W., Raghunathan S., Wolf W., Labahn J., Kahmann R. and Saenger W. (1991). Three-dimensional structure of the *E. coli* DNA binding protein FIS. Nature 349, 178-180.

Krapf F., Herrmann M., Leitmann W. and Kalden J.R. (1989). Antibody binding of macromolecular DNA and RNA in the plasma of SLE patients. Clin. Exp. Immunol. 75, 336-342.

Larrick J.W., Danielson L., Brenner A., Abrahamson M., Fry K.E. and Borrebaeck C.A. (1989). Rapid cloning of rearranged immunoglobulin genes from human hybridoma cells using mixed primers and the polymerase chain reaction. Biochem. Biophys. Res. Commun. 160, 1250-1256.

Li T.Z. and Steinmann C.R. (1989). Plasma DNA in systemic lupus erythematosus. Characterization of cloned base sequences. Arthitis Rheum. 32, 726-733.

Mohan C., Adams S. Stanik V. and Datta S.K. (1993). Nucleosome: a major immunogen for pathogenic autoantibody-inducing T cells of lupus. J. Exp. Med. 177, 1367-1381.

Naiki M., Chiang B-L, Cawley D., Ansari A., Rozzo S.J., Kotzin B.L., Zotnik A. and Gershwin M.E. (1992). Generation and characterisation of cloned T helper cell lines for anti-DNA responses in NZB.H-2bm12 mice. J. Immunol. 149, 4109-4115.

Paul E., Iliev A., Livneh A. and Diamond B. (1992). The anti-DNA-associated idiotype 8.12 is encoded by the VlII gene family and maps to the vicinity of L chain CDR1. J. Exp. Med. 149, 3588-3595.

Radic M. Z., Mackle J., Erikson J., Mol C., Anderson W.F., and Weigert M. (1993). Residues that mediate DNA binding of autoimmune antibodies. J. Immunol. 150, 4966-4977.

Rajagopalan S., Zordan T., Tsokos G. and Datta S.K. (1990). Pathogenic anti-DNA autoantibody-inducing T helper cell lines from patients with active lupus nephritis: Isolation of CD4-8- T helper cell lines that express the gd T-cell antigen receptor. Proc. Natl. Acad. Sci. USA 87, 720-7024.

Randen I., Pascual V., Victor K., Thompson K.M., Førre Ø., Capra J.D. and Natvig J.B. (1993). Synovial IgG rheumatiod factors show evidence of an antigen-driven immune response and a shift in the V gene repertoire compared to IgM rheumatoid factors. Eur. J. Immunol. 23, 1220-1225.

Rumoe P.M. and Steinmann C.R. (1990). Endogenous circulating DNA in systemic lupus erythematosus. Occurence as multimeric complexes bound to histone. J. Clin. Invest. 86, 69-74.

Sano H. and Morimoto C. (1982). DNA isolated from DNA/anti-DNA immune complexes in systemic lupus erythematosus is rich in guanine-cytosine content. J.Immunol. 128, 1341-1345.

Silvestris F., Searles R.A., Williams R.C. Rodriguez M.A. and Schwartz R. S. (1989). Distribution of anti $F(ab')_2$ antibodies and the 16/6 idiotype in systemic lupus erythematosus (SLE) probands and kindreds. J. Clin Immunol 9, 462-468.

Stewart A.K., Huang C., Stollar B. D. and Schwartz R.S. (1993). High-frequency representation of a single VH gene in the expressed human B cell repertiore. J. Exp. Med. 177, 409-418.

Tan E.M. (1989). Antinuclear antibodies: diagnostic markers for autoimmune diseases and probes for cell biology. Adv. Immunol. 44, 93-151.

Van Es J.H., Gmelig Meyling F.H., van de Akker W.R., Aanstoot H., Deksen R.H. and Logtenberg T. (1991). Somatic mutations in the V-regions of a human IgG anti-dsDNA autoantibody suggest a role for antigen in the induction of SLE. J. Exp. Med. 173, 461-470.

Vlahakos D.V., Foster M.H., Adams S., Katz M., Ucci A.A.,Barrett K.J., Datta S.K. and Madaio M.P. (1992). Anti-DNA antibodies form immune deposits at disitinkt glomerular and vascular sites. Kidney Int. 41, 1690-1700.

Winkler T.H., Jahn S. and Kalden J.R. (1991). IgG human monoclonal anti-DNA autoantibodies from patients with systemic lupus erythematosus. Clin. Exp. Immunol. 85, 379-385.

Winkler T.H., Fehr H. and Kalden J.R.. (1992). Analysis of immunoglobulin variable region genes from human IgG anti-DNA hybridomas. Eur. J. Immunol. 22, 1719-1728.

Differential usage of an autoantibody-associated V_H gene by human B-cell tumors

Freda K. Stevenson,[1] Myfanwy B. Spellerberg,[1] Joanna Treasure,[2] Caroline J. Chapman,[1] Leslie E. Silberstein,[3] Terry J. Hamblin,[1] and David B. Jones[2]
[1]Molecular Immunology Group, Tenovus Laboratory, and [2]Department of Pathology, Southampton University Hospitals, Tremona Rd., Southampton SO9 4XY UK; and [3]Department of Pathology and Laboratory Medicine, Hospital of the University of Pennsylvania, Philadelphia, Pennsylvania, USA

INTRODUCTION

During maturation of a normal B lymphocyte, selection and recombination of a particular V_H gene with other genetic elements occurs, giving rise to a functional V_H-D-J_H unit. Selection is from among the six known V_H families, which together comprise 100-200 genes (Berman et al, 1988; Schroeder et al, 1989), and the resulting recombinations in the B-cell population encode the functional repertoire. Since this repertoire can include autoantibodies, there has been considerable interest in the nature of the V_H genes selected to encode these potentially pathological molecules.

In the case of monoclonal IgM autoantibodies, such as rheumatoid factor and anti-DNA antibodies, it appears that there is some preferential selection of certain V_H genes to encode particular specificities (reviewed in Stewart et al, 1992). However, preferential gene usage is extreme in autoimmune hemolytic anaemia, where the monoclonal IgM anti-red cell antibodies of anti-Ii specificity appear always to be encoded by a single gene, V_H4-21, a member of the V_H4 family (Pascual et al, 1992). Such restriction of an antibody specificity to a single V_H gene is highly unusual, and suggests involvement of a conserved sequence of the

NATO ASI Series, Vol. H 80
Autoimmunity: Experimental Aspects
Edited by Moncef Zouali

variable region in red cell binding (Stevenson et al, 1991). Autoanti-red cell antibodies synthesized by normal B cells in patients with infectious mononucleosis are similarly encoded by the V_H4-21 gene (Chapman et al, 1993).

Although the gene appears mandatory for anti-red cell antibodies, it is not confined to these. Analysis of its involvement in encoding other antibodies has been facilitated by the raising of a MoAb (9G4) specific for Igs encoded by theV_H4-21 gene (Stevenson et al, 1986; Thompson et al, 1991). Using the MoAb, and confirmatory nucleotide sequencing, usage of the gene by anti-DNA antibodies in patients with SLE has been demonstrated (Stevenson et al, 1993a). However, a panel of control monoclonal IgM, IgG and IgA paraproteins was negative (Stevenson et al, 1989). In summary therefore, the V_H4-21 gene appears to be strongly associated with autoantibodies, particularly those with specificity for red cells or DNA.

Linkage between autoimmunity and B-cell tumors has been indicated by the fact that Igs secreted by neoplastic B cells frequently have autoantibody activity (Dighiero et al, 1983). Also, in the B-cell tumor chronic lymphocytic leukemia (CLL), the Ig expressed at the cell surface often has rheumatoid factor (RF) activity (Casali et al, 1988). Since RF has been found to be encoded by a restricted set of V_H and V_L genes, this results in an asymmetric V-gene usage in CLL (Logtenberg et al, 1989). In the case of the V_H4-21 gene, it is clearly involved in the B-cell tumors which secrete anti-red cell antibodies, and these comprise ~12% of the IgM-secreting neoplasms known as Waldenstrom`s macroglobulinemia (Crisp and Pruzanski, 1982). It was of interest therefore to use the 9G4 MoAb to trace involvement of this autoantibody-associated V_H gene in a range of B-cell tumors which express Ig at the cell surface.

RESULTS

Specificity of the 9G4 MoAb.

The correlation between expression of the 9G4 idiotope and usage of the V_H4-21 gene, based on published data (Thompson et al, 1991; Pascual et al, 1992; Pascual et al, 1993; Logtenberg et al, 1993) and additional unpublished data from our laboratory is shown in Table 1. As indicated, reactivity with the MoAb is highly associated with usage of the V_H4-21 gene segment. Recently, mutational analysis has located the motif recognized by the MoAb to amino acids 23-25 in the first framework region of V_H4-21 (Potter et al, 1993).

Table 1. Expression of the 9G4 idiotope by sequenced immunoglobulins of various V_H families and individual members of V_H4

	9G4 idiotope expression	
V_H family	Negative	Positive
V_H1	27	0
V_H2	3	0
V_H3	54	0
V_H5	2	0
V_H6	2	0
V_H1,2,3,5,6	159	0
V_H4 family		
V_{2-1}	4	0
V_{11}	1	0
V_{71-2}	3	0
V_{71-4}	6	0
V_{79}	2	0
V_{4-21}	1	63

Expression of V_H4-21-associated idiotope by B-cell tumors.

Expression of the V_H4-21-associated idiotope by B-cell tumors of various categories (Table 2) indicates some asymmetry. Thus in CLL, 3/81 (3.7%) of cases were positive, whereas in the lymphomas (all categories), 9/62 (14.5%) were positive. This difference is statistically significant ($0.1>P>0.05$ by χ^2 analysis), and becomes more significant if the lymphocytic lymphomas, which are very similar in cell morphology and clinical behaviour to CLL, are included with the CLL cases.

The data suggest therefore that there is preferential usage of the V_H4-21 gene by tumor cells in B-cell lymphoma but not in CLL (Stevenson et al, 1993b). Within the lymphoma group there are insufficient cases to detect significant differences among different histological categories, but the gene apparently can be used by mucosa-associated and splenic marginal zone lymphomas. The phenotype of the positive tumors (Table 3) shows that the surface Ig was generally IgM with either κ or λ light chains. The CLL cases were all CD5-positive, and the lymphoma cases CD5-negative, indicating that the V_H4-21 gene can be used by either CD5-positive or negative neoplastic B cells.

Table 2. Expression of V_H4-21-associated idiotope by B-cell Tumors

Tumor category	Number of cases	Positive cases
Chronic lymphocytic leukemia	81	3
Lymphocytic lymphoma	9	0
Marginal zone lymphoma	1	1
Follicle centre lymphoma:		
Follicular, low grade	18	2
Intermediate and high grade	28	5
Lymphoma of mucosa-associated lymphoid tissue	6	1*

* This single positive case showed morphological transformation to a high grade tumor.

Nucleotide sequence analysis of selected tumors.

In order to confirm involvement of the V_H4-21 gene, and to investigate the pattern of mutations in the gene in the various neoplasms, representative cases of 9G4-positive tumors were taken at random for nucleotide sequence analysis from the CLL and follicular lymphoma categories. The single case of marginal zone lymphoma (MZL) was also investigated.

For all three tumors, the nucleotide sequence confirmed usage of the V_H4-21 gene. For the case of CLL, the nucleotide sequence was identical to the germline sequence with no mutations; Figure 1 shows the deduced amino acid sequence. For the MZL, there were only three nucleotide differences from the germline sequence, but two of these gave rise to amino acid substitutions (Fig.1).

Table 3. Phenotypic Analysis of Idiotope-Positive Cases

Tumour category	Tissue	Ig expression	CD5
Chronic lymphocytic leukemia	Blood	IgMλ	+
	Blood	IgMκ	+
	Blood*	IgMκ	+
Marginal zone lymphoma	Blood*, spleen	IgMDλ	-
Follicle centre cell lymphoma:			
Follicular, low grade	Lymph node, *blood	IgMκ	-
	Lymph node	IgMλ	-
Intermediate and high grade	" "	IgMλ	-
	" "	IgMκ	-
	" "	IgMκ	-
	" "	IgMλ	-
	" "	λξ	-
Lymphoma of mucosa-associated lymphoid tissue	Lung	IgMλ	-

* Nucleotide sequence analysis of V_H genes was carried out on the tissues indicated.
ξ The heavy chain was not identified in this case.

In contrast, the nucleotide sequence from the case of follicular lymphoma revealed 29 mutations scattered among complementarity-determining regions (CDRs) and framework regions, and giving rise to 14 amino acid substitutions. In one case (MZL), the tumour-derived IgM was available and was found to have anti-red cell activity.

DISCUSSION

Analysis of the V_H-genes used to encode autoantibodies provides information at several levels: first, preferential usage of certain V-genes to encode particular antibody specificities may be revealed; second, mutational patterns of the expressed gene sequences may indicate whether the cell of origin has undergone hypermutation and antigen-selection. The latter point is important since it is possible that autoantigen could have a role in "driving" B cells, even following neoplastic transformation (Friedman et al, 1991). The third point is that there appears to be a tantalizing connection between autoimmunity and B-cell tumors which may be accessible via a study of the V_H genes involved. For all three points, analysis of the behaviour of the V_H4-21 gene has added useful information.

		CDR1	
V_H4-21 (GL)	QVQLQQWGAGLLKPSETLSLTCAVYGGS	FSGYYWS	WIRQPPGKGLEWIG
CLL (IgM)	--	------------	----------------------
MZL (IgM)	--	--D---------	----------------------
FCCL (IgM)	--	---D------N	-------S----------T----

	CDR2	
V_H4-21 (GL)	EINHSGSTNYNPSLKS	RVTISVDTSKNQFSLKLSSVTAADTAVYYCAR
CLL (IgM)	--------------------------	--
MZL (IgM)	--------------------------	-------------------------R---------------------------
FCCL (IgM)	------R--TA-------------	----A---I--------T-------R--T--L-----------L---------

	CDR3		
V_H4-21 (GL)			
CLL (IgM)	VGWYYDFWSGYSVSD	YYYMDV	JH6
MZL (IgM)	SSAPPIVVVTAIPLTAQPG	YYGMDV	JH6
FCCL (IgM)	NGTSGD	FDY	JH4

Figure 1. Deduced amino acid sequence comparison of the V_H4-21 genes expressed by three neoplastic B-cell populations, with the germline gene. CLL: chronic lymphocytic leukemia; MZL: marginal zone lymphoma; FCCL: follicular centre cell lymphoma.

The V_H4-21 gene is switched on early in development and remains active in adult lymphoid tissue, where B cells using the gene comprise ~3% of the total B-cell population of the lymph node (Stevenson et al, 1989). Restricted usage of the gene by anti-red cell antibodies leads to a rise in serum levels of immunoglobulin encoded by the V_H4-21 gene in certain infections such as EBV, where these autoantibodies are synthesized (Chapman et al, 1993). A further possible involvement of the V_H4-21 gene in response to infection has been indicated recently by the finding that certain IgM monoclonal antibodies which recognize lipid A, a common constituent of gram-negative bacteria, also recognize red cells, and are encoded by the V_H4-21 gene (Bhat et al, 1993). It appears therefore that this V_H gene may represent an important constituent of immunity, but that it can be involved in potentially damaging autoantibodies such as anti-red cell and anti-DNA antibodies (Pascual et al, 1992; Stevenson et al, 1993a), possibly due to molecular mimicry between infective organisms and autoantigens (Sinha et al, 1990).

Since the gene is expressed in ~3% of normal lymph node B cells, it might be expected that, on a random basis, its incidence in B-cell tumors would be similar. In fact the incidence in CLL (3.7%) is close to this figure. However, the incidence in B-cell lymphoma (14.5%) is higher than expected, consistent with preferential expression in this group of tumors (Stevenson et al, 1993b). The autoantibody reactivity of only one of these non-secreting tumors has been investigated so far, and for this case (MZL) the IgM was found to have anti-red cell activity (Stevenson et al, 1993b).

Analysis of nucleotide sequences of the V_H4-21 gene, which is facilitated by its low level of polymorphism (Sanz et al, 1989), revealed that the CLL case is unmutated, as has been reported for other V_H genes in CLL (Kuppers et al, 1991). In the case of marginal zone lymphoma, V_H gene is also relatively unmutated, and this would suggest that the cell of origin has not been exposed to the hypermutation mechanism. In contrast, in the case of follicular lymphoma the V_H gene is highly mutated, consistent with the fact that normal B cells undergo hypermutation in the follicle centre (Jacob et al, 1991). This degree of mutation is similar to that reported for another case of follicular lymphoma involving the V_H4-21 gene (Bahler and Levy, 1992). However, the pattern of the mutations differs between the two cases, with the previously reported case having clusters of replacement mutations in the CDRs suggestive of antigen-selection (Bahler and Levy, 1992). Clearly, more information is required to gain further insight into the origin of tumors, their relationship to infection and autoimmunity, and into a possible role for antigen in pathogenesis.

Concluding remarks

The V_H4-21 gene is rearranged early in development and encodes antibodies produced following certain infections. It also selectively encodes autoantibodies against red cells and DNA. Expression of the gene in B-cell tumors has revealed an asymmetry, with no preferential usage in CLL, but an increased usage in B-cell lymphoma. This suggests that certain autoantibodies may be sythesized by B cells involved in lymphoma which differ from those known to be associated with CLL. Sequence analysis of this relatively non-polymorphic gene has indicated that the cell of origin in the case of follicular lymphoma had been exposed to the hypermutation mechanism.

ACKNOWLEDGEMENTS

This work has been supported by Tenovus, the Arthritis and the Rheumatism Council and the Wessex Medical Trust, U.K. FKS is a recipient of a NATO collaborative grant.

REFERENCES

Bahler, D.W., and Levy, R. (1992). Clonal evolution of a follicular lymphoma: evidence for antigen selection. Proc.Natl.Acad.Sci. USA 89,6770-6774.

Berman, J.E., Mellis, S.J., Pollock, R., Smith, C.L., Suh, H., Heinke, B., Kowal, C., Surti, U., Chess, L., Cantor, C.R., Alt, F.W. (1988). Content and organization of the human Ig V_H locus: definition of three new V_H families and linkage to the Ig C_H locus. EMBO J. 7,727-738.

Bhat, N.M., Bieber, M.M., Chapman, C.J., Stevenson, F.K., Teng, N.N.H. (1993). Human anti-lipid A monoclonal antibodies bind to human B cells and the "i" antigen on cord red blood cells. J.Immunol in press.

Casali, P., Prabhakar, B.S., Notkins, A.L. (1988). Characterization of multireactive autoantibodies and identification of LEU-1+ B lymphocytes as cells making antibodies binding multiple self and exogenous molecules. Int.Rev.Immunology 3,17-45.

Chapman, C.J., Spellerberg, M.B., Smith, G.A., Carter, S.J., Hamblin, T.J., Stevenson, F.K. (1993). Autoanti-red cell antibodies synthesized by patients with infectious mononucleosis utilize the V_H4-21 gene segment. J.Immunology 151,105-1061.

Crisp, D. and Pruzanski, W. (1982). B-cell neoplasms with homologous cold reacting antibodies. Amer.J.Med. 72,915-922.

Dighiero, G., Guilbert, B., Fermant, J.P., Lymberi, P., Danon, F., Avrameas, S. (1983). Thirtysix human monoclonal immunoglobulins with antibody activity against cytoskeleton proteins, thyroglobulin and native DNA: immunologic studies and clinical correlations. Blood 62,264-269.

Friedman, D.F., Cho, E.A., Goldman, J., Carmack, C.E., Besa, E.C., Hardy, R.R., Silberstein, L.E. (1991). The role of clonal selection in the pathogenesis of an autoreactive human B cell lymphoma. J.Exp.Med. 174,525-537.

Jacob, J., Kelsoe, G., Rajewsky, K., Weiss, U. (1991). Intraclonal generation of antibody mutants in germinal centers. Nature 254,389-392.

Kuppers, R., Ganse, A., Rajewsky, K. (1991). B cells of chronic lymphatic leukemia express V genes in unmutated form. Leuk.Res. 15,487-496.

Logtenberg, T., Schutte, M.E.H., Inghirami, G., Berman, J.E., Gemlig-Meyling, F.H.J., Insel, R.A., Knowles, D.M., Alt, F.W. (1989). Immunoglobulin V_H gene expression in human B cell lines and tumors. Biased V_H gene expression in chronic lymphocytic leukemia. Int. Immunol. 1,362-366.

Pascual, V., Victor, K., Lelsz, D., Spellerberg, M.B., Hamblin, T.J., Thompson, K.M., Randen, I., Natvig, J. Capra, J.D., Stevenson, F.K. (1991). Nucleotide sequence analysis of the V regions of two IgM cold agglutinins. Evidence that the V_H4-21 gene segment is responsible for the major cross-reactive idiotype. J.Immunology 146,4385-4391.

Pascual, V., Victor, K., Spellerberg, M., Hamblin, T.J., Stevenson, F.K., Capra, J.D. (1992). V_H restriction among human cold agglutinins. The V_H4-21 gene segment is required to encode anti-I and anti-i specificities. J. Immunology 149,2337-2344.

Potter, K.N., Li, Y., Pascual, V., Williams, R.C. Jr., Byres, L.C., Spellerberg, M., Stevenson, F.K., Capra, J.D. (1993). Molecular characterization of a cross-reactive idiotope on human immunoglobulins utilizing the V_H4-21 gene segment. J.Exp.Med. in press.

Schroeder, H.W., Hillson, J.L., Perlmutter, R.M. (1990). Structure and evolution of mammalian V_H families. Int. Immunol. 2,41-50.

Sanz, I., Kelly, P., Williams, C., Scholl, S., Tucker, P., Capra, J.D. (1989). The smaller human V_H gene families display remarkably little polymorphism. EMBO J. 8,3741-3748.

Sinha, A.A., Lopez, M.T., McDevitt, H.O. (1990). Autoimmune diseases: the failure of self tolerance. Science 248,1380-1388.

Stevenson, F.K., Wrightham, M., Glennie, M.J., Jones, D.B., Cattan, A.R. Feizi, T., Hamblin, T.J., Stevenson, G.T. (1986). Antibodies to shared idiotypes as agents for analysis and therapy for human B cell tumors. Blood 68,430-436.

Stevenson, F.K., Smith, G.J., North, J., Hamblin, T.J., Glennie, M.J. (1989). Identification of normal B-cell counterparts of neoplastic cells which secrete cold agglutinins of anti-I and anti-i specificity. Brit.J.Haem. 72, 9-15.

Stevenson, F.K., Spellerberg, M.B., Jefferies, L.C., Goldman, J., Hamblin, T.J., Pascual, V., Capra, J.D., Silberstein, L.E. (1991). Monoclonal gammopathies and autoimmunity. Topics in ageing research in Europe. Monoclonal gammopathies III Vol.14. Proceedings of the EURAGE meeting, Brussels, Belgium.

Stevenson, F.K., Longhurst, C., Chapman, C.J. Ehrenstein, M., Spellerberg, M.B., Hamblin, T.J., Ravirajan, C.T., Latchman, D., Isenberg, D. (1993a). Utilizaton of the V_H4-21 gene segment by anti-DNA antibodies from patients with systemic lupus erythematosus. J. Autoimmunity, in press.

Stevenson, F.K., Spellerberg, M.B., Treasure, J., Chapman, C.J., Silberstein, L.E., Hamblin, T.J., Jones, D.B. (1993). Differential usage of an Ig heavy chain variable region gene by human B-cell tumors. Blood 82,224-230.

Stewart, A.K., Huang, C., Long, A.A., Stollar, B.D., Schwartz, R.S. (1992). V_H-gene representation in autoantibodies reflects the normal human B-cell repertoire. Immunol. Rev. 128,101-122.

Thompson, K.M., Sutherland, J., Barden, G., Melamed, M.D., Randen, I., Natvig, J.B., Pascual, V., Capra, J.D., Stevenson, F.K. (1991). Human monoclonal antibodies against blood group antigens preferentially express a V_H4-21 variable region gene-associated epitope. Scand.J.Immunol. 34,509-518.

Idiotypes of DNA-Binding Autoantibodies:Mapping by Epitope Scanning and Molecular Modelling

Norman A. Staines, Francis J. Ward, Paul Hobby[1], Andrew N. Denbury, Sukhvinder Bansal[2], Brian J. Sutton[3], D. Gwyn Williams[1], David A. Isenberg[4], Jatinderpal Kalsi[4] and Samuel R. Oleinick[5]

Infection & Immunity Group, King's College London, Campden Hill Road, London W8 7AH, UK,
[1]Renal Unit, Division of Medicine, UMDS, Guy's Hospital, London SE1 9RT, UK,
[2]School of Pharmacy, King's College London, Manresa Road, London SW3 6LX,
[3]Department of Biophysics, King's College London, 26-29 Drury Lane, London WC2B 5RL,
[4]Bloomsbury Rheumatology Unit, University College London, 40-50 Tottenham Street, London W1P 9PG, UK, and [5]Department of Medicine, VA Medical Center (111F), 921 N. E. 13th Street, Oklahoma City, OK 73104, USA.

SLE and DNA-binding antibodies

One of the cardinal features of systemic lupus erythematosus (SLE) is the production of autoantibodies against a range of intracellular autoantigens. The spectrum of antibodies, and their fine specificities, may vary between individuals with SLE reflecting the varied nature of the aetiologial and pathogenic processes that contribute to the SLE syndrome. One relatively constant feature of the serology of SLE, however, is the production of antibodies that react with DNA and these are found both in humans and also in mice that develop the disease. It is generally accepted that these antibodies play a significant role in the developing pathology, especially that associated with nephritis. For example, transfusion of some DNA-binding antibodies into healthy BALB/c mice (Vlahakos et al., 1992) or young MRL/*lpr* mice in early stages of disease (Lake & Staines 1988) will cause significant changes in kidney function. It is quite likely, if not certain, that other autoantibodies also contribute significantly to lupus nephritis and the dominant role ascribed to the DNA antibodies may to an extent be a result of the concentration of research effort directed at them.

NATO ASI Series, Vol. H 80
Autoimmunity: Experimental Aspects
Edited by Moncef Zouali

DNA-reactive antibodies are not unique to SLE, although those with specificity for double stranded (ds) DNA are more likely to be associated with pathology, whereas antibodies against single stranded (ss) DNA appear as a natural consequence of many infections and traumatic insults (see Shoenfeld & Isenberg, 1993). It is understandable that such antibodies should be produced as a result of tissue damage that would release nuclear material from cells. However, there is some hesitancy in assuming that the DNA-binding antibodies in SLE, and other situations, are always produced as a result of an antigen-driven immune response simply because DNA itself is an unimpressive experimental immunogen. It is immunogenic under some circumstances, for example, nucleosomes which are naturally released from cells as part of the apoptotic process are able to stimulate anti-DNA antibody production. Further, the way in which mutations accumulate in the variable heavy chain (V_H) and light chain (V_L) regions of DNA-binding antibodies argues for an antigen driven response in which high affinity clones are selected by DNA itself (Shlomchik et al., 1990; Radic et al., 1993), and this is the best proof that the antibodies are stimulated by DNA itself.

The idiotypes of DNA antibodies

Antibody idiotypes (Ids) are structural attributes of antibody V regions that can be detected by anti-Id antibodies. Extensive studies of DNA antibodies from individuals with SLE, other autoimmune and infectious diseases, and from normal individuals have shown, in general, that they carry Ids that are widely distributed on both DNA and other antibodies. They are public or cross-reactive idiotypes, and there is some evidence for inheritance of Ids in families (Isenberg et al., 1990). There are also clear demonstrations that certain Ids associate especially with lupus disease. These come from studies at the population level (Isenberg et al., 1990) or at the level of Id expression in immune deposits in biopsy material (Ebling & Hahn 1980; Isenberg & Collins, 1985; Watts et al., 1991) and from experiments on antibody pathogenicity (Hahn & Ebling 1984; Vlahakos et al., 1992; Katz et al., 1993).

Mostly, the Ids of DNA antibodies have been defined by heterologous polyclonal antisera raised against purified serum antibodies, or monoclonal antibodies (mAbs), but a smaller number of studies with monoclonal anti-Id antibodies also show that the Ids are widespread in their distribution. Thus, we might assume that each Id is composed of a set of idiotopes, each identified by a single anti-Id antibody, and each distributed in an idiosyncratic pattern on different immunoglobulins according to their usage of individual V genes. Because V genes originate in families of homologous members, and because families are also related, we might also expect individual idiotopes to be encoded by more than one V gene, or indeed by other members of the immunoglobulin superfamily such as T cell receptor (TCR) genes.

In many systems, anti-Id antibodies are immunosuppressive, or can induce the production of the corresponding Id+ antibodies themselves. SLE is no exception, and there have been several demonstrations that anti-DNA anti-Id antibodies will suppress DNA antibody production and the progression of lupic disease in mice. However, the original observations from Hahn and Ebling (1984) showed that permanent suppression could not be achieved and that there was a late rebound of antibody production involving antibodies expressing Ids different from those originally suppressed, and fulminating disease eventually developed in mice treated in this way. Some subsequent studies have produced results in accord with this (Zouali et al., 1985) and others have not (Teitelbaum et al., 1984; Jacob & Tron 1984) although it is possible that a combined treatment with anti-Id antibodies and cyclosporin may be more effective in BWF1 mice than the antibodies on their own (Morland et al, 1991). A related observation is that immunization of normal BALB/c mice, that do not develop idiopathic lupus disease, will do so if immunized with a DNA-binding mAb carrying the 16/6 Id administered in Freund's Complete Adjuvant (Mendlovic et al., 1988), and disease in MRL/Mp mice can be accelerated by a similar immunization with DNA-reactive mAbs (Ravirajan & Staines, 1991).

Thus, the study of the Ids of DNA-reactive antibodies has potential value in the clinical assessment of disease, in the analysis of immunopathological mechanisms, in unravelling the origins and relationships of autoantibodies and in identifying targets for immune-based anti-Id therapy of disease. Against the background of the information about the DNA antibody Ids described above, we know very little about the structural basis of the idiotypes. Knowledge of this would contribute to understanding the nature and significance of the biological and clinical experiments.

Analysis of idiotype structure

There are no reasons to think that antibodies are any different from other proteins in the way their epitopes are formed. The various approaches that have been used to analyse the structure of the Ids of DNA-binding antibodies are summarised in Table 1. By analogy with other proteins, it is to be expected that many B-cell epitopes (those detected by anti-Id antibodies) would depend upon conformational structures that would be lost upon degradation of the molecule. Some Ids do not survive denaturation of the antibody H and L chains in Western blotting, and others are dependent upon the correct relationship of the H and L chains for their expression. This appears to be the structural basis of private Ids which are limited in their expression to one antibody. Typically, private Ids are found on anti-protein antibodies produced as a result of prolonged immunization where these antibodies are the products of mutated V genes. So private Ids are likely to define mutated genes. The elegant studies of the structure of co-crystals of an anti-lysozyme antibody and an anti-Id antibody against it provide clear support for this idea (Bentley et al., 1990). In this case the private idiotype is revealed to be centred on the paratope of the antibody: the Id and its anti-Id dock end-to-end.

Table 1. Approaches to Analysis of Antibody Idiotypes

Serological analysis of Id distribution on different antibodies
Expression of Ids by denatured antibody chains
Comparison of V gene sequences of Id+ and Id- antibodies
Synthetic V region sequence peptides as immunogens/antigens
Structural analysis of Id-anti-Id co-crystals

Antibodies against public Ids on the other hand do not always compete effectively with antigen for the antibody. From this it is concluded that some idiotopes reside outside the paratope. This information collectively implies that a public Id may include structures that are (a) relatively conserved or identical between different immunoglobulins, (b) not necessarily related to the specificity of the antibody, and (c) encoded by heritable sequences in V genes. Other observations support this definition of a public Id. For example, there are cases where synthetic peptides corresponding to sequences in an antibody V region have been used to induce the production of anti-peptide antibodies that react with the intact antibody (Mazza et al., 1987; Meek et al., 1990). Further, heritable Ids may be conserved on antibodies of changing fine specificity produced in A/Jax mice as they respond to antigenic stimulation by arsonate (Ars-KLH and V_H-

Id^{CR}) (Naparstek et al., 1986) or in lupus mice as they age and make DNA antibodies of progressively higher affinity (Shlomchik et al., 1991).

A new approach to idiotype structure

Against this background we have recently used epitope scanning methods to identify epitopes in the V regions of DNA autoantibodies that are recognised by antibodies in anti-Id sera. These continuous linear epitopes are putative idiotopes, and their location allows the construction of a linear idiotope map for the V_H and V_L regions. In parallel, the structure of the antibody can be modelled by computer molecular modelling methods. When the linear idiotope map is superimposed on the hypothesised structure, a 3-dimensional topological model of the antibody V region and its idiotopes is formed (Figure 1).

Figure 1 Strategy to produce a three-dimensional idiotope map of DNA-binding antibodies.

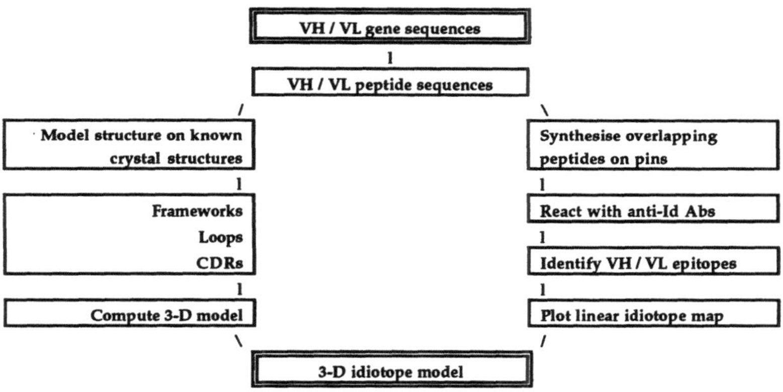

Choice of antibodies. Four mAbs, two murine and two human, have been studied, all bind DNA and their properties are summarised in Table 2. Antibodies V-88 and WRI-176 were chosen because they are members of the 16/6 Id family, and antibodies F-423 and BEG-2 because they were derived from foetal tissues and might be expected to be coded by genes that had not mutated away from the germ line sequences.

Table 2 DNA-binding mAbs analysed with overlapping peptides

mAb	Origin	VH	VL	Specificity	References
V-88 (IgG1k)	BWF1 adult female	7183 member	k1A	ssDNA>dsDNA >RNA	Morgan et al., 1985a
F-423 (IgG3k)	MRL/lpr 16-day foetus	VH (?)	k (?)	ssDNA>dsDNA >RNA	Staines et al., 1993a Ravirajan & Staines 1993
BEG-2 (IgMl)	Human 12 week foetal liver	VH4	l (?)	ssDNA>dsDNA >p(ADP ribose)	Watts et al., 1991
WRI-176 (IgMk)	Human polymyositis	VH3	kIIIA	ssDNA=dsDNA	Submitted

Peptide synthesis and assay methods. From the inferred peptide sequences of the antibody V region genes, sets of overlapping peptides were made, using Fmoc chemistry, according to techniques described by Geysen et al. (1987). In all cases, individual duplicate hexameric or octameric peptides were made commencing at each amino acid in the V_H and V_L regions respectively. The peptides, with acetylated amino termini and anchored by their carboxy termini on polypropylene pins, were used (repeatedly after regeneration) in an enzyme linked immunosorbent assay (ELISA) to determine with which of them antibodies in various sera reacted. The procedures have been described in detail (Staines et al., 1993b). Sera used have included heterologous antisera raised in rabbits against the individual mAbs that are designed, through absorption on normal serum immunoglobulin, to be specific for the immunizing mAb - these are conventional heterologous anti-Id polyclonal reagents. In many cases these antisera have been tested against V region peptides without absorption because any antibodies against non-V region epitopes should be undetectable in the peptide ELISA system. Emphasis here will be placed on the analysis of the murine antibody, V-88, for which we currently have the most information.

Sera from human subjects with a variety of different autoimmune and infectious diseases and from healthy and lupic mice have also been examined for V region peptide-reactive antibodies.

V region epitopes recognised by heterologous anti-Id antisera

Antisera raised in rabbits, against purified mAbs, reacted with V_H and V_L region peptides in all the human and mouse antibodies studied. The example of V-88 is typical, and four major epitopes in each of the V_H and V_L regions were detected: reactions with these were detectable by antibodies in the original unabsorbed antiserum and the same antiserum after absorption with pooled normal mouse immunoglobulin. In the case of the absorbed antiserum the background reactions were reduced, and some significant peaks of reactivity were

Figure 2: Epitope scanning of the V_H region of the V-88 mAb with rabbit anti idiotype antibodies and with antibodies occurring naturally in sera from MRL/*lpr* mice

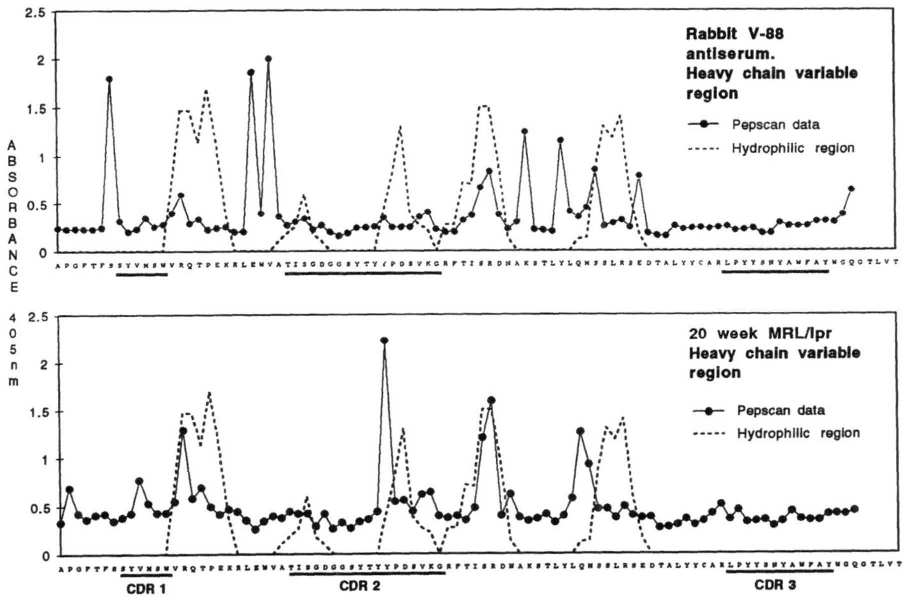

Shown are the pepscan results from using a rabbit antiserum raised against mAb V-88 (upper panel) and a serum from a 20 week-old MRL/lpr mouse, each tested against V_H region hexapeptides. The amino terminal residue for each peptide is indicated. CDRs are underlined. The ELISA readings are compared with a hydrophilicity plot.

lost in the absorption process. It was quite clear, however, that the dominant anti-peptide antibodies were easily detectable in the original antiserum. The specificity of the reactions of these polyclonal antisera is not easy to define; for example, when tested against the peptides of a heterologous, but structurally related, antibody (WRI-176) there were reactions indicating that one anti-mAb antiserum contains antibodies that react with V region epitopes not limited to the immunizing antibody. Formal adsorption-elution experiments have not yet been conducted to determine to what extent the reactions with individual peptides are specific for the reference antibody.

Figure 2 shows a typical linear idiotope map generated from such scans. The rabbit antibodies reacted with epitopes in both CDR and FR regions of the V_H region. The reactive V-88 hexapeptides did not conform to a particular set of physical properties. Algorithms based on the properties of hydrophilicity, amphipathicity or mobility failed to identify any obvious trend that could predict epitopes. However, most epitopes do have predicted large areas of solvent accessibility suggesting that many of the antibodies reactive with the peptides would also be reactive with the native molecule. As Figure 2 indicates, the epitopes recognized by antibodies in MRL/*lpr* mouse sera tended to be in more hydrophilic regions than those epitopes recognized by rabbit antibodies. It appears that antibodies reactive with hydrophilic peptides may be induced by a response against the whole antibody molecule and those against more hydrophobic peptides by a reaction against degradation products of the antibody. This would be consistent with the fact that the rabbit was immunized with mAb emusified in adjuvant, whereas the mice make antibodies as a natural part of the disease process.

V region epitopes recognised by murine lupic sera

Using the mAb V-88 peptides, a number of mouse sera have been examined. In these experiments, and in others like them, the strategy has been to test pools of sera (usually five) rather than individual sera. This is because the peptides have a limited life-span and this permits more sera to be screened. It was found that lupus mice have antibodies in their sera that bind to V region peptides (Table 3). It was notable that MRL/*lpr* mice had more antibody than BWF1 mice. In the former, IgM antibodies preceded IgG antibodies; in the latter there was only a weak reaction of IgM antibodies with little switching to IgG, and the patterns of reactivity were very similar to those of sera from BALB/c mice. From these experiments it appears that MRL mice develop antibodies against V region epitopes as their disease progresses. The reasons for this are not known, and it was also notable that BWF1 mice apparently had lower titres of antibodies of this type than MRL mice. The appearance of IgG antibodies implies the involvement of T cells. The response of B cells against surface-accesible epitopes of antibodies in lupus mice is in general accord with responses to many other proteins. The appearance of antibodies against peptide epitopes that are relatively inaccessible in the intact immunoglobulin molecules implies a degree of protein degradation or processing. Comparing the responses of immunized rabbits and lupus mice (not intentionally immunized) indicates that this occurs more as a result of intentional immunization than as part of the lupus disease process.

Table 3. Antibodies in lupus-prone and normal mice that bind V region peptides derived from mouse mAb V-88

Strain	Age					
	weeks		weeks		weeks	
	IgM	IgG	IgM	IgG	IgM	IgG
MRL/lpr	++	+	+++	++	++	+++
BWF1	+	-	++	-	++	-
BALB/c	+	-	+	-	-	+

V region epitopes recognized by human sera

A large number of sera from normal humans and those with SLE and other autoimmune rheumatic diseases have been examined for antibody reactivity against V region peptides from both human and mouse antibodies. Human sera generally gave higher background readings than mouse sera with antibody V region peptides, so that the interpretation of the pepscans was more difficult. However, the results were analogous to those found with mouse sera in that there were obvious areas of epitopic activity indicating that some sera, or pools of sera, contained antibodies absent from others. Using peptides of the V_H region of mAb BEG-2 (Table 2) we examined reactions of antibodies in polyclonal rabbit anti-mAb BEG-2 antisera, rabbit anti-V-88 antiserum and numerous human sera from patients with SLE and other autoimmune conditions. Many of the human sera expressed Id.BEG-2 (Table 2). No distinct hexamer epitopes specific for BEG-2 were identified with these antisera. The rabbit anti-BEG-2 antiserum reacted with two hexamers in FR1 and FR3. The rabbit anti-V-88 antiserum reacted with two overlapping hexamers in FR1, one in FR2, two sets of overlapping hexamers in FR3 and one in the terminal CDR3. The human sera which were Id.BEG-2+ gave inconclusive results, identifying multiple hexamers with each serum, but as other sera lacking the BEG-2 idiotype also reacted with various peptides, no distinct patterns which separated the two sets were seen.

Limitations of the pepscan method

The pepscan method is designed primarily to identify candidate sequences that may be epitopes for particular antibodies. It is essentially a screening method, and verification of a particular epitope usually follows by preparing synthetic peptides in large amounts and high purity for formal immunochemical analysis. Such experiments are in progress. There are two major limitations to the pepscan method. First, the purity and amount of each peptide on its pin cannot be determined and different pins may have different peptide loadings. We deal with this to a certain extent by making replicate peptides, and the fact that different sera give different reaction patterns also argues against this being a major problem. Second, some peptides may non-specifically adsorb immunoglobulins. It is commonplace that this happens with peptides made in this way, but it is circumvented by diluting sera/antibodies to give low background values. This is effective with mouse sera, but less so with human sera, but on the other hand this may reflect the presence of many more anti-peptide antibodies in humans than in mice. However, the absorbance values in a typical scan do not fall in a normal distribution (Figure 3), which might be expected if immunoglobulin binding occurred at random. There has been a V region binding motif identified that is important for the binding of monomeric or aggregated IgG to rheumatoid factors (Hay et al., 1991), and self-association sites have also been identified (Kaveri et al., 1991) in antibody V regions. These findings mean that antibodies can associate with peptides irrespective of their antigen-combining specificity.

The fact that sera from different individuals and from different disease groups react differently argues strongly against non-specific immunoglobulin adsorption being a major factor. It raises the question, however, of the specificity of the anti-peptide antibodies. From absorption-elution experiments some anti-peptide antibodies also react with the native immunoglobulin, and we know that rabbit antisera against mAbs such as those used here tend to react with denatured antibody H and L chains in Western blots. As antibodies are generally conserved molecules, it is not surprising to find that antisera against one mAb also react with others, both in their native form and as peptides as used here. Peptides larger than the hexamers or octamers employed might limit cross-reactions, but still a degree of latitude is anticipated in the critical sequences of V region peptides.

The 16/6 idiotype

The mouse antibody V-88 was previously found to express an idiotype related to the human 16/6 Id system (Morgan et al., 1985b). In comparing theequence of mAb V-88 with other published sequences of 16/6+ human mAbs (Young et al., 1990) (Table 4) we identified a region in the FR2/CDR2 area of the H

Figure 3. Epitope scanning results reveal positive reactions and do not show a normal distribution

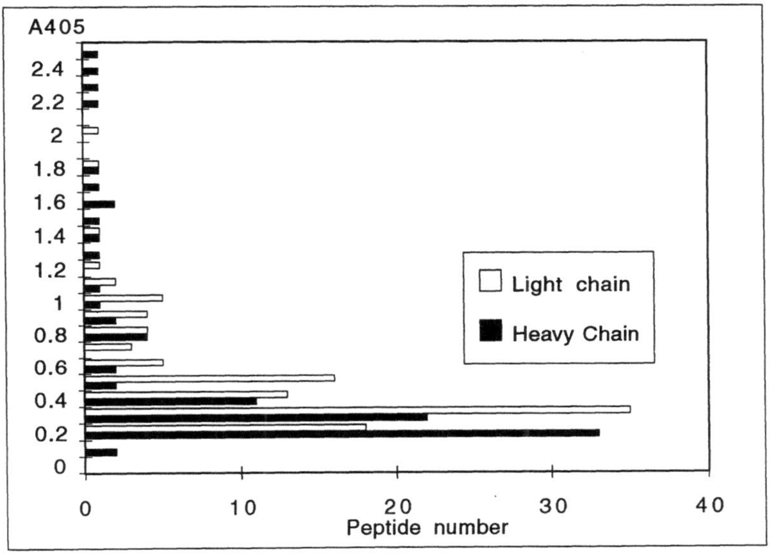

The figure shows the frequency distribution of reactivity in the ELISA of different peptides of mAb V-88 when reacted with antibodies in a pool of MRL/lpr sera.

chain that was relatively conserved (Staines et al., 1993b). This happened to be a region that was especially active when the peptides were scanned with the rabbit-anti-V-88 antiserum, but it was also recognized by antibodies in the mouse and human lupus sera. The sequence LEWVATISG, spanning FR2 and CDR2, is a putative idiotope of the 16/6 Id.

In another approach (Kalsi et al., to be published), pools of human SLE sera were made according to their expression of Id.16/6 and these were tested against the peptides derived from the sequence of WRI-176 which binds to a monoclonal anti-Id.16/6 reagent (Tucker L, Watt RA, Isenberg DA, unpublished observations). The rationale being that an anti-Id response is likely to be found in individuals also expressing the Id. It was found that in an Id.16/6+ pool there were antibodies against peptides in the centre of the VH CDR2 region, and that these antibodies were absent from a pool of Id.16/6- sera. Studies by Schwartz et al. (personal communication) have identified a gene sequence in the VH CDR2 that marks all Id.16/6+ human antibodies that have been tested. Thus from both the protein and gene approaches, the same area of the molecule is implicated as part of the 16/6 Id.

Table 4. Sequence homologies of FW2/CDR2 of DNA-binding mAb VH88 with other molecules

Id.16/6+ antibodies											
	FW3		CDR3								
L	E	W	V	A	T	I	S	G	D	VH88	Mouse
-	-	-	-	S	A	-	-	-	S	VH26	Human
-	-	-	-	S	A	-	-	-	S	VH18/2	Human
-	-	-	-	S	A	-	-	D	S	VHA73	Human
-	-	-	-	S	A	-	-	P	S	VHA85	Human
-	-	-	-	S	S	-	-	-	S	VHGL18	Human
-	-	-	-	S	V	-	Y	.	I	VHA57	Human
Hsp 60 family											
I	A	Q	-	-	-	-	-	A	N	Rat Hsp 60 $^{142-152}$ (GroEl)	
I	A	Q	-	-	-	-	-	A	N	Human GroEl $^{145-155}$	
I	A	Q	-	-	S	-	-	A	N	*burgdorferi*	
I	A	Q	-	-	-	-	-	A	N	*psittaci*	
I	A	Q	-	-	-	-	-	A	N	*cerevisiae*	
I	A	Q	-	G	-	-	-	A	N	*coli*	
I	A	Q	-	G	-	-	-	A	N	*burnetti*	
I	A	A	T	-	A	-	-	A	G	*bovis* TB65k144-153	
I	A	Q	-	-	A	-	-	A	A	*B.subtilis*	
I	A	Q	-	G	T	-	-	A	N	*Legionella pneumophilis*	

Sequence homologies of other proteins with antibody V regions

The same area of the V-88 H chain also contains a sequence that is homologous to a conserved part of the Hsp 60 sequence (Table 4). Limited absorption/elution experiments suggest that antibodies that bind the VATISG peptide of V-88 also cross-react with the VATISA peptide of mammalian Hsp 60 (GroEl), the homologue of the bacterial Hsp 60. It was interesting that a rabbit anti-V-88 antiserum and a human SLE serum (Re), with reactivity for the antibody sequence, also reacted with homologous sequences from a variety of bacteria (Figure 4).

Figure 4. Rabbit anti-V-88 and human SLE serum (Re) have antibody activity against Hsp peptide sequences

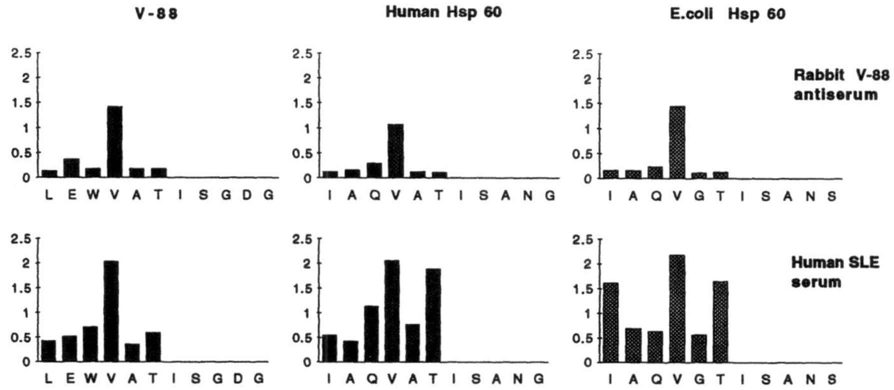

The rabbit anti-V-88 antiserum and a serum from a SLE patient (Re) were reacted in the peptide ELISA with sets of overlapping hexamers corresponding to limited sequences of the Hsp 60 from humans and from *E. coli* (as shown in Table 4) and with the relevant homologous sequence from the V_H of mAb V-88.

These cross reactions identify sequences in antibody V regions that occur also in non-antibody molecules. There are precedents for this (Swanson et al., 1991; Puccetti et al., 1990), but the fact that immune reactivity to Hsps is commonplace in arthritic diseases suggests a functional link such that an immune reaction to Hsp 60 would produce antibodies that would in turn be anti-idiotypic for antibodies bearing a correct V region motif. This could explain the appearance of antibodies bearing Id.16/6 in many situations, such as infection with *Klebsiella* (37% of patients) and in tuberculosis (60%), filariasis (25%) and schistosomiasis (24%) (as reviewed by Watts & Isenberg 1990) in which immune responses against bacterial or human Hsp 60 are known, or might reasonably be expected, to occur. The functional link could be mediated by either B cells or T cells or both. The involvement of T cells might, as implied above, account for the large amounts of antibody found that react with V region peptides in both humans and MRL mice.

The linear idiotype maps may be converted to topological maps by computer molecular modelling software. We have used the INSIGHT II and HOMOLOGY programmes for this purpose. In the case of mAb V-88, a suitable known structure was available: myeloma protein MCPc603 has been crystallised and its structure solved to 2.7Å resolution (Satow et al., 1986). The two antibodies have V_H and V_L framework regions that are ~60% homologous at the protein level; loop structures were based upon the canonical forms of Chothia and Lesk (1989), and parts of the FR3 and CDR3 of VH were found by conformational search of the data base for homologous structures. From these, an energy-minimalized structure was modelled to represent mAb V-88 and the solvent accessibility of its amino acids determined by the DSSP programme.

Superimposition of the linear idiotope maps on this model showed that the epitopes recognised by the antibodies in the heterologous rabbit antisera resided in both b sheet and loop regions, whereas the antibodies from the (allogeneic) MRL/lpr mice reacted predominantly with epitopes in the loops. Both rabbit and mouse antibodies identified epitopes in the H chain FR3 region that are on a protrusive structure on the modelled antibody. This region has been shown to characterise the third V_H clan of antibodies in the classification of Kirkham et al. (1992), and it also influences the conformation of the CDR2 loop which was found, as described earlier, to contribute to the make-up of Id.16/6. This reaction clearly shows that the idiotopes on the protrusive structure identify not the product of a single V_H gene, but rather products of a related family of genes. It explains clearly why particular Ids may associate with antibodies of greatly diverse specificities and it also suggests that rationally-directed anti-Id therapy may effectively eliminate or control many clones of B cells of widely diverse specificities.

Acknowledgements
The authors gratefully acknowledge the support of The Arthritis & Rheumatism Council, Lupus UK and Boots Research

References

Bentley GA, Boulot G, Riottot MM, Poljak RJ. (1990) Three-dimensional structure of an idiotype-anti-idiotype complex. *Nature* **348,** 254-257.

Chothia C, Lesk AM. (1987) Canonical structures for the hypervariable regions of immunoglobulins. *J Mol Biol* **196,** 901-917

Ebling F, Hahn BH. (1980) Restricted subpopulations of DNA antibodies inkidneys of mice with systemic lupus: comparison of antibodies in serum and renal eluates. *Arhritis Rheum* **23,** 392

Geysen HM, Rodda SJ, Mason TJ, Tribbick G. Schoofs PG. (1987) Strategies for epitope analysis using peptide synthesis. *J Immunol Methods* **102,** 259-274

Hahn BH, Ebling FM. (1984) Suppression of murine lupus nephritis by administration of an anti-idiotypic antibody to anti-DNA. *J Immunol* **132,** 187-190

Hay FC, Soltys AJ, Tribbick G, Geysen HM. (1991) Framework peptides from KIIIb rheumatoid factor light chains with binding activity for aggregated IgG. *Eur J Immunol* **21,** 1837-1841

Isenberg DA, Collins C. (1985) Detection of cross-reactive anti-DNA antibody idiotypes on renal tissue bound immunoglobulins from lupus patients. *J Clin Invest* **76,** 287-294

Isenberg DSA, Williams W, Axford J, Bakimer R, Bell D, Casaseca-Grayson T, Diamond B, Ebling F, Hahn B, Harkiss G, Mackworth-Young C, Le Page S, Massicotte H, Rauch J, Ravirajan C, Schwartz R, Shoenfeld Y, Staines Y, Todd-Pokropek A, Tucker L, Watts R, Zouali M. (1990) Comparison of DNA antibody idiotypes in human sera: an international collaborative study of 19 idiotypes from 11 different laboratories. *J Autoimmunity* **3,** 393-414

Jacob J, Tron F. (1984) Induction of anti-DNA auto-anti-idiotypic antibodies in (NZBxNZW)F1 mice: possible role for specific immune suppression. *Clin Exp Immunol* **58,** 293

Katz MS, Foster MH, Madaio MP. (1993) Independently derived murine glomerular immune deposit forming anti-DNA antibodies are encoded by near identical VH gene sequences. *J Clin Invest* **91,** 402-408

Kaveri S-V, Halpern R, Kang, C-Y, Kohler H. (1991) Antibodies of different specificities are self-binding: implication for antibody diversity. *Mol Immunol* **28,** 773-778

Kirkham PM, Mortari F, Newton JA, Schroeder HW. (1992) Immunoglobulin VH clan and family identity predicts variable domain structure and may influence antigen binding. *EMBO J* **11,** 603-609

Lake RA, Staines NA. (1988) A monoclonal DNA-binding autoantibody causes a deterioration in renal function in MRL mice with lupus disease. *Clin Exp Immunol* **73,** 103-110

Mazza G, Guigou V, Moinier D, Corbet S, Ollier P, Fougerau M. (1987) Molecular interactions in the 'GAT' idiotypic network. An approach using synthetic peptides. *Ann Inst Pasteur/Immunol Paris* **136D,** 259-269

Meek K, Takei M, Dang H, Sanz I, Dauphinee MJ, Capra JD, Talal N. (1990) Anti-peptide antibodies detect a lupus-related idiotype that maps to the H chain CDR2. *J Immunol* **144,** 1375-1381

Mendlovich S, Brocke S, Shoenfeld Y, Ben-Bassat M, Meshorer A, Bakimer P, Moses E. (1988) Induction of a systemic lupus erythematosus-like disease in mice by a common anti-DNA idiotype. *Proc Natl Acad Sci USA* **85,** 2260-2264

Mahana W, Guilbert B, Avrameas S. (1987) Suppression of anti-DNA antibody production in MRL mice by treatment with anti-idiotypic antibodies. *Clin Exp Immunol* **70,** 538-545

Morgan A, Buchanan RRC, Lew AM, Olsen I, Staines NA. (1985a) Five groups of antigenic determinants on DNA identified by monolconal antibodies from (NZBxNZW)F_1 and MRL/*lpr/lpr* mice. *Immunology* **55,** 75-83

Morgan A, Isenberg DA, Naparstek Y, Rauch J, Duggan D, Khiroya R, Staines NA, Schattner A. (1985b) Shared idiotypes are expressed on mouse and human anti-DNA autoantibodies. *Immunology* **56,** 393-399

Morland C, Michael J, Adu D, Kizaki T, Howie AJ, Morgan A, Staines NA. (1991) Anti-idiotype and immunosuppressant treatment of murine lupus. *Clin Exp Immunol* **83,** 126-132

Naparstek Y, Andre-Schwartz J, Manser T, Wysocki LJ, Breitman L, Stollar BD, Gefter M, Schwartz RS. (1986) A single germline VH gene segment of normal A/J mice encodes autoantibodies characteristic of systemic lupus erythematosus. *J Exp Med***164,** 614-626

Pucetti A, Koizumi T, Migliorini P, André-Schwartz J, Barrett KJ, Schwartz RS. (1990) An immunoglobulin light chain from a lupus-prone mouse induces autoantibodies in normal mice. *J Exp Med* **171,** 1919-1930

Radic MZ, Mackie J, Erikson J, Mol C, Anderson WF, Weigert M. (1993) Residues that mediate DNA binding of autoimmune antibodies. *J Immunol* **150,** 4966-4977

Ravirajan CT, Staines NA. (1991) Involvement in lupus disease of idiotypes Id.F-423 and Id.IV-228 defined, respectively, upon foetal and adult MRL/*Mp/lpr/lpr* DNA-binding monoclonal autoantibodies. *Immunology* **74,** 342-347

Satow Y, Cohen GH, Padlan EA, Davies DR. (1986) Phosphocholine binding immunoglobulin Fab McPC603. An X ray diffraction study at 2.7 Å. *J Mol Biol* **190,** 593-604

Shlomchik MJ, Sascelli M, Shan H, Radic MZ, Pisetsky D, Marshak-Rothstein A, Weigert M. (1990) Anti-DNA antibodies from autoimmune mice arise by clonal expansion and somatic mutation. *J Exp Med* **171,** 265-292.

Shoenfeld Y, Isenberg DA (eds). (1993) *Natural autoantibodies - Their physiological role and regulatory significance.* CRC Press, Boca Raton, Florida

Staines NA, Ravirajan CT, Morgan A, Belcher AJ, Henry AJ, Lake RA, Smith DA, Hamblin AS, Hara M, Adu D, Morland C, Isenberg DA (1993a) Expression and relationships of sèven public idiotypes of DNA-binding autoantibodies on monoclonal antibodies and serum immunoglobulins. *Lupus* **2,** 25-33

Staines NA, Ward FJ, Denbury AN, Mitchiner J, Hartley O, Eilat D, Isenberg DA, Bansal S. (1993b) Primary sequence and location of the idiotopes of V-88, a DNA-binding monoclonal autoantibody, determined by idiotope scanning with synthetic peptides on pins. *Immunology* **78,** 371-378

Swanson SJ, Lin B-F, Mullenix MC, Mortensen RF. (1991) A syntheitc peptide correspondng to the phosphorylcholine (PC)-binding region of human C-recative protein possesses the TEPC-15 myeloma PC-idiotype. *J Immunol* **146,** 1596-1601

Teitelbaum D, Rauch J, Stollar BD, Schwartz RS. (1984) *In vivo* effects of antibodies against a high frequency idiotype of anti-DNA antibodioes in MRL mice. *J Immunol* **132,** 1282-1285

Vlahakos DV, Foster MH, Adams S, Ucci AA, Barrett KJ, Datta SD, Madaio MP. (1992) Monoclonal anti-DNA antibodies derived from lupus-prone mice form immune deposits at distinct glomerular locations and induce different structural and functional abnormalities in vivo. *Kidney Intl* **41,**1690-1700

Watts RA, Isenberg DA. (1990) Autoantibodies and anti-bacterial antibodies: from both sides now. *Ann Rheum Dis* **49,** 961-965

Watts RA, Hillson JL, Oppliger IR, Mackenzie L, Lydyard PM, Mackworth-Young CG, Brown C, Staines NA, Isenberg DA. (1991) Sequence analysis and idiotypic relationships of BEG-2, a human fetal antibody reactive with DNA. *Lupus* **1,** 9-17

Young F, Tucker L, Rubinstein D, Guillaume T, André-Schwartz J, Barrett KJ, Schwartz RS, Logtenberg T. (1990) Molecular analysis of a germ line-encoded idiotypic marker of pathogenic human lupus autoantibodies. *J Immunol* **145,** 2545-2553

Zouali M, Jolivet M, Leclerk C, Ravisse P, Audibert F, Eyquem A, Chedio L. (1985) Suppression of murine lupus autoantibodies to DNA by administration of muramyl dipeptide and syngeneic anti-DNA. *J Immunol* **135,** 1091-1096

T cell receptor variable gene repertoire in rheumatoid synovial T cells responding to BCG

[1]Jacob B. Natvig, [2]Kay B. Wilson, [1]Alison Quayle, [1]Sel Suleyman, [1]Jens Kjeldsen-Kragh, [1]Øystein Førre, [1]Mouldy Sioud, and [2]J.Donald Capra
[1]Institute of Immunology and Rheumatology and the Rhematism Hospital, The University of Oslo, Oslo, Norway; and [2]Department of Microbiology, University of Texas, Southwestern Medical Center, Dallas, USA.

INTRODUCTION

Rheumatoid arthritis (RA) is a progressive disease that results in destruction of the joint tissues. RA tissues show infiltration with B and T lymphocytes, macrophages and dendritic cells as well as proliferation of synovial cells. These cells and their products appear to be responsible for the chronic inflammatory processes in the joint. T cells are the major inflammatory cells in the rheumatoid inflammatory tissues (Abrahamsen et al. 1975) . Since we do not know the etiology of this destructive disease, we have approached the question of the T cell receptor repertoire (TCR) by studying antigen-specific T cell clones responding to BCG (Quayle et al. 1992, Wilson et al. 1993) that represent frequent responses in these local T cells (Abrahamsen et al. 1978).

RESULTS

The methods utilized have been described in detail (Quayle et al. 1992, Wilson et al. 1993). The results showed the following. First, as shown in Table 1, there is a specific expansion of BCG-specific T cell clones that express the Vß8 gene segment in the synovial fluid of one patient (US) with early rheumatoid arthritis who was extensively studied (Wilson et al. 1993) .

NATO ASI Series, Vol. H 80
Autoimmunity: Experimental Aspects
Edited by Moncef Zouali

Table 1: Cloning of BCG-specific SF and PB lines.
(Results of FACS staining)

Number of clones positive for	Synovial fluid (SF)	Peripheral blood (PB)
αβ*	26	32
Vβ8	9**	2
Vβ6	3	4
Vβ5	0	0
Vβ12	0	0

* All clones produced were αβ+, CD4+ and CD8-
** p 0.001 (Fisher's tes)

Second, T cell clones responding to BCG reacted with a variety of different BCG antigens. The results are shown in Table 2 (Wilson et al. 1993). None of the BCG driven clones appeared to recognize the 65 Kd heat shock protein (HSP) of mycobacteria while a series of them identified antigens in the range of 19-60 Kd..

Table 2: Specificity of the clones

Clone	Specificity (MW, Kd)	Vβ gene expressed
SF2F9	15-20	8
SF3C2	55-60	8
SF1B10	24-30	8
SF3B7	UKN	8
PB1G3	UKN	8
SF3D3	15-19	6
SF2B9	24-30	6
PB1D8	21-24	6
PB1D6	15-19	6
SF2G4	UKN	2

SDS-separated BCG antigen blotted onto Nitrocellulose was used.
UKN: The antigenic specificity could not be determined.

Third, when TCR usage was analyzed in BCG-specific clones, it became apparent that there was a wide range of Vß families utilized and also different Dß and Jß gene segments (Table 3). Even within the group of Vß8-expressing BCG-specific clones, the CDR3 region was very divergent with different N segment additions as shown in Table 4 (Wilson et al 1993).

Table 3: TCR Vβ gene usage and amino acid sequences of the third hypervariable region of BCG-specific clones

	3'Vβ	5'N	Dβ	3'N	5'Jβ	Vβ-Dβ-Jβ
USSF2F9	CASS	PI	G	AG	NTEAFFGQG	Vβ8-Dβ1.1-Jβ1.1
USSF3C2	CASS	LT	RG	A	EAFFGQG	Vβ8-Dβ1.1-Jβ1.1
USSF1B10	CASS	Q	TG	IAP	EAFFGQG	Vβ8-Dβ1.1-Jβ1.1
USSF3B7	CASS	PW	G	-	QPQHFGDG	Vβ8-Dβ1.1-JB1.5
USSF2B9	CASSL*	R	D	K	NYQYTFGSG	Vβ6-Dβ1.1-Jβ1.2
USSF2G4	CSAR	E	TG	SG	TGELFFGEG	Vβ2-Dβ1.1-Jβ2.2
USSF1G2	CSA	N	GA	GSL	NEQFFGPG	Vβ2-Dβ1.1-Jβ2.1
USPB1D8	CASS	AC	T	TP	NYGYTFGSG	Vβ6-Dβ1.1-Jβ1.2
USPB1D6	CASSL	VG	GG	AGL	EQYFGPG	Vβ6-Dβ2.1-Jβ2.7
USPB1G3	CASS	-	S	QG	TEAFGQG	Vβ8-Dβ1.1-Jβ1.1

	3'Vβ	5'N	Dβ	3'N	Dβ	5'Jβ	Vβ-Dβ-Jβ
USSF3D3	CASS	S	LS	GGW	D	TQYFGPG	Vβ6-Dβ1.1inv-Dβ2.1-Jβ2.5

* The underlined sequences are contributed by the Vβ gene segment preceeding the heptamer-nonamer sequence

Fourth, by special cloning with HSP, we were able to identify some heat-shock protein-specific clones in patient US. These clones (Quayle et al. 1992) were studied with overlapping peptides. Two of the clones reacted in a region very homologous in human and bovine HSP. By testing overlapping peptides the minimal sequence was found to be localized within the stretch 243-251. The MHC restriction was also studied and turned out to be HLA-DQ for the clone USSF2C6 and HLA-DR for clones USSF1F11 and USSF1C6.

Table 4: Nucleotide sequence of the TCR Vβ chain third hypervariable region of BCG-stimulated SF clones

	3'Vβ8	5'N	Dβ	3'N
SF7	AGCAGT	--	AGGG	AGACA
SF9	AGCAGTTTAG*	--	CAG	CTT
SF16	AGCAGT	CTTCAG	GGGA	G
SF26	AGCAGTTTA	AAGTTCCC	CAG	CTCAATAG
SF28	AGCAGTTTA	CCC	GGGACAGGGG	TCGAA
SF39	AGCAGTTTA	AC	GAC	GGGCC
SF42	AGCAGT	ACCATTCAAGCAGG	TGGT	--
SF55	AGCAGTTT	CGTAA	ACAG	GTGG
SF57	AGCAGTTT	C	GGGG	CCGGATCG
SF58	AGCAGTTTA	--	CA	AGGA
SF63	AGCAGT	CTCCC	GGGG	--
SF66	AGCAGTTTA	CC	GGGA	ACTCGAAGC
SF101	AGCAGT	GGTTC	GGG	TGT
SF1028	AGCAGT	GGTCA	GACTA	A
SF1032	AGCAGTTT	--	GACT	GCAAG

* The underlined sequences are contributed by the 3' end of the Vβ gene segment preceeding the heptamer-nonamer sequence

DISCUSSION

The data presented here suggest that the T cell responses in early rheumatoid arthritis to BCG antigens in the local inflammatory tissues are rather heterogenous at least in the patient US we have studied extensively (Quayle et al. 1992, Wilson et al. 1993). Several approaches have been used to address the extent of heterogenity of the TCR repertoire in rheumatoid arthritis. One approach involves a strategy in which T cells are isolated in bulk from the synovium or the blood and their TCR's analyzed by molecular approaches, generally by PCR (Capra and Natvig 1993). These studies have given variable results both in different studies and in different patients (Sioud et al 1991, Uematzu et al. 1991, Paliard et al. 1991).

Here, we have used a more specific method in that a candidate antigen was chosen in order to clone certain synovial fluid T cells. The antigen-specific approach, however, has a major limitation in that the specific antigen chosen may not be relevant to the disease. In spite of these limitations we examined the T cell receptor repertoire to BCG antigens in rheumatoid arthritis. BCG appears to be a good candidate because T cells from the synovium react particularly strongly to it (Abrahamsen et al 1978). The data suggest that the T cell response to BCG in terms of Vß gene segment usage is heterogenous. This is not surprising as there are probably hundreds of potential antigenic epitopes contained within BCG. Although there is a certain preference for the Vß8 gene family, our results show that within the Vß8 response there is a considerable heterogeneity. This diversity can best be noted in several Vß8-Jß1.1 rearrangements which differ widely in nucleotide number and sequence (Table 4).

When looking at specific 65 Kd heat-shock protein responses, we also find a heterogenous picture of TCR V gene usage together with the interesting observation that 2 of 3 clones react with a region with striking homology between human and mycobacterial HSP (Quayle et al. 1992). In one case, direct crossreaction with a human HSP was demonstrated.

Evidence that there is an oligoclonal response in terms of Vß gene restriction in RA has lead other groups (Paliard et al. 1991) to postulate that the inciting antigen for RA could actually be a superantigen. Superantigens have been well characterized in the mouse and have been shown to produce a limited Vß gene response but have no effect on Dß and Jß gene segment utilization. However, our results differ from the current concept of superantigens in that other Vß gene segments are found to be expressed in response to BCG. It must be kept in mind, however, that superantigens have been largely defined in inbred strains of mice whereas humans are obviously outbred. Therefore, the concept of a superantigen invoking a single Vß gene response in the mouse may be too stringent a definition for the human population.

REFERENCES

Abrahamsen, T.G., Frøland, S.S., Natvig, J.B., Pahle, J. (1975). Elution and characterization of lymphocytes from rheumatoid inflammatory tissue. Scand. J. Immunol. 4,823-830.

Abrahamsen, T.G., Frøland, S.S., Natvig, J.B. (1978). In vitro mitogen stimulation of synovial fluid lymphocytes from rheumatoid arthritis and juvenile rheumatoid arthritis patients: dissociation between the responses to antigens and polyclonal mitogens. Scand. J. Immunol. 7,81-90.

Capra, J.D., Natvig, J.B. (1993). Is there V region restriction in autoimmune disease. Immunologist 1,16-19.

Paliard, X., West, S.G., Lafferty, J.A., Clements, J.R., Kappler, J.W., Marrack, P., Kotzin, B.L. (1991). Evidence for the Effects of a superantigen in rheumatoid arthritis. Science 253,325-329.

Quayle, A.J., Wilson, K.B., Li, S.G., Kjeldsen-Kragh, J., Oftung, F., Shinnick, T., Sioud, M., Førre, Ø., Capra, J.D., Natvig J.B. (1992). Peptide recognition, T cell receptor usage and HLA restriction elements of human heat-shock protein (hsp) 60 and mycobacterial 65-kDa hsp-reactive T cell clones from rheumatoid synovial fluid. Eur. J. Immunol. 22,1315-1322.

Sioud, M., Kjeldsen-Kragh, J., Quayle, A.J., Wiker, H.G., Sørskaar, D., Natvig, J.B., Førre, Ø. (1991). Immune responses to 18.6 and 30-kDa mycobacterial antigens in rheumatoid patients, and Vß usage by specific synovial T-cell lines and fresh T cells. Scand. J. Immunol. 34,803-812.

Uematsu, Y., Wege, H., Straus, A., Ott, M., Bannwarth, W., Lanchbury, J., Panayi, G., Steinmetz, M. (1991). The T-cell-receptor repertoire in the synovial fluid of a patient with rheumatoid arthritis is polyclonal. Proc. Natl. Acad. Sci. 88, 8534-8538.

Wilson, K.B., Quayle, A.J., Suleyman, S., Kjeldsen-Kragh, J., Førre, Ø., Natvig, J.B., Capra, J.D. (1993). Heterogeneity of the TCR repertoire in synovial fluid T lymphocytes responding to BCG in a patient with early rheumatoid arthritis. Scand. J. Immunol. 38,102-112.

An Analysis of T cell receptor diversity in the NOD mouse: What can it tell us about the autoimmune process?

Adelaida Sarukhan, Henri-Jean Garchon and Claude Carnaud
INSERM U 25
Hôpital Necker
161 rue de Sèvres
75743 PARIS Cedex 15
France

INTRODUCTION

Little is known concerning the features of the T cell receptor (TcR) repertoires which emerge in the course of cell-mediated autoimmune diseases. The finding of a highly restricted variable (V)α or Vβ usage by T cells from animals experimentally induced to respond to self antigens, such as myelin basic protein (MBP) (Acha-Orbea H., 1988) or type 2 collagen (David C.S., 1989, Chiocchia G., 1991), may not necessarily be true for spontaneously-triggered autoimmune reactions. In these reactions, the multiplicity of natural epitopes, the sequential release of sequestered antigens from damaged tissues and the aberrant expression of major histocompatibility complex (MHC) or costimulatory molecules on non-hematopietic cells may contribute to a rapid diversification of V gene usage. The question is important, not only from a theoretical, but also from a practical point of view since many strategies for the prevention of autoimmune diseases by modification of TcR expression are based on the assumption that, for a given individual and a given autoimmune disorder, the T cell repertoire is of sufficiently limited heterogeneity to warrant manipulation.

The non-obese-diabetic (NOD) mouse (Makino S., 1980) provides a remarkable opportunity to explore this issue in an animal model. NOD mice

NATO ASI Series, Vol. H 80
Autoimmunity: Experimental Aspects
Edited by Moncef Zouali

become overtly diabetic at 12 to 18 weeks of age as a result of an autoimmune, cell-mediated process which appears spontaneously, in 4 to 6 week-old animals. The features of the murine disease are strikingly similar to those observed in patients suffering from insulin-dependent type 1 diabetes (IDDM). Both disorders are under a polygenic control which involves MHC genes (Leiter E., 1989, Nepom, G.T., 1990); both are mediated by T cells (Miyazaki A., 1985, Bottazzo G.F., 1985) and both seem to require the coordinate intervention of CD4- and CD8-positive lymphocytes (Bendelac A., 1987, Bottazzo G.F., 1985). Finally, both involve an aberrant activation of T cells by putative β cell antigens which are still in the process of characterization (Elias D., 1990, Baekkeskov S., 1990).

There is still no clear answer as to whether anti-pancreatic activated T cells in NOD mice use a limited set of V genes. On the one hand, histochemical (O'Reilly L.A., 1991, Nakano N., 1991) and PCR analyses (Maeda T., 1991, Waters S.H., 1992) of infilrating T cells reveal heterogeneity of TCR expression. Similarly, sequencing of the TcR α and β chains from several pathogenic T cell clones did not result in the identification of recurrent V or J genes (Nakano N., 1991, Candeias S., 1991, Prud'homme G.J., 1991). On the other hand, recent experimental evidence from our laboratory indicates that the T cells which transfer the disease into naive recipients are biased towards the Vβ6 element (Edouard P., 1993). Similar biases for Vβ8.1 or Vβ8.2 have been reported by other groups (Fukuda M., 1989, Bacelj A., 1989, Bacelj A., 1990). The absence of an unequivocal answer is due, amongst other reasons, to the difficulty in delineating precisely the subset of effector cells involved in the pathogenesis of diabetes and in defining the stage of the disease at which the analysis is performed. Furthermore, no valid assertion can be made about possible biases of V gene usage by infiltrating T cells until there is an accurate and complete description of the basic repertoire emerging first in the thymus and secondarily in the peripheral lymphoid tissues of the NOD mouse. The present paper addresses two of these issues. First, we provide an almost complete account of the Vβ repertoire found in the spleen of young prediabetic NOD mice, both in terms of gene family distribution and nucleotide sequence. Second, we explore the question of clonal heterogeneity at the inception of inflammatory foci formation, i.e. at 4 weeks of age.

An Analysis of the Peripheral Vβ Repertoire in NOD Mice

A thorough analysis of Vβ distribution in the peripheral lymphoid tissues of NOD mice may disclose unexpected biases due to positive or negative selection of T cell precursors by endogenous peptides and superantigens. Such biases could account for the persistence of undesirable autoreactive T cells or, alternatively, for the disappearance of suppressor T cells in the peripheral lymphoid organs of autoimmune mice. Even though the question of when and where the T-cell repertoire becomes pathogenic is poorly understood, it is clear that a detailed knowledge of the evolution of this repertoire in primary and secondary lymphoid organs, as well as in the inflammatory sites, is critical. In order to describe the complete Vβ repertoire in the spleen of young prediabetic mice, we have set up an anchored PCR (An-PCR) procedure (Loh E.Y., 1989) which permits the amplification of all TCR β chains, irrespective of the variable 5' end. The main steps of the experimental protocol are described in Figure 1.

Total RNA extraction
↓
Retrotranscription into cDNA
↓
cDNA dG-tailing
↓
PCR # 1 : 25 cycles
(primers: oligo-dC and Cβ-external primer)
↓
Size selection by gel electrophoresis
↓
PCR # 2 : 20 cycles
(primers: oligo-dC and Cβ-internal primer)
↓
Subcloning of PCR products
↓
Colony screening with Cβ and Vβ-specific cDNA probes

Figure 1. Main steps involved in the semi-quantitative analysis of Vβ repertoires following amplification by An-PCR.

As a prerequisite, we had to show that this technique provided a faithful and reproducible representation of the Vβ distribution, compatible with that obtained by immunofluorescence staining with anti-Vβ monoclonal antibodies (mAb). Two precautions appeared to be important: a limited number of cycles in each round of amplification and the use of nested primers at the 3' end to increase the specificity of the PCR reaction. Under such conditions, the percentages obtained by fluorocytometry and An-PCR were in good agreement for practically all the Vβ families for which an antibody was available, with the noticeable exception of Vβ6 and, to a lesser extent, of Vβ11. The ratio between the values estimated by cell surface staining and PCR amplification was, however, constant, irrespective of the initial concentration of Vβ6-positive cells, so that a correction factor of 4-fold for Vβ6 and of 2-fold for Vβ11 could be applied to the data generated by An-PCR. It is probable that Vβ6-positive lymphocytes express, on a per cell basis, lower amounts of Vβ transcripts than most other T cell subsets.

The Vβ distribution found in the spleen of an 8 week-old prediabetic mouse, as determined by An-PCR, is shown in Figure 2. For the sake of comparison, we present in the same Figure the data available from the literature for C57BL/6 mice, which are, like NOD mice, IE negative.

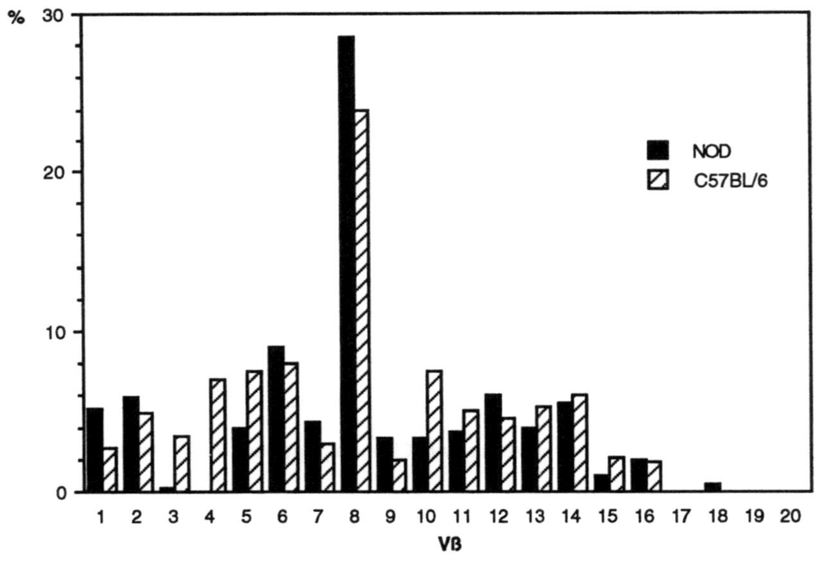

Figure 2. Comparison of the Vβ distribution in NOD and C57/BL6 mice. Data for C57BL6 are those reported by McDuffie using mAb staining (McDuffie M., 1992) and by A.P. Singer for Vβ1,10,12,15 and 16, using RNase protection assay (Singer A.P., 1990).

As has been observed for all mouse strains, the usage of Vβ segments is non-random and is, in great part, shaped by MHC-regulated positive and negative selection events occurring in the thymus (Kappler J., 1987). Another major factor influencing the Vβ usage between different mouse strains is the presence of endogenous retroviral superantigens, in particular endogenous mouse mammary tumor viruses (Mtv), which have the capacity to expand or delete specific Vβ subsets by directly bridging the lateral sides of the TCR β chain to the MHC class II molecules (Janeway C.A., 1989, Dellabona P., 1990). Superantigens can also expand, delete or anergize mature T cells in the periphery (Ochi A., 1992, Webb S.R., 1990, Kawabe Y., 1991, Rammensee H.G., 1989). Being $Mls1^b$, NOD mice do not delete the Vβ family members which react with Mtv 7, namely Vβ6, 7, 8.1 and 9 (reviewed in Tomonari K., 1993). Vβ5 and Vβ11, which react with IE-associated superantigens, are also not deleted. On the other hand, our data confirm the previously-reported deletion of Vβ3 associated with the $Mls2^a$ phenotype (Fairchild S., 1991). Mtv-3 which maps to chromosome 11 and which causes the deletion of Vβ3-expressing T cells, is clearly harbored in the NOD genome.

Two other Vβs, Vβ20 and Vβ15, could represent good candidates for a superantigen-mediated deletion. Vβ20 deletions have been already reported in several strains, but not in NOD mice. They have been associated to the expression of IE (Six A., 1991). In the case of the NOD mouse, which is IE negative, it must be postulated that the deletion involves the bridging with IA instead of IE. A similar mechanism has been proposed in order to explain the IE-independent deletion of Vβ17a cosegregating with Mtv-3 in (NOD x C57L)F1 x C57L backsross mice (McDuffie M., 1992). The percentage of Vβ15-positive T cells as determined by An-PCR appears also to be low, in particular when it is compared to that of other Mls-2a strains such as C3H/HeJ (5.1% in C3H vs 1.0% in NOD) (Singer P.A., 1990). It is also lower than in most other strains examined, including C57BL/6, AKR/J and CBA/J (Table 1). This decrease in Vβ15 expression is intriguing since no endogenous deleting element has been described for this segment. Vβ15-positive cells are capable however of reacting, at least *in vitro,* to Mtv-encoded antigens. They can for instance be stimulated by an exogenous viral superantigen such as ORF-9 (Choi Y., 1991). Similar stimulation by an endogenous ligand could lead ultimately to deletion *in vivo.*

	NOD	C3HBL/6	C57BL/6	AKR/J	CBA/J
Vβ15 (%)	1.0	5.1	2.1	6.6	8.7

Table 1. A comparison of Vβ15 expression in various inbred strains of mice. For NOD mice, Vβ15 expression was determined by An-PCR. Data for the other strains were obtained by the RNase protection assay (Singer P.A., 1990)

The absence of Vβ17 and Vβ19 transcripts was not totally unexpected since in the $V\beta^b$ haplotype to which the NOD mouse belongs, these two gene segments are not functional (Wade T. 1988, Louie M.C., 1987). Occasional transcripts resulting from first rearrangement attempts are probably too rare to be detected under the present experimental conditions. Finally, comparison of Vβ16 levels in IE positive and negative strains confirmed the view that the deletion of this TcR segment requires specifically the presence of an IE molecule (Singer P.A., 1990).

Most laboratory mouse strains segregate into two haplotypes: $V\beta^a$ and $V\beta^b$ (Behlke M., 1986). The $V\beta^a$ haplotype which carries a large deletion encompassing Vβ genes 5, 8, 9, 11, 12 and 13, presents several sequence differences with the $V\beta^b$ haplotype. A majority of strains, including NOD mice, belong to the latter haplotype. Sequencing of examples from all Vβ families isolated in the NOD strain revealed no particular polymorphism. All sequences were strictly identical to the consensus sequences published in the literature for the $V\beta^b$ haplotype. The aberrant presence of anti-pancreatic autoreactive clones in the NOD strain can therefore not be ascribed to an atypic polymorphism encoded by a Vβ segment. If faulty T-cell selection or activation occurs, it is likely to be due to extrinsic factors such as the conformation of MHC class II products (Acha-Orbea H., 1987) or to superantigens.

Oligoclonality of Infiltrating T Cells In The Pancreas Of NOD Mice

Nucleotide sequencing of TcR α and β chains from various origins has shown that most of the diversity is concentrated at the junction of V(D)J segments which is designated, by analogy with immunogloblins, the third complementarity determining-region or CDR3 (Kabat E.A., 1987). The diversity encountered in this region results from imprecise joining of the segments due to the activity of two enzymes: an exonuclease which causes nucleotide deletions, also called nibbling, and the terminal deoxynucleotidyl transferase (TdT) which, on the contrary, inserts new nucleotides at the 3' ends of the segments giving rise to what is referred to as N additions. In view of its remarkable diversity, many authors have postulated that the CDR3 loop is responsible for the fine antigenic specificity of the TcR by establishing contacts with the peptide inserted into the MHC groove (Davis M.M., 1988). Mutations resulting in the substitution of a single amino acid in the inserted peptide have indeed been shown to induce selection of T cells with corresponding changes in their CDR3 (Davis M.M., 1992).

In the case of an autoimmune reaction, the analysis of CDR3 diversity may provide useful insights into the degree of oligoclonality of the cell-mediated response, its plasticity and its variability with respect to the age and to the individual history of the subject. With these aims in mind, we have undertaken a large-scale analysis of sequences from CDR3 regions associated with TcR β-chains in NOD mice. For both practical and theoretical reasons, we have focused on early T cell infiltrates from 4 week-old mice, and we have selected Vβ6-positive transcripts. Firstly, we established that the earliest age at which a Vβ6-positive signal could be generated by PCR from isolated islets of an individual mouse was, indeed, 4 weeks. Islets from 3 week-old donors were reproducebly negative. Eight independent batches of cDNA from purified islets of 4 week-old mice were then processed for Vβ6-specific amplification. Five of these batches were from the islets of individual mice, and three were from pools of ten mice each. Spleen RNA from two of these donors was treated according to the same protocol and sequenced. A total of 187 randomly selected Vβ6-positive clones of pancreatic origin were sequenced, an average of 20 clones per batch. Remarkably, all sequences obtained were different from one batch to the other, but, within a single batch, we found a considerable number of repeated transcripts. This suggested an overall diversity of the infiltrating CDR3 repertoire, but at the same time a marked oligoclonality of the invading Vβ6 cells within a single batch. In contrast to pancreatic transcripts, the 48 Vβ6 transcripts from the spleen of two donors, were all different.

A careful analysis of the CDR3 regions with respect to Jβ usage, CDR3 length and N region additions, revealed several interesting differences between intrapancreatic and splenic T lymphocytes. It has been previously shown that Jβ segments are not randomly used. The reasons for such bias are still not fully understood, but they may reflect both mechanistic and selective processes (Candeias, S. 1991). A non-random usage of Jβ segments is also found in the case of peripheral T lymphocytes from the spleen of 4 week-old NOD mice. As can be seen in Figure 3, which compares Jβ usage by Vβ6-positive T splenocytes of NOD mice to that of Vβ17a in SJL mice (Candeias S., 1991) and Vβ5 and 8 in BALB/c mice (Feeney A.J., 1991), the Jβ2 cluster is notably more frequently used than the Jβ1 cluster (86.6% vs 13.4% in NOD transcripts).

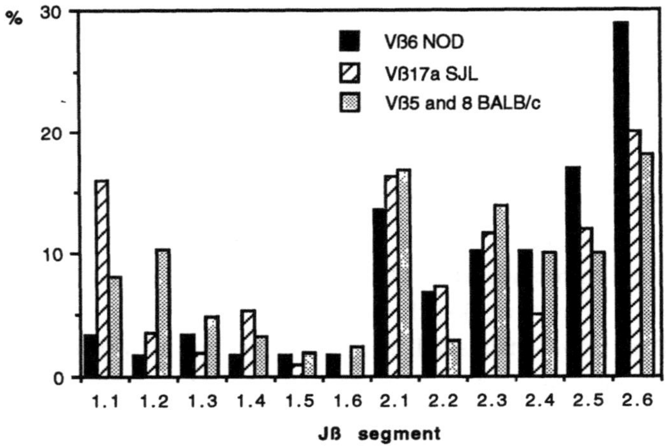

Figure 3. Jβ usage by peripheral Vβ6-positive T cells in NOD mice, compared to that of Vβ17a and Vβ5 and 8 in the SJL and BALB/c strains, respectively.

On the other hand, when we compare, within the NOD strain, the Jβ usage of infiltrating T cells versus that of T cells from the spleen (Table 2), we notice dramatic differences. Most remarkable is the fact that the Jβ1 cluster is used with a higher frequency (23.1% versus 13.4% in the spleen). Such a difference is likely to reflect a bias in the selection of the T lymphocytes which colonize the pancreas.

Jβ	**1.1**	**1.2**	**1.3**	**1.4**	**1.5**	**1.6**	**Total**
Islets	7.7	1.9	0.0	5.8	0.0	7.7	23.1%
Spleen	3.4	1.5	3.4	1.7	1.7	1.7	13.4%

JB	**2.1**	**2.2**	**2.3**	**2.4**	**2.5**	**2.7**	**Total**
Islets	7.7	11.5	15.4	3.8	5.8	32.7	76.9%
Spleen	13.6	6.8	10.2	10.2	17.0	28.8	86.6%

Table 2. Comparison of Jβ usage by Vβ6-positive T lymphocytes from the spleen and the pancreas of NOD mice at 4 weeks of age

The length of the CDR3 loop results from the opposite effects of nucleotide nibbling and N sequence addition. Even though the CDR3 length varies to a much lesser extent in TCR α and β chains than in H and L chains of immunoglobulins, the number of residues encoded within the loop nevertheless seems to play an important role in determining the specificity of cell-mediated responses (Casanova J.L., 1992). The number of N insertions reflects also the maturity of the repertoire since TdT becomes functional only after birth (reviewed in Feeney A.J., 1991). Direct comparison, at 4 weeks of age, of CDR3 lengths in intrapancreatic versus splenic Vβ6-positive T cells reveals a shift in the average number of amino acids (Figure 4).

CDR3 length (amino acids)

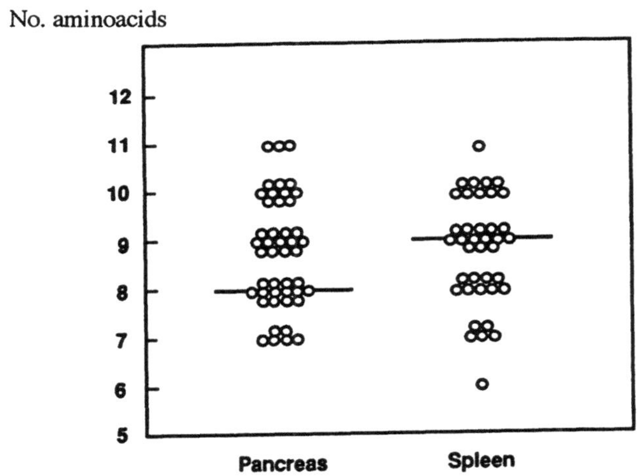

Figure 4. CDR3 lengths measured in transcripts from splenocytes and intrapancreatic T cells of NOD mice.

The shorter CDR3 regions displayed by infiltrating T cells could be due either to greater exonucleotide nibbling or to less N sequence addition. The average number of N additions per sequence (2.4 in pancreatic T cells versus 3.2 in the spleen) is in support of the latter alternative. Thus, at 4 weeks of age, the early inflammatory T cells appear to express a less mature repertoire than splenic T lymphocytes. An interesting parallel can be drawn with the results reported on encephalitogenic T cell clones (Zhang X.M., 1992). There too, autoreactive T cells show fewer N additions than the rest of the T cell pool. As already suggested (Howe M.L., 1970, Schneider R., 1989 and Riggs J.E., 1991), T cells generated during fetal life could have a higher propensity for self-reactivity than those generated later in life. It is thus not entirely surprising that the first wave of inflammatory cells, responsible for the incipient lesions of insulitis, exhibits ontogenic characteristics of early T cells. The evolution of the pathogenic process could be largely influenced by this population of "early settlers" as suggested by the fact that experimental manipulations performed at birth, such as injections of anti-CD3 mAb (Hayward, A.R., 1989) or induction of neonatal tolerance (Bendelac A., 1989), delay to a considerable extent the onset of overt diabetes.

Concluding Remarks

The analysis by An-PCR of the Vβ repertoire expressed in the spleen of NOD mice reveals limited, but noticeable alterations. The first and most conspicuous of those, concerns the well-known deletion of Vβ3-positive T cells which has been shown to be mediated by Mtv-3. In addition to this deletion, we have detected a marked decrease of T cells using Vβ15 and Vβ20. Deletion of Vβ20-positive cells has been described in several strains and has been associated with IE expression. In the case of NOD mice which are negative for IE, one must assume that IA can be used as a substitute by the putative superantigen. Concerning Vβ15, it is interesting to note that no deleting agent has been identified in any of the inbred strains examined. The fact that Vβ15-positive T cells can be activated in vitro by the ORF sequence of Mtv 9 supports, however, the idea of an endogenous retroviral sequence being responsible for the deletion observed.

One can only speculate on the possible impact of such deletions upon the autoimmune process developing in NOD mice. Attempts at explaining protection against diabetes as resulting from deletions of specific Vβ subsets have not been conclusive. It is for instance clear that the protective effect afforded by transgenic expression of IE (Nishimoto H.,1987, Lund T., 1990, Bohme J., 1990) is not due to a deletion of Vβ5 or 11 subsets as originally postulated. On the other hand, the possibility that the absence of certain Vβ elements might indirectly favor the emergence of antipancreatic autoimmunity deserves exploration. Regulatory or suppressive T cells present in non-autoimmune strains of mice and protecting them from anti-pancreatic attack, could conceivably use a restricted set of Vβ genes including Vβ15 and 20.

Having isolated a complete panel of NOD Vβ transcripts, we have also determined their nucleotidic sequences. No differences from published sequences of $V\beta^{b}$ mouse strains were noticed, thus ruling out a protein polymorphism as the source of T cell autoreactivity in the NOD strain.

The analysis of CDR3 sequences, which constitutes the second part of this paper, reveals some interesting features of the pathogenic T cell repertoire. In particular, this analysis discloses a limited diversity of sequences within the pancreas of a single individual and, at the same time, a relative heterogeneity of the CDR3 when sequences are compared from individual to individual. This suggests that pathogenic T cells can accomodate a large array of CDR3 sequences, but that the initial colonization of the pancreas is achieved by a limited number of precursors which expand clonally, presumably upon encounter with self antigens. That this repertoire is the object of a local selection process is further supported by the presence of differences in Jβ usage, CDR3 length and N sequence additions, from the splenic T cell repertoire of the same mice. The intra-pancreatic T cell repertoire at 4 weeks of age appears ontogenically less mature than the splenic repertoire, especially with respect to the number of N sequence additions. The influence of this "early" T cell repertoire will have to be properly evaluated in future experiments, but indirect evidence already suggests that the presence of T cells expressing such a repertoire may be crucial for the evolution of the pathogenic process.

Acknowledgements

We would like to thank Y.A. Vassitch and S. Delaire for excellent technical assistance. This work was supported by funds from the Institut National de la Santé et de la Recherche Médicale (INSERM). A. Sarukhan is the recipient of a scholarship of the Universidad Nacional Autonoma de Mexico, Mexico.

References

Acha-Orbea H, McDevitt HO (1987) The first external domain of the nonobese diabetic mouse class II I-A beta chain is unique. Proc. Natl. Acad. Sci. U. S. A. 84:2435-2439

Acha-Orbea H, Mitchell DJ, Timmermann L, Wraith DC, Tausch GS, Waldor MK, Zamvil SS, McDevitt HO, Steinman L (1988) Limited heterogeneity of T cell receptors from lymphocytes mediating autoimmune encephalomyelitis allows specific immune intervention. Cell 54:263-273

Bacelj A, Charlton B, Koulmanda M, Mandel TE (1990) The role of V beta 8 cells in disease recurrence in isografts in diabetic NOD mice. Transplant. Proc. 22:2167-2168

Bacelj A, Charlton B, Mandel TE (1989) Prevention of cyclophosphamide-induced diabetes by anti-V beta 8 T-lymphocyte-receptor monoclonal antibody therapy in NOD/Wehi mice. Diabetes 38:1492-1495

Baekkeskov S, Aanstoot HJ, Christgau S, Reetz A, Solimena M, Cascalho M, Folli F, Richter-Olesen H, Camilli P (1990) Identification of the 64K autoantigen in insulin-dependent diabetes as the GABA-synthesizing enzyme glutamic acid decarboxylase. Nature 347:151-156

Behlke M A, Chou HS, Huppi K, Loh DY (1986) Murine T-cell receptor mutants with deletions of beta-chain variable region genes. Proc. Natl. Acad. Sci. U. S. A. 83:767-771

Bendelac A, Boitard C, Bach JF, Carnaud C (1989) Neonatal induction of allogeneic tolerance prevents T cell-mediated autoimmunity in NOD mice. Eur. J. Immunol. 19:611-616

Bendelac A, Carnaud C, Boitard C, Bach JF (1987) Syngeneic transfer of autoimmune diabetes from diabetic NOD mice to healthy neonates. Requirement for both L3T4+ and Lyt-2+ T cells. J. Exp. Med. 166:823-832

Bohme J, Schuhbaur B, Kanagawa O, Benoist C, Mathis D (1990) MHC-linked protection from diabetes dissociated from clonal deletion of T cells. Science 249:293-295

GF, Dean BM, McNally JM, Mackay EH, Swift PG, Gamble DR (1985) In situ characterization of autoimmune phenomena and expression of HLA molecules in the pancreas in diabetic insulitis. N. Engl. J. Med. 313:353-360

Candeias S, Katz J, Benoist C, Mathis D, Haskins K (1991) Islet-specific T-cell clones from nonobese diabetic mice express heterogeneous T-cell receptors. Proc. Natl. Acad. Sci. U. S. A. 88:6167-6170

Candeias S, Waltzinger C, Benoist C, Mathis D (1991) The Vbeta 17+ T cell repertoire: skewed J beta usage after thymic selection: dissimilar CDR3s in CD4+ versus CD8+ cells J. Exp. Med. 174:989-1000

Casanova JL, Cerottini JC, Matthes M, Necker A, Gournier H, Barra C, Widmann C, Macdonald R, Lemonnier F, Malissen B, Maryanski JL (1992) H-2-restricted cytolytic T lymphocytes specific for HLA display T cell receptors of limited diversity. J. Exp. Med. 176:439-447

Chiocchia G, Boissier MC, Fournier C (1991) Therapy against murine collagen-induced arthritis with T cell receptor V beta-specific antibodies. Eur. J. Immunol. 12:2899-2905

Choi Y, Kappler JW, Marrack P (1991) A superantigen encoded in the open reading frame of the 3' long terminal repeat of mouse mammary tumour virus. Nature 350:203-207

David CS, Banerjee S (1989) T cell receptor genes and disease susceptibility. Arthritis. Rheum. 32:105-107

Davis MM, Bjorkman PJ (1988) T-cell antigen receptor genes and T-cell recognition. Nature 334:395-402

Dellabona P, Peccoud J, Kappler J, Marrack P, Benoist C, Mathis D (1990) Superantigens interact with MHC class II molecules outside of the antigen groove. Cell 62:1115-1121

Edouard P, Thivolet C, Bedossa P, Olivi M, Legrand B, Bendelac A, Bach JF, Carnaud C (1993) Evidence for a preferential V beta usage by the T cells which adoptively transfer diabetes in NOD mice. Eur. J. Immunol. 23:727-733

Elias D, Markovits D, Reshel T, Van der Zee R, Cohen IR (1990) Induction and therapy of autoimmune diabetes in the non-obese diabetic (NOD/Lt) mouse by a 65kDa heat shock protein. Proc. Natl. Acad. Sci. U.S.A. 4:1576-1580

Feeney AJ (1991) Junctional sequences of fetal T cell receptor b chains have few N regions. J. Exp. Med.174:115-124.

Fairchild S, Knight AM, Dyson PJ, Tomonari K (1991) Co-segregation of a gene encoding a deletion ligand for Tcrb-V3+ T cells with Mtv-3. Immunogenetics 34:227-230

Fukuda M, Horio F, Kubo R, Hattori M (1989) Monoclonal antibody F23.1 against T cell bearing Vß8 TCR elements can prevent the development of insulitis in NOD mice. Diabetes 38 (suppl):12 A

Hayward AR, Shreiber M (1989) Neonatal injection of CD3 antibody into nonobese diabetic mice reduces the incidence of insulitis and diabetes. J. Immunol. 143:1555-1559

Howe ML, Goldstein AL, Battisto JR (1970) Isogeneic lymphocyteinteractions: recognition of self antigens by cells of the neonqtal thymus. Proc. Natl. Acad. Sci. USA 67:613-619.

Janeway CA JR, Yagi J, Contrad PJ, Katz ME, Jones B, Uroegop S, Buxser S (1989) T-cell responses to Mls and to bacterial proteins that mimic its behavior. Immunol. Rev. 6:61-88

Jorgensen JL, Esser U, Fazekas de St. Groth B, Reay PA, Davis MM (1992) Mapping T-cell receptor-peptide contacts by variant peptide immunization of single-chain transgenics. Nature 355:224-230.

Kabat EA, Wu TT, Reid Miller M, Perry HM, Gottesman KS (1987) Sequences of proteins of immunological interest. Department of Health and Human Services, Bethesda, MD, USA, 4th edition.

Kappler JW, Roehm N, Marrack P (1987) T cell tolerance by clonal elimination in the thymus. Cell 49:273-280

Kawabe Y, Ochi A (1991) Programmed cell death and extrathymic reduction of Vbeta8+ CD4+ T cells in mice tolerant to Staphylococcus aureus enterotoxin B. Nature 349:245-248

Leiter E H (1989) The genetics of diabetes susceptibility in mice. FASEB. J. 3:2231-2241

Loh E Y, Elliott J F, Cwirla S, Lanier L L, Davis M M (1989) Polymerase chain reaction with single-sided specificity: analysis of T cell receptor delta chain. Science 243:217-220

Louie M C, Nelson C A, Loh D Y (1989) Identification and characterization of new murine T cell receptor beta chain variable region (V beta) genes. J. Exp. Med. 170:1987-1998

Lund T, O'Reilly L, Hutchings P, Kanagawa O, Simpson E, Gravely R, Chandler P, Dyson J, Picard JK, Edwards A, Kioussis D, Cooke A (1990) Prevention of insulin-dependent diabetes mellitus in non-obese diabetic mice by transgenes encoding modified I-A beta-chain or normal I-E alpha-chain. Nature 345:727-729

Maeda T, Sumida T, Kurasawa K, Tomioka H, Itoh I, Yoshida S, Koike T (1991) T-lymphocyte-receptor repertoire of infiltrating T lymphocytes into NOD mouse pancreas. Diabetes 40:1580-1585

Makino S, Kunimoto K, Muraoka Y, Mizushima Y, Kahagiri K, Tochino Y (1980) Breeding of a non-obese, diabetic strain of mice. Exp. Anim. 1:1-13

McDuffie M, Schweiger D, Reitz B, Ostrowska A, Knight AM, Dyson PJ (1992) I-E-independent deletion of V beta 17a+ T cells by Mtv-3 from the nonobese diabetic mouse. J. Immunol. 148:2097-2102

Miyazaki A, Hanafusa T, Yamada K, Miyagawa J, Fujino-Kurihara H, Nakajima H, Nonaka K, Tarui S (1985) Predominance of T lymphocytes in pancreatic islets and spleen of pre-diabetic non-obese diabetic (NOD) mice: a longitudinal study. Clin. Exp. Immunol. 60:622-630

Nakano N, Kikutani H, Nishimoto H, Kishimoto T (1991) T cell receptor V gene usage of islet beta cell-reactive T cells is not restricted in non-obese diabetic mice. J. Exp. Med. 173:1091-1097

Nepom GT (1990) HLA and type I diabetes. Immunol. Today 9:314-315

Nishimoto H, Kikutani H, Yamamura K, Kishimoto T (1987) Prevention of autoimmune insulitis by expression of I-E molecules in NOD mice. Nature 328:432-434

O'Reilly L A, Hutchings P R, Crocker P R, Simpson E, Lund T, Kioussis D, Takei F, Baird J, Cooke A (1991) Characterization of pancreatic islet cell infiltrates in NOD mice: effect of cell transfer and transgene expression. Eur. J. Immunol. 21:1171-1180

Oshi A, Migita K, Yuh K (1992) Do anergic T cells live or die? Res. Immunol. 143:304-306

Prud'homme GJ, Long TY, Bocarro DC, Balderas RS, Theofilopoulos AN (1991) Analysis of pancreas-infiltrating T cells in diabetic NOD mice: fusion with BW5147 yields a high frequency of islet-reactive hybridomas. Autoimmunity 10:285-289

Rammensee HG, Kroschewski R, Frangoulis B (1989) Clonal anergy induced in mature V beta 6+ T lymphocytes on immunizing Mls-1b mice with Mls-1a expressing cells. Nature 339:541-544

Riggs JE; Stowers RS Mosier DE (1991) Adoptive transfer of neonatal T lymphocytes rescues immunoglobulin production in mice with severe combined immune deficiency. J. Exp. Med. 173:265-268

Schneider R, Lees RK, Pedrazzini T, Zinkernagel RM, Hengartner H, MacDonald HR (1989) Postnatal disappearance of self-reactive (V beta 6+) cells from the thymus of Mlsa mice. Implications for T cell development and autoimmunity. J. Exp. Med.169:2149-2158

Singer PA, Theofilopoulos AN (1990) T-cell receptor V beta repertoire expression in murine models of SLE. Immunol. Rev. 144:103-127

Six A, Jouvin-Marche E, Loh DY, Cazenave PA, Marche PN (1991) Identification of a T cell receptor beta chain variable region, V beta 20, that is differentially expressed in various strains of mice. J. Exp. Med. 174:1263-12

Tomonari K, Fairchild S, Rosenwasser OA (1993) Influence of viral superantigens on Vβ- and Vα-specific positive and negative selection. Immunol Rev 131:131-168

Wade T, Bill J, Marrack PC, Palmer E, Kappler JW (1988) Molecular basis for the nonexpression of V beta 17 in some strains of mice. J. Immunol. 141:2165-2167

Waters S H, O'Neil JJ, Melican DT, Appel MC (1992) Multiple TCR V beta usage by infiltrates of young NOD mouse islets of Langerhans. A polymerase chain reaction analysis. Diabetes 41:308-312Webb SR, Sprent J (1990) Response of mature unprimed CD8+ T cells to Mlsa determinants. J. Exp. Med. 171:953-958

Zhang XM, Heber-Katz E (1992) T cell receptor sequences from encephalitogenic T cells in adult Lewis rats suggest an early ontogenic origin. J. Immunol. 148:746-752

Possible Linkage of Myasthenia Gravis Susceptibility in R111 S/J Mice to a Genomic Deletion of T Cell Receptor Vβ Genes

Mohan Shenoy*, Bo Wu*, Rashmi Kaul*, Elzbieta Goluszko*, Chella David+, and Premkumar Christadoss*
*The Department of Microbiology, University of Texas Medical Branch, Galveston, Texas 77555-1019; and the +Department of Immunology, Mayo Clinic, Rochester, MN 55905, USA.

INTRODUCTION

Myasthenia gravis (MG) is an autoimmune disease characterized by muscle weakness and fatigue. The defective neuromuscular transmission in MG is due to a functional loss of the nicotinic acetylcholine receptor (AChR) in the postsynaptic membrane, which is the result of the effects of anti-AChR antibodies and complement (Lindstrom *et al*, 1988). Many caucasian MG patients possess HLA-B8, and those having a subtype of the DR3 haplotype polymorphic at the DQβ locus, run the greatest risk of developing MG (Bell, *et al*, 1986). Moreover, HLA-DR3-restricted, AChR-specific T cell lines were generated from a patient with MG (Bell, *et al*, 1986). Experimental autoimmune myasthenia gravis (EAMG) susceptibility has also been linked to MHC genes (Christadoss, 1989; Christadoss and Shenoy, 1992). A gene conversion event between the immune response genes $E\beta^b$ and $A\beta^b$ in the B6.CH-2^{bm12} (bm12) strain, which altered three amino acids (Ile-67 to Phe; Arg-70 to Gln; Thr-71 to Lys) of the Aβ chain, suppressed EAMG development (Christadoss *et al*, 1985), and thus, the Aβ chain was implicated in the pathogenesis. Further, the immune response to AChR could be suppressed by *in vivo* administration of anti-IA antibodies (Waldor *et al*, 1983). EAMG in mice can either be prevented, or clinical remission induced, by *in vivo* administration of monoclonal antibodies (mAb) to CD4 (Christadoss and Dauphinee, 1986), suggesting a requirement for $CD4^+$ helper T cells in the production of pathogenic anti-AChR antibodies by B cells. In a recent study, MHC class II and $CD4^+$ cell deficiency, completely prevented class II-deficient C57BL6 mice from developing EAMG (Kau *et al.*, Submitted). In contrast, β2

NATO ASI Series, Vol. H 80
Autoimmunity: Experimental Aspects
Edited by Moncef Zouali

to Theiler's virus-induced demyelination (Rodriguez *et al*, 1992). Further, restricted use of Vβ genes has been observed in the autoimmune T cell response to the dominant encephalitogenic NH2-terminal peptide of the myelin basic protein, and a mAb specific for the TCR-Vβ8 family was found to be effective in suppressing autoimmune encephalomyelitis (Zamvil *et al*, 1988; Urban *et al*, 1988). The exact cellular and molecular nature of the TCR interaction with MHC and peptide in the evolution of autoimmunity and autoimmune diseases is still obscure. We have utilized TCR-Vβ mutants and congenic mice to study the cellular and molecular mechanisms of EAMG pathogenesis. TCR-Vβ mutant strains SJL, C57BR, C57L and SWR (TCR-$V\beta^a$ haplotype) have a genomic deletion of Vβ 5.2, 8.3, 5.1, 8.2, 5.3, 8.1, 13, 12, 11 and 9 genes (Behlke *et al*, 1986), in contrast to the wild type strains with normal complementation of the TCR-Vβ genes (TCR-$V\beta^b$ haplotype). Another TCR-Vβ mutant strain, R111/SJ (TCR-$V\beta^c$ haplotype), has a genomic deletion of approximately 130 Kb of DNA encoding 13 Vβ genes (Vβ 5.2, 8.3, 5.1, 8.2, 5.3, 8.1, 13, 12, 11, 9, 6, 15 and 17) of the 21 known Vβ genes including those deleted in the $V\beta^a$ haplotype strain, are deleted (Haqqi *et al*, 1989). The data presented here suggest a possible linkage of EAMG susceptibility to the genomic deletion of chromosomal segment bearing the TCR-Vβ genes.

RESULTS

Augmentation of EAMG in R111/SJ mice:

B10.R111 (H-2^r; $V\beta^b$) and R111/SJ (H-2^r; $V\beta^c$) mice were injected with 20ug of AChR in complete Freund adjuvant (CFA). They were reinoculated on day 34 with 20ug of AChR in CFA. Animals were evaluated for muscle weakness by M.S. and P.C. in a blind study. Muscle weakness was graded as described (Christadoss and Dauphine, 1986). It was confirmed to be due to MG by observing improvement in muscle strength after anti-cholinesterase medication (Christadoss *et al*, 1986). A significant decremental response to repetitive nerve stimulation was observed in most of the mice with muscle weakness. Only one of nine (12%) B10.R111 mice developed muscle weakness, while a dramatic augmentation in the incidence of clinical EAMG was seen in R111/SJ mice, with 8 of 10 (80%) mice developing muscle weakness. The time of onset and the clinical course of the disease in B10.R111 and R111/SJ mice are illustrated in Figure 1. The difference in susceptibility observed between R111/SJ and B10.R111 mice was statistically significant at $p=0.004$ (Fisher exact test). Among several inbred strains tested (Christadoss, 1989), the R111/SJ strain demonstrated the highest incidence of clinical EAMG after two immunizations with AChR in CFA. Thus, it can serve as an excellent model to study MG at the cellular and molecular levels. The augmented susceptibility observed in R111/SJ mice may be due to one or more of the following: (i) the compensatory overexpression of one or more non-deleted Vβ genes; (ii) the genomic deletion of Vβ 5.2, 8.3, 5.1, 8.2, 5.3, 8.1, 13, 12, 11, 9, 6, 15 and 17 genes, with one or more of these deleted genes involved in the regulation/suppression of an autoimmune response to AChR; (iii) the deletion of genes linked to those coding for TCR Vβ genes; and (iv) the presence of non-H-2 genes, other than the TCR genes in the R111/SJ background.

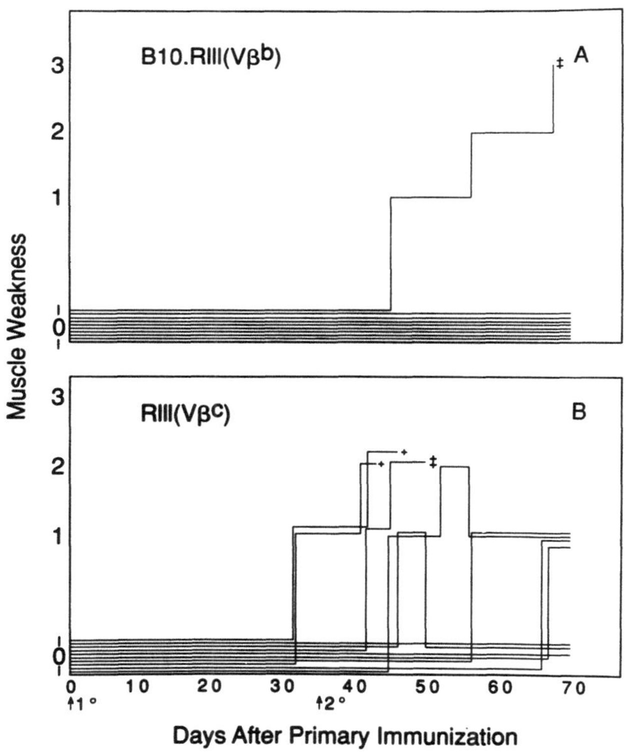

Fig 1: The onset of clinical MG and its course in B10.R111 and R111/SJ mice.
The + sign indicates the death of the mouse due to the severity of myasthenia. The + + signs indicate that the mouse was sacrified because of severe disease. Fisher's exact test was used to evaluate the statistical significance. B10.R111 versus R111/SJ, $p = 0.004$.

Possible linkage of EAMG susceptibility to the genomic deletion of Vβ genes

To rule out the role of non-H-2 genes other than the deleted TCR Vβ genes in the R111/SJ background in EAMG susceptibility, congenic B10 (H-2^b; $V\beta^b$), B10.TCR^a (H-2^b; $V\beta^a$) and partially congenic B10.TCR^c ($V\beta^c$) mice were immunized with AChR in CFA, as described earlier, and screened for EAMG susceptibility. B10.TCR^a congenic mice were produced by making an F1 between C57BL10 (B10) ($V\beta^b$) and C57L ($V\beta^a$), and backcrossing to B10 for 10 generations while selecting for the $V\beta^a$ haplotype. The selection was based on FACS analysis with anti-Vβ8 positive cells in homozygous mice. At the end of the 10th generation, the mice were intercrossed and those showing 0% Vβ8-positive T cells were selected for further inbreeding. The B10.TCR^c strain was generated by crossing B10 and R111 ($V\beta^c$) and backcrossing to B10, while selecting for the $V\beta^c$ haplotype by monitoring Vβ8- and Vβ6-positive T cells. B10.TCR^c mice belonged to the 7th generation. Nine of ten (90%) B10.TCR^c (H-2^b; $V\beta^c$) mice developed muscle weakness, an incidence equivalent to its $V\beta^c$ counterpart R111/SJ (H-2^r; $V\beta^c$), while only 4 of 14 (28%) B10 and 1 of 7 (14%) B10.TCR^a mice developed muscle weakness (Table 1). The differences in EAMG susceptibility between B10.TCR^c and B10, and between B10.TCR^c and B10.TCR^a were statistically significant, with p values of 0.016 and 0.012, respectively (Fisher exact test). On the contrary, the difference in the clinical incidence between B10 and B10.TCR^a did not attain statistical significance (p=1.75). The augmented susceptibility observed in B10.TCR^c mice when compared to B10.TCR^b or B10.TCR^a mice suggests a possible linkage of EAMG susceptibility to the deleted chromosomal segment bearing the TCR-Vβ genes or to the linked genes.

Augmented anti-AChR antibody response in B10.TCR^c:

The kinetics of the autoantibody response against mouse muscle AChR in B10, B10.TCR^c and B10.TCR^a mice was also determined. An augmented anti-AChR antibody response was evident in the serum of B10.TCR^c mice on days 47 and 57 after the first immunization with AChR in CFA, when compared with the antibody response of B10 or B10.TCR^a mice. On day 57 after the primary immunization with AChR in CFA, the antibody response in B10.TCR^c mice was significantly higher than that of B10 or B10.TCR^a mice (Table 1; B10.TCR^c vs B10, p=0.033; B10.TCR^c vs B10.TCR^a, p=0.06; Student's t test). Thus, the augmented anti-AChR antibody response correlated with the increased susceptibility of B10.TCR^c mice.

Strain*	TCR Vβ Haplotype	Muscle Weakness (Grade)				Total Incidence (%)	Anti-AChR Ab#±SEM (10^{-10}M)
		0	1	2	3		
B10.TCR^c	c	1	4	3	2	9/10 (90)	12.3 ± 3.1
B10.TCR^a	a	6	0	0	1	1/7 (14)	1.6 ± 1.2
B10	b	10	2	0	2	4/14 (28)	2.6±1.3

Table 1: Clinical and Immunological Manifestations of EAMG in B10.TCR Vβ Congenic Mice. *Mice were immunized twice at monthly intervals with *Torpedo* AChR (20 ug) in CFA. +Muscle weakness was graded between 0 to 3 according to the method described earlier (Christadoss, 1986). 0, animal with normal strength. Grade 1, Normal at rest but weak with chin on the floor, flaccid tail and inability to raise the head, after exercise consisting of 20-30 paw grips on cage top steel grids. Grade 2, Grade 1 weakness even at rest with paralysis of forelimbs. Grade 3, severe weakness with dehydration and quadriplegia. #Auto-Ab to mouse AChR was determined at the termination of the experiment on d57 after the primary immunization with AChR in CFA. Auto-Ab expressed as alpha bungarotoxin binding sites precipitated per liter of serum. SEM, standard error of the mean.

DISCUSSION

The B10.TCR^c mice used in this study belonged to the N7 generation, and is therefore, not totally congenic with B10 or B10.TCR^a mice. Studies on B10.TCR^c mice after the N10 generation are being conducted in order to definitively link EAMG susceptibility in R111/SJ mice to the deleted chromosomal segment bearing the Vβ genes. Overexpressed non-deleted Vβ genes in R111/SJ and B10.TCR^c mice could have contributed to the augmented EAMG susceptibility. Thus, overexpressed, non-deleted, Vβ gene bearing T cells could be the effectors involved in helping B cells to produce pathogenic anti-AChR antibodies. In preliminary studies 49.7% of lymph node T cells of R111/SJ mice expressed the Vβ2 gene (by FACS), while in B10.R111 (Vβ^b) and C57L (Vβ^a) mice only 6.9% and 1.7% lymph node T cells expressed Vβ2 gene, respectively. The incidence of disease in C57L mice is equivalent to that of B10.TCR^a (Vβ^a) mice. In another study, an R111/SJ T cell line reactive to AChR and the dominant AChR α182-198 epitope, expressed the Vβ2 gene product (by FACS and

PCR). The expression of other Vβ gene products is being evaluated by utilizing the other Vβ primers in a PCR analysis. Therefore, overexpressed Vβ2 gene in R111/SJ mice may be involved in the development of EAMG because of its affinity for the class II AChR epitope complex. The Vβ2 cells are being depleted in R111/SJ mice to study the effector role of Vβ2 gene in EAMG pathogenesis.

One of the primary genetic differences between B10.TCR^{a} and B10.TCR^{c} mice is the presence of Vβ6, Vβ15 and Vβ17 genes in the former. The Vβ17 gene is non-functional in B10.R111 and B10 mice. Therefore, deletion of Vβ6 and/or Vβ15 gene(s) in mice of the Vβ^{c} haplotype could also have augmented EAMG susceptibility. Also, non-functional Vβ17 gene in B10 mice did not augment EAMG. It is quite plausible that T cells bearing Vβ6 and or Vβ15 gene products could be involved in the regulation/suppression of an autoimmune response to AChR. In another study , *in vivo* depletion of Vβ6 cells has augmented EAMG in C57L (Vβ^{a}) mice, suggesting a possible regulator role for Vβ6 cells in the development of EAMG (Bo Wu *et al*, in preparation). Infante *et al* (1992) reported that T cell clones reactive to AChR-α146-162 preferentially expressed TCR-Vβ6 gene, and therefore, Vβ6^{+} cells were implicated as the effector cells helping B cells to produce pathogenic antibodies in B6 (H-2^{b}) mice with EAMG. In our hands, depletion of Vβ6 cells failed to prevent EAMG in C57L (Vβ^{a}; H-2^{b}) and B6 (Vβ^{b}; H-2^{b}) mice. The *in vivo* Vβ6 depletion and the genomic Vβ deletion studies reported here failed to show any effector role of Vβ6 cells in EAMG development. On the contrary, depletion of Vβ6 cells augmented disease incidence in C57L mice. Thus our data bend towards a suppressor/regulator role for Vβ6 cells in EAMG development. Because the immunopathology of EAMG mimics that of human MG (Christadoss, 1989, Lindstrom *et al* 1988), it is conceivable that specific TCR-Vβ gene-expressing cells may be involved in the effector phase of an autoimmune response to AChR. Thus, overexpression of those Vβ cells which have the affinity for AChR peptide-Class II complex may contribute to the development of MG. Also, clonal elimination, anergy or decreased frequency of specific Vβ gene bearing regulator/suppressor cells may predispose or augment MG susceptibility in individuals genetically prone to develop MG. Similar TCR effector/regulatory mechanisms might operate in the development of other autoimmune diseases.

ACKNOWLEDGEMENTS

We thank Drs. Bellur Prabhakar, Teh-sheng Chan and Rolf Konig for their critical review of the manuscript, and Mardelle Susman for editing. This work was supported by the MDA, Sealy Smith Endowment and NIH. R. Kaul is supported by a James W. McLaughlin Foundation postdoctoral fellowship.

REFERENCES

Banerjee, S., Haqqui, T.M., Luthra, H.S., Stuart, J.M. and David, C.S. (1988). Possible role of Vβ T cell receptor genes in susceptibility to collagen induced arthritis in mice. J Exp Med 167,832-839.

Beall, S.S., Concannon, P., Biddison, W.A., McFarland, H., Hood, L., and McFarlin, D.E. (1987). Analysis of germline Vα and Vβ genes in patients with multiple sclerosis. J Cell Biochem 11D,223.

Behlke, M.A., Hubert, J.S., Huqqi, K. and Loh, D.Y. (1986). Murine TCR mutants with deletions of β-chain variable region genes. Proc Natl Acad Sci (USA) 83,767-771.

Bell, J., Smoot, S., Newley, C., Toyka, K., Rassenti, L., Smith, D., Hohlfeld, R., McDevitt, H., and Steinman, L. (1986). HLA-DQ beta-chain polymorphism linked to MG. Lancet 1058-1060.

Christadoss, P. (1989). Immunogenetics of experimental autoimmune myasthenia gravis. Crit. Rev. in Immunol. 9,247-278.

Christadoss, P, and Dauphinee, M.J. (1986). Immunotherapy for myasthenia gravis: A murine model. J. Immunol. 136,2437-2440.

Christadoss, P, Lindstrom, J.M., Melvold, R.W. and Talal, N. (1985). Mutation at I-Aβ chain prevents experimental autoimmune myasthenia gravis. Immunogenetics 21,33-38.

Christadoss, P., Lindstrom, J., Munro, S. and Talal, N. (1985). Muscle acetylcholine receptor loss in murine experimental autoimmune MG: Correlated with cellular, humoral and clinical responses. J. Neuroimmunol. 8,29-34.

Christadoss, P., Shenoy, M. (1992). Molecular immunopathogenesis of experimental autoimmune myasthenia gravis. Regional Immunol. 4,314-320.

Christadoss, P., Lindstrom, J.M., Talal, N., Duvic, C.R., Kalantri, A. and Shenoy, M. (1986). Immune response gene control of lymphocyte proliferation induced by acetylcholine receptor specific helper factor derived from lymphocytes of myasthenic mice. J. Immunol. 137,1845-1849.

Demaine, A.G., Millward, B.A., Vaughan, R.W. and Welsh, K.I. (1987). T cell receptor beta chain gene polymorphisms are associated with certain autoimmune diseases. J Cell Biochem. 11D,255.0

Demaine, A.G., Vaughan, R.W., Taube, D.H. and Welsh, K.I. (1988). Association of membranous nephropathy with T cell receptor constant β chain and Ig heavy chain with region polymorphisms. Immunogenetics 27,19-23.

Haqqi, T., Banerjee, S., Anderson, G.D. and David, C.S. (1989). An inbred mouse strain with a massive deletion of T cell receptor Vβ genes.J Exp Med 169,1903-1909.

Hohlfeld, R., Conti-Tronconi, B., Kalies, I., Bertrams, Jr. and Toyka, K.V. (1985). Genetic restriction of autoreactive acetylcholine receptor specific T lymphocytes in myasthenia gravis. J. Immunol. 135,2393-2399.

Hoover, M.L., Angelini, G., Ball, E., Stastny, P., Marks, J., Rosenstock, J., Raskin, P., Ferrara, G.B., Tosi, R. and Capra, J.D. (1986). HLA-DQ and T cell receptor genes in insulin-dependent diabetes mellitus. Cold Spring Harbor Symp, Quant Biol 51,803-809.

Lindstrom, J., Shelton, D. and Fujii, Y. (1988). Myasthenia gravis. Adv. Immunol. 42,233-284.

Lindstrom, J.M., Einarson, B.L. and Tzartos, S. (1981). Production and assay of antibodies to acetylcholine receptor. Methods Enzymol 74,432-460.

Rodriguez, M., Patrick, A.K., Peese, L.R. and David, C.S. (1992). Role of T cell receptor Vβ genes in Theiler's virus-induced demyelination of mice. J Immunol. 148,921-927.

Shenoy, M., Oshima, M., Atassi, M.Z. and Christadoss, P. (1993). Suppression of experimental autoimmune myasthenia gravis by epitope-specific neonatal tolerance to synthetic region α146-162 of acetylcholine receptor. Clinic. Immunol. Immunopath. 66,230-238.

Urban, J, Kumar, V., Kono, D.H., Gomez, C., Horvath, S.J., Clayton, J., Ando, D.G., Sercarz, E.E. and Hood, L. (1988). Restricted use of T cell receptor V genes in murine autoimmune encephalomyelitis raises possibilities for antibody therapy. Cell 54,577-592.

Waldor, M.K., Sriram, S., McDevitt, H.D., Steinman, L. (1983) *In vivo* therapy with monoclonal anti-Ia antibody suppresses immune response to acetylcholine receptor. Pro. Natl. Acad. Sci. USA. 80:2713-2717.

Weetman, A.P., So, A.K., Roe, C., Walport, M.J. and Foroni, L. (1987). T cell receptor Vβ chain polymorphism linked to primary autoimmune hypothyroidism but not Graves disease. Hum Immunol 20,167-173.

Zamvil, S., Mitchell, D.J., Lee, N.E., Moore, A.C., Waldor, M.K., Sakai, K., Rothbard, J.B., McDevit, HO, Steinman, L. and Acha-Orbea, H. (1988). Predominant expression of a T cell receptor Vβ gene subfamily in autoimmune encephalomyelitis. J Exp Med 167,1586-1596.

Genetic and environmental factors in Systemic lupus erythematosus

Pierre Youinou*, Gilbert Semana[1], Jean-Charles Piette[2], Sylviane Muller[3], Jatinderpal Kalsi[4], David A Isenberg[4], Loic Guillevin[5], Denise Salmon[6] Jean-Paul Viard[7] and Jean-François Bach[7]
*Laboratory of Immunology,Brest University Medical School Hospital, BP824, F29609 Brest cedex, France; [1]Laboratory of Immunology, Rennes University, Medical School, Rennes, France; [2]Department of Internal Medicine, La Pitié-Salpêtrière Hospital, Paris, France; [3]CNRS, Strasbourg, France; [4]Department of Immunology, University College London Medical School, London, UK; [5]Department of Internal Medicine, Avicenne Hospital, Bobigny, France;[6]CNRS, Paris, France; and [7]Department of Clinical Immunology, Necker Hospital, Paris, France

Systemic lupus erythematosus (SLE) is a multifaceted autoimmune disorder (Shoenfeld and Isenberg 1989). Its diagnosis relies on a set of criteria established by the American College of Rheumatology (Tan *et al* 1982). There is compelling evidence that the genetic background plays an important role, not only in the susceptibility to the disease, but also in its clinical and serological expression (Arnett and Reveille 1992). The first association of SLE with a genetic trait was reported in 1971 with the HLA class I allele B8 (Grumet *et al 1971*). Subsequently a large number of family and population studies have confirmed this finding. However it is also known that environmental factors play an important role.

Thirty-two families were enrolled in the present study, with one patient in each of 30 families, two patients in another and three patients in the remaining

NATO ASI Series, Vol. H 80
Autoimmunity: Experimental Aspects
Edited by Moncef Zouali

family. The diagnosis of definite SLE required the presence of at least four criteria from the revised classification list (Tan *et al.*, 1982). A total of 222 individuals were evaluated (32 probants, 175 relatives and 15 spouses). Twenty five to over 500 normal controls, depending on the test under investigation, were also tested. A variety of autoantibodies were sought. They can be classified into five groups: antinuclear antibodies (ANA), including antibodies to synthetic peptides encompassing Sm protein D, U1-RNP protein A and Ro/SSA 60 kiloDalton (kDa) protein; anti-cytoskeleton and anti-organelle antibodies; rheumatoid factors (RF) directed to hüman or rabbit IgG; anti-tissue and anti-phospholipid (APL) antibodies. Six genetic marker systems were also analysed: HLA class II genes, gamma and kappa immunoglobulin chains and alpha-1 anti-trypsin allotypes and T-cell receptor (TCR) germ-line genes. In addition, ross-reactive idiotypes (CRI) were tested. This report will focus on selected data that can be divided into three sections: autoantibodies, idiotypic markers, and HLA-DR/DQ and TCR germ-line genes. Based upon these results, our conclusions address some of the discrepancies previously reported, but are in full agreement with the disease heterogeneity concept proposed by Harley et al (1989).

AUTOANTIBODIES

Using the indirect immunofluorescence (IIF) technique with HEp-2 cells as the substrate, ANA were detectable in 91% of the patients, 34% of the relatives and 20% of the spouses, compared with only 3% of the controls (Table 1).

Table 1: Antinuclear antibodies (ANA) in patients with SLE and their family members. Shown are the positive/total numbers of patients tested (% positive).

Specificity	Patients	Relatives	Spouses	Controls
ANA	29/32 (90.6)	59/175 (33.7)	3/15 (20.0)	7/205 (3.4)
dsDNA	18/32 (56.3)	17/175 (9.7)	1/14 (7.1)	2/205 (1.0)
Histone	0/30	11/165 (6.7)	0/13	0/105
Sm	0/32	0/175	0/13	0/85
U1-RNP	2/32 (6.3)	0/175	0/14	0/85
Ro/SSA	3/32 (9.4)	0/175	0/14	0/85
La/SSB	0/32	0/175	0/14	0/85

Anti-double stranded (ds) DNA antibodies were positive in 56% of the patients and 10% of their family members, whereas anti-Sm, -U1-RNP, -Ro/SSA, -La/SSB antibodies were only occasionally found. However, some of the corresponding peptides appeared to be recognized. For Sm D1-20, tests were positive in 13% of the patients, 20% of the relatives and 17% of the spouses; for U1-RNP A 35-58, tests were positive in 53%, 26% and 20% of the cases respectively, and for Ro/SSA 60kDa 304-324, tests were positive in 13%, 7% and 1% of the cases respectively. These findings confirm previous results (Barakat *et al* 1990, Barakat *et al* 1991), except the ability of the third peptide to be recognized by SLE patients and their family members (Barakat *et al* 1992).Additionally, various combinations of APL antibodies were observed in patients and their family members. For example, anti-cardiolipin (ACL) antibodies were found in five patients, 27 relatives and two spouses, ACL combined with anti-phosphatidyl-ethanolamine (APE) antibodies were found in four patients, 17 relatives and one spouse, while anti-phosphatidyl ethanolamine (PE) antibodies were found alone in nine patients, 22 relatives and four spouses. This is in accordance with another report (Karmochkine *et al* 1992). Some patients and family members also had antibodies to 2-glycoprotein I, with no evidence of ACL activity, thus confirming and extending recent observations (Viard *et al* 1992).Antibodies to cytoskeleton and organelles were tested by IIF on gerbil and Alexander cells. Antibodies to vimentin were frequently detected as previously reported by Blaschek *et al* (1988a). The frequencies of anti-Golgi apparatus antibodies were surprisingly higher in the relatives (25% positive) and spouses (40% positive) than in the patients (9% positive). These findings may be related to the finding of such antibodies in patients with rubella or cytomegalovirus infections (Blasckek *et al* 1988b), and suggests that an as yet unknown environmental factor could play a role in the pathophysiology of the disease, as previously suggested for patients with systemic sclerosis (Maddison *et al* 1986). RF directed to human (RFh) and rabbit (RFr) IgG were detected by enzyme-linked immunosorbant assays. IgM-RFh was present in 53% of the patients, 25% of the relatives, 7% of the spouses and in 4% of the controls. IgM-RFr was present in 56%, 29% 20% and 2% of the cases, respectively. In summary, non-organ-specific autoantibodies predominate largely in SLE families in agreement with other studies (Shoenfeld *et al* in press). Some peptides appear to be far more immunogenic than others. Autoantibodies are clearly interconnected.

IDIOTYPIC MARKERS

We have employed four anti-CRI antibodies: 16/6, with an original antibody specificity directed to single stranded (ss)DNA, poly(dT) and poly(I); WRI-176, with an original antibody specificity directed to ssDNA, dsDNA, poly(dT) and poly(ADP ribose); RT84, with an original antibody specificity directed to ssDNA, dsDNA and CL; and RT72 with an original antibody specificity directed to ss DNA, dsDNA, CL, histones H1 and H4 and Sm D44-67 peptide.As shown by Zouali and Eyquem (1984), Isenberg *et al* (1985), El-Roeiy *et al* (1987) and Dang *et al* (1988) high frequency of idiotypic markers exists among SLE relatives, as well as patients. The 16/6, WRI-176, RT84 and RT72 CRI were found in 28%, 32%, 41% and 11% of the patients, respectively; 5%, 14%, 33%, and 4% of the relatives ; 0, 0, 31, and 0% of the spouses; and 1%, 0%, 3% and 0% of the controls. All combinations of CRI were observed. There were families that expressed only one of the four, others with none and other families with two or three idiotypes. This observation agrees with the positive correlation between 16/6 and WRI-176 ($p=0.0416$) and negative correlations between 16/6 and RT84 ($P=0.0329$) and between WRI-176 and RT84 ($p=0.0427$). The latest CRI was associated with the production of antibodies to Sm D1-20 ($p=0.0267$) and U1-RNP A 35-58 ($p=0.0035$) and to IgM anti-PE antibodies ($p=0.0124$). None of the idiotypes were associated with anti-dsDNA activity. In summary, SLE patients do express CRI, of which some correlate positively or negatively with each other, and in addition are associated with auto-antibody production.

HLA-DR/DQ AND TCR-V GERM-LINE GENES

HLA class II genes were amplified by a polymerase chain reaction and identified by hybridization with sequence-specific oligonucleotides (the primers and probes have been available since the 11th Histo-compatibility Workshop). We failed to find any significant differences in haplotype frequencies between the patients and controls (DR2, i.e. DR15/DQ6 or DR16/DQ5, present in 25% and 24%; DR3/DQ2 in 36% and 26%; and DR4/DQ8 or DR1/DQ5 or DR BON/DQ5 in 32% and 21%). However, DR3/DQ2 was associated with anti-dsDNA antibodies, DR4/DQ8 with IgM-RF, and antibodies to the U1-RNP A 35-58 peptide. Numerous studies utilizing serologic HLA typing demonstrated that DR3, and to a lesser degree DR2 are more strongly associated with anti-Ro/SSA and La/SSB antibodies than with SLE itself (Arnett *et al* 1989, Harley *et al* 1989). Specific amino acid residues in the second hypervariable region of HLA-DQA1 and DQB1 chain genes have been shown to promote the Ro/SSA and La/SSB autoantibody responses (Reveille *et al* 1991).

The TCR Vβ locus was studied by restriction fragment length polymorphism analysis. The Vβ8, Vβ11 and Cβ loci are bi-allelic (23 and 2 Kb, for Vb8; Vb20 and 25 Kb for Vβ11 ; and Vβ9 and 10 Kb for Cβ). By tracing the traits through the family trees, it is possible to deduce the Vβ8-Vβ11-Cβ haplotype of the individuals. Again, we could find no difference in the haplotype frequencies between patients and controls. Neither could we show associations between these haplotypes and the production of any given autoantibody. Wong *et al* (1988) reported also that none of three TCR chain Vβ genes co-segregated with the disease in families with at least two patients.Interestingly, there was a selection of TCR Vβ elements by HLA class II determinants predisposing to autoantibody production. An example is given in Table 2.

Table 2: .Selection of TCR 23-20-9 haplotype by HLA-DR13/DQ6 to elicit antibodies to U1-RNP A 35-58 peptide (% of responders).

Presence of TCR 23-20-9 haplotypes	HLA haplotype DR13/DQ6	DR4/DQ8	DR3/DQ2	DR15/DQ6
yes	52	4	22	38
no	13	44	33	41
Difference	$p<0.02$	$p<0.001$	$p>0.05$	$p>0.05$

Sixty-three individuals responded to the peptide U1-RNP A 35-58. Thirty-eight of them were DR13/DQ6 positive. Among these, the proportion of responders was higher in those with the 23-20-9 TCR Vβ haplotype than in the remainder. Conversely, 51 were DR4/DQ8 positive, the proportion of responders being lower in those with the 23-20-9 TCR Vβ haplotype than in the remainder. This example suggests that there are positive and negative selections. In summary, there were no differences in terms of HLA class II and TCR Vβ germ-line gene frequencies between the patients and the controls. DR3/DQ2 and DR4/DQ8 were associated with autoantibody activity. These two sets of genes may act in concert to promote autoantibody responses.

CONCLUSION

Thus, the genetics of SLE proves to be complex. A number of loci contribute to a disease which is clinically and serologically heterogeneous. Various combinations of genes together with environmental agents could account for this polymorphism.

Acknowledgements:

This work was supported by the INSERM which is gratefully acknowledged. Thanks are also due to the Société Nationale Française de Médecine Interne.

REFERENCES

Arnett FC, Hamilton RG, Reveille JD, Bias WB, Harley JB, Reichlin M (1989) Genetic studies of Ro (SSA) and La (SSB) autoantibodies in families with systemic lupus erythematosus and primary Sjögren's syndrome.Arthritis Rheum 32 : 413-419

Arnett FC, Reveille JD (1992) Genetics of systemic lupus erythematosus. Rheum Dis Clin North America 18 : 865-892

Barakat S, Briand JP, Weber JC, van Regenmortel MHV, Muller S (1990) Recognition of synthetic peptides of Sm-D autoantigen by lupus sera. Clin Exp Immunol 81 : 256-262

Barakat S, Briand JP, Abuaf N, van Regenmortel MHV, Muller S (1991) Mapping of U1 snRNP polypeptide A with synthetic peptides and autoimmune sera. Clin Exp Immunol 86 : 71-78

Barakat S, Meyer O, Torterotot F, Youinou P, Briand JP, Kahn MF, Muller S (1992) IgG antibodies from patients with primary Sjögren's syndrome and systemic lupus erythematosus recognize different epitopes in 60-kD SSA/Ro protein. Clin Exp Immunol 89 : 38-45

Blaschek MA, Boehme M, Jouquan J, Simitzis AM, Fifas S, Le Goff P, Youinou P (1988a) The relationship of anti-vimentin antibody to cardiolipin antibody in systemic lupus erythematosus. Ann Rheum Dis 47 : 708-716

Blaschek MA, Pennec YL, Simitzis AM, Le Goff P, Kerdraon G, Jouquan J, Youinou P (1988 b) Anti-Golgi complex autoantibodies in patients with primary Sjögren's syndrome. Scand J Rheumatol 17 : 291-296

Dang H, Takei M, Isenberg DA, Shoenfeld Y, Backimer R, Rauch J, Talal N (1988) Expression of an interspecies idiotypes in sera of SLE patients and their first-degree relatives. Clin Exp Immunol 71 : 445-450

El-Roeiy A, Gleicher N, Isenberg DA, Kennedy RC, Shoenfeld Y (1987) A common anti-DNA idiotype and other autoantibodies in sera of offspring of mothers with systemic lupus erythematosus. Clin Exp Immunol 68 : 528-534

Grumet FC, Coukel A, Bodmer JC, Bodmer WF, McDevitt HO (1971) Histo-compatibility (HL-A) antigens associated with systemic lupus erythematosus. N Engl J Med 285 : 193-196

Harley JB, Sestak AL, Willis LG, Man Fu S, Hansen JA, Reichlin M (1989) A model for disease heterogeneity in systemic lupus erythematosus. Arthritis Rheum 32 : 826-836

Isenberg DA, Shoenfeld Y, Walport M, Mackworth-Young C, Dudeney C, Todd-Pokropek A, Brill S, Weinberger A, Pinkas J (1985) Detection of cross-reactive anti-DNA idiotypes in the serum of systemic lupus erythematosus patients and of their relatives. Arthritis Rheum 28 : 999-1007

Karmochkine M, Cacoub P, Piette JC, Godeau P, Boffa MC (1992) Antiphosphatidylethanolamine antibody as the sole antiphospholipid antibody in systemic lupus erythematosus. Clin Exp Rheumatol 10 : 603-605

Maddison PJ, Skinner RP, Perevia RS, Black CM, Ansell BM, Jayson MIV, Rowell NR, Welsh KI (1986) Antinuclear antibodies in the relatives and spouses of patients with systemic sclerosis. Ann Rheum Dis 45 : 793-799

Reveille JD, MacLeod MJ, Whittington K, Arnett FC (1991) Specific amino acid residues in the second hypervariable region of HLA-DQA1 and DQB1 chain genes promote the Ro(SSA)/La(SSB) autoantibody responses. J Immunol 146 : 3871-3876

Shoenfeld Y, Isenberg DA (1989) The mosaic of autoimmunity. Elsevier, Amsterdam

Shoenfeld Y, Slor H, Shafrir S, Krause I, Granados J, Villarreal GM, Alarcón-Segovia D (1993) The diversity and pattern of inheritance of autoantibodies in multiple case families of systemic lupus erythematosus. Ann Rheum Dis in press

Tan EM, Cohen AS, Fries JF, Masi AT, McShane DJ, Rothfield NF, Green Schaller J, Talal N, Winchester RJ (1982) The 1982 revised criteria for the classification of systemic lupus erythematosus. Arthritis Rheum 25 : 1271-1277

Viard JP, Amoura Z, Bach JF (1992) Association of anti-ß2 glycoprotein I antibodies with lupus-type circulating anticoagulant and thrombosis in systemic lupus erythematosus. Am J Med 93 : 181-186

Wong DW, Bentwich Z, Martinez-Tarquino C, Seidman JG, Duby AD, Quertermous T, Schur PH (1988) Nonlinkage of the T cell receptor α, ß and γ genes to systemic lupus erythematosus in multiplex families. Arthritis Rheum 31 : 1371-1376

Zouali M, Eyquem A (1984) Idiotype restriction in human auto-antibodies to DNA in systemic lupus erythematosus. Immunol letters 7 : 187-190

Role of Immunocompetent Cells in Skin Sclerosis and Autoimmunity in Tight Skin Mice

Constantin Bona, Shinobu Shibata, Christian Daian and Kuppuswamy Kasturi
Mount Sinai School of Medicine
Depatment of Microbiology
1 Gustave L. Levy Place
New York, NY 10029
USA

1. Introduction

Scleroderma is a connective tissue disease which affects the skin and internal organs. It is a complex and clinically heterogenous disease with three major clinical forms ranging from limited skin involvement (morphea) to diffuse skin sclerosis and severe internal organ involvement (progressive skin sclerosis, SSc) and CREST syndrome in which skin sclerosis is associated with calcinosis, Raynaud phenomenon, esophageal densitometry and telangiectasia. Although, the exact mechanisms involved in the pathogenesis of SSc are not known, the following three major alterations might play a role in the pathogenesis:

a) excessive deposition of collagen within the skin and internal organs.
b) vascular lesions of capillaries and small arteries.
c) autoimmune manifestations. (LeRoy, 1989)

The autoimmune manifestations in scleroderma are characterized by the presence of autoantibodies as well as by abnormalities of cellular immune response.

The antigens that are the specific target of autoantibodies associated with scleroderma have been identified and characterized. Some of the autoantibodies are present only in scleroderma, while others are associated with various systemic autoimmune diseases. Table 1 illustrates the specificity and clinical association of various autoantibodies .

NATO ASI Series, Vol. H 80
Autoimmunity: Experimental Aspects
Edited by Moncef Zouali

Table I. Specificity and clinical association of autoantibodies in SSC.

Autoantibody	**Antigen**	**Clinical Association**
Anti-Scl70	Topoisomerase I	Diffuse SSc
Anti-fibrillarin	Fibrillarin-a34 kD protein associated with nucleoli	Diffuse SSc
Anti-Nor-90	90kD nucleolar protein	Diffuse SSc with lung involvement
Anti-centromeric	Kinetochore protein (CENP A,B, C)	CREST
Anti-FcR	FcγRII	SSC, lupus
Anti-RNA polymerase I	RNA polymerase I	SSc, lupus, MCD
Anti-PM-Scl	Complex of 20-110kD	SSc, polymyositis
Anti-Ku	70-80kD nucleolar	SSc, polymyositis, lupus
Anti-mitochondrial	M2 complex	CREST, primary biliary cirrhosis

Experimental models of human diseases represent excellent tools for studying the pathogenesis of autoimmune diseases.

Many features of human scleroderma are found in tight skin (TSK) mouse. Cutaneous hyperplasia with increased collagen synthesis in skin fibroblasts and the characteristic histopathological changes of skin are the main features of tight skin disease (Green et al 1976) . Abnormalities in collagen metabolism, or similar to those present in patients with scleroderma, have also been demonstrated in TSK mice (Osborn et al 1983, Jimenez et al 1986, and Peltonen et al 1990). This disorder appeared spontaneously in B10D2 mice as a result of a mutation on chromosome 2 (Green et al 1976). The TSK/+ mutant is currently maintained by sister-brother breeding with C57BL/6 pa/pa mice as the offspring homozygous for TSK mutation dies *in utero*.

2. Resemblance and differences between skin alterations in TSK mice and scleroderma

Major histopathological alterations in TSK mice are observed in skin connective tissue, lung, heart and skeleton. The skin is characterized by an increased thickness of dermis, increase in the number of fibroblasts and a firm attachment of hypodermis to deep tissue. The increase of collagen is striking compared to normal mice. The hyperplasia extends around adnexa, panniculus carnosus and replaces subcutaneous fat. Despite this extensive proliferation, there is little inflammatory infiltrate. (Fig.1).

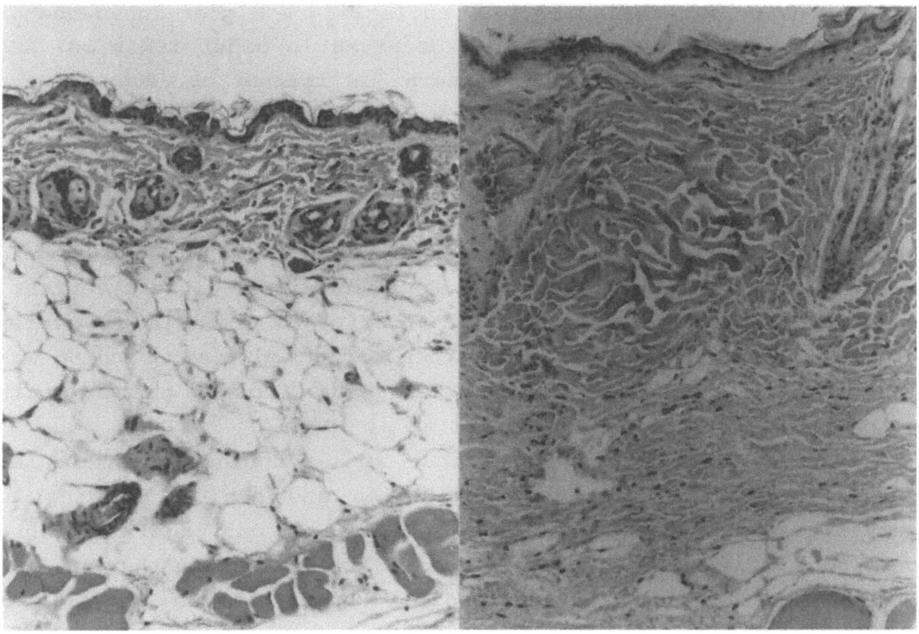

Figure 1. Histopathologic aspect of skin of 4 month old C57BL/6 pa/pa mice (left) and TSK mice (right).

Therefore, the hyperplasia observed in the deep dermis is somewhat analogous to human SSc. However, differences are observed between human SSc and TSK mouse skin. In the latter, there is preservation of adnexa, absence of an apparent inflammatory phase and hyperplasia in all layers of the dermis, rather than in the subcutaneous fat (Phelps et al submitted).

3. Resemblance of autoantibodies of TSK mice and human scleroderma

Studies conducted during the past five years in our laboratory demonstrated that TSK mice spontaneously produce autoantibodies specific for target autoantigens as those detected in human scleroderma. The presence of autoantibodies to a variety of nuclear antigens constitutes a common feature of connective tissue diseases, including scleroderma. These autoantibodies react with various components of nucleus or nucleoli. A higher incidence of antinuclear antibodies (ANA) was observed in older TSK mice. ANA giving a speckled staining pattern in immunofluorescence were observed in 44% of the animals. By contrast, anti-DNA autoantibodies were observed only in 10% of TSK mice (Bocchieri et al 1991). Some sera also contained antibodies which stained nucleoli or nucleoli and cytoskeleton. The analysis of a large panel of hybridomas obtained from TSK mice showed that some of them bind to cytoskeleton proteins. Among 11 mAb which stained the cytoplasm, one bound to actin, one to keratin and five to vimentin. A series of mAbs bound to nucleoli and mitochondria and their binding was inhibited by the anti-mitochondria serum from a patient with primary biliary cirrhosis (Muryoi et al 1992).

Previously, it was shown that anti-cytoplasmic antibodies generally belong to natural antibodies and exhibit binding to multiple antigens (Naparstek et al 1986). In our panel, we found only four mAbs exhibiting multispecificity indicating that TSK ANA and anti-cytoplasmic autoantibodies do not belong to natural antibodies.

Diffuse scleroderma, the most severe form of SSc is associated with the presence of anti-Scl70 antigen antibodies which are actually specific for topoisomerase I (Shero et al, 1986). Anti-topo I autoantibodies are found in SSc but not in other systemic autoimmune diseases. Our studies showed that there is an increase in the level of serum anti-topoisomerase I autoantibodies in aged TSK mice (Muryoi et al 1991). The monoclonal antibodies isolated from TSK mice bind to epitopes which interact with autoantibodies from scleroderma patients (Muryoi et al 1991). The epitope specificity of these antibodies has been determined using a series of fusion proteins containing contiguous portions of topo I polypeptide. Western blot analysis demonstrated that both human and mouse mAbs bound strongly to fusion protein C encompassing the NH_2- terminal portion of the enzyme, and weakly to fusion proteins F and G containing regions close to the COOH-terminal end of the molecule. This crossreactivity is related to a tripeptide sequence homology in F,G and C fusion proteins. It is interesting that a pentapeptide sequence homologous to that found in fusion protein C was identified in the UL70 protein of cytomegalovirus, suggesting that activation of autoreactive B cell clones by molecular mimicry is possible. Both human and mouse mAbs exhibiting the same antigen specificity, also share an interspecies cross-reactive idiotope (Muryoi et al 1992). These data suggest that B cell clones producing anti-topo autoantibodies present in human and mouse repertoire are conserved during phylogeny, and are activated during the development of scleroderma disease.

We have also observed that TSK mice spontaneously produce anti-RNA polymerase I (RNA, pol I) autoantibodies (Shibata et al 1993). Such autoantibodies have been described in humans afflicted with rheumatoid arthritis, mixed connective diseases and scleroderma (Stetler et al 1982). Human anti-RNA polymerase I autoantibodies are characterized by binding to the native enzyme localized in nucleoli and particularly to fibrillar elements seggregated nucleoli, inhibition of enzymatic activity and increased binding to phosphorylated enzyme (Stetler et al 1992).

From TSK mice, we have prepared a large panel of anti-RNA-pol I monoclonal antibodies. The analysis of the immunochemical properties of these mAbs showed similarities and differences with human autoantibodies. Compared to human autoantibodies, TSK mAbs bound not only to rat hepatoma enzyme, but also to murine, bacterial and viral RNA polymerase I (Shibata et al, 1993). TSK antibodies resemble human antibodies with respect to: a) binding of large enzyme subunits, b) increased binding to phosphorylated enzyme, c) inhibition of enzymatic activity and d) A speckled immunofluorescence pattern of nucleoli. Some mAbs differ from human antibodies as they enhance enzymatic activity and lack the "necklace" immunofluorescence staining pattern of actinomycin D treated Vero cells (Shibata et al, 1993).

In early studies, we have found that TSK mice like other animals prone to systemic autoimmune diseases (NZB, $\underline{me}^{v}$, MRL) spontaneously produce anti-FcR autoantibodies (Boros et al, 1990). Anti-FcR mAbs obtained from TSK mice exhibit the following characteristics: a) they bind to both FcγR-II and -III expressed on neutrophils and their binding is inhibited by Fab fragments of 3G8 and IV.3 mAbs. b) they cause degranulation of neutrophils accompanied by release of hydrolase from both azurophilic and specific granules (Boros et al 1991).

Because these observations indicated spontaneous occurrence of anti-FcR autoantibodies in TSK mice, we further studied the presence of such antibodies in scleroderma patients. A high percentage of patients with SSc and to a lesser extent SLE and Sjogren's syndrome, have anti-FcR Ig in their serum. These autoantibodies were specific for FcγRI (CD664) and FcγRII (CD32). These naturally occurring anti-FcγR antibodies recognized various cells including neutrophils, B cells, NK cells and macrophages (Boros et al, 1993). From a scleroderma patient we obtained anti-FcR mAbs. One of these antibodies was extensively studied, It bound to FcγRIII A and B expressed on NK cells, activated macrophages and neutrophils (Szegedi et al, 1993).

These data clearly demonstrate that TSK mice which develop a scleroderma-like syndrome spontaneously produce autoantibodies which exhibit similar immunochemical properties with autoantibodies present in scleroderma patients.

4. Induction of skin fibrosis and autoantibodies by TSK bone marrow cells

One of the postulates to define pathogenic lymphocytes is the ability to transfer disease to normal animals by lymphocytes from animals prone to autoimmune diseases (Bona 1991). Indeed, transfer of T cells from autoimmune animals to normal mice can result in the induction of autoimmune diseases such as experimental allergic enephalomyelitis,(Lublin et al 1985). With bone marrow cells, it can result in the induction of diabetes (Serezze et al 1988). Thus, we investigated whether the transfer of bone marrow cells of aged TSK mice can induce skin fibrosis and the production of autoantibodies in C57BL pa/pa mice. C57BL/6 pa/pa mice differ from TSK mice only by the tsk mutation and do not reject skin grafts (Phelps et al, submitted). These genetic characteristics rule out eventual scleroderma-like skin alterations accompanying graft vs host reaction (Jaffee et al 1983).

a) Histopatholic alterations in the skin of C57BL pa/pa mice infused with TSK bone marrow cells.

Injection of $25x10^6$ TSK bone marrow cells into C57BL/6 pa/pa mice lead to skin fibrosis, increased transcription of α_1 collagen type I gene and production of anti-topoisomerase I and RNA polymerase I autoantibodies after a 3 month prodromal period (Phelps et al, submitted). This long prodromal period was also required for the occurrence of overt diabetes in F_1 mice infused with NOD bone marrow cells (Serreze et al, 1988). C57BL/6 pa/pa mice infused with TSK bone marrow cells showed several abnormalities including skin alterations characterized by an initial inflammatory phase followed by fibrosis, occurring throughout the dermis and most evident at the dermal subcutaneous junction (Figure 2).

The occurrence of fibrosis was also illustrated by increased thickness of the dermis (0.6μm in TSK, 0.19μm in C57BL/6 pa/pa and 0.32μm in C57BL/6 pa/pa infused with TSK bone marrow cells). Electron microscopy confirmed the histopathological observations namely, lack of apparent changes in the superficial dermis, deep cellular infiltration of the dermis by macrophages, activated lymphocytes, mastocytes and plasma cells.

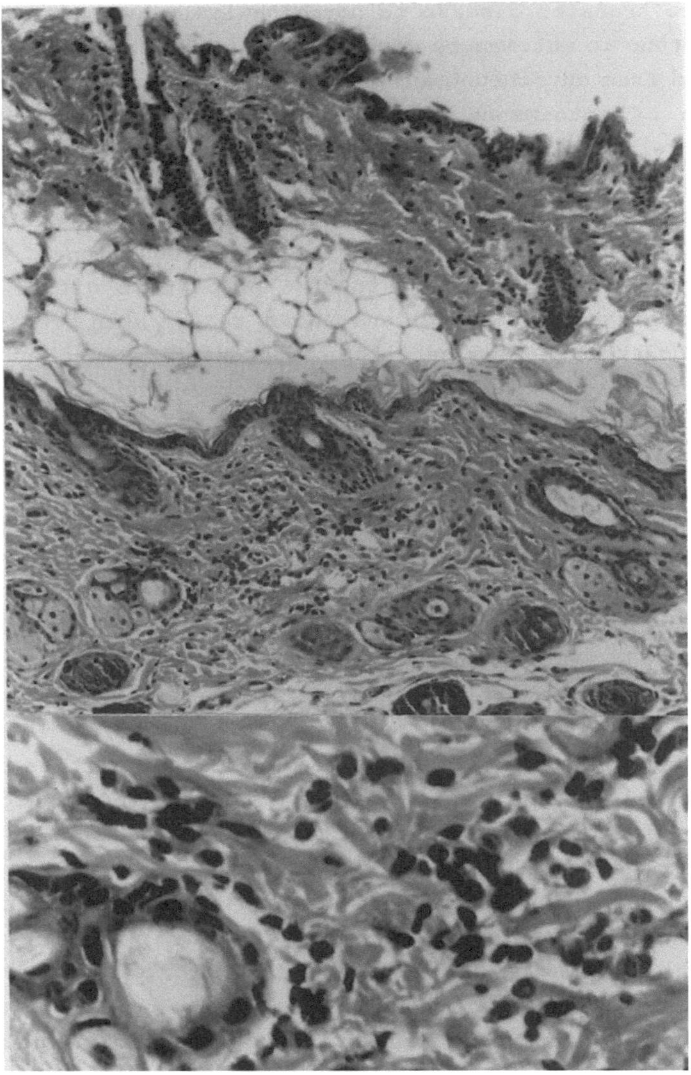

Figure 2. Histopatholic aspect of the skin of C57BL/6 pa/pa mice infused with C57BL/6 pa/pa bone marrow cells(upper), TSK bone marrow (middle)x 10 and bottom (x30) HE staining.

These alterations are very similar to those observed in the inflammatory phase of morphea or scleroderma (Fleischmajer et al 1977). Collagen fibers showed an extreme variation in diameter with small fibers, admixed with large fibers. Some of the fibers had irregular transverse or wavy appearance (Figure 3).

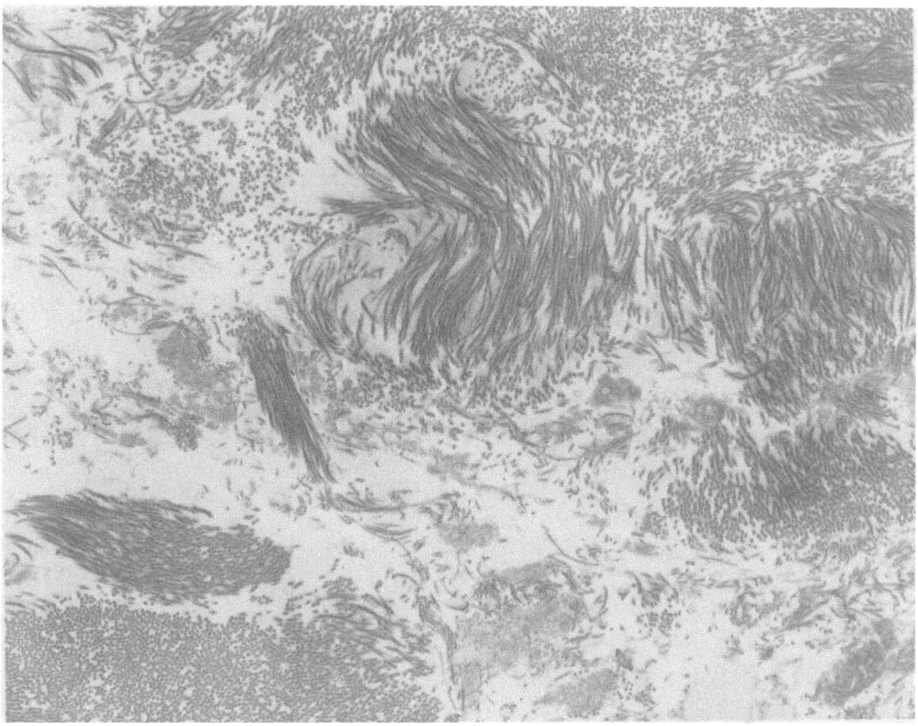

Figure 3. Electron microscopy of skin of 4 month old TSK mice. Collagen bundles in mid dermis x10.000

A histometric analysis of collagen fibers in mid-dermis showed that while in pa/pa mice collagen fibers are preponderant (100-200μm) in TSK mice higher numbers of fibers with a smaller diameter (30-100nm) are observed (Fleischmajer et al 1990). Pa/pa mice infused with TSK bone marrow cells showed also fibers of smaller diameter, but in a smaller proportion than in TSK/+ mice. A similar pattern is observed in scleroderma patients where small diameter fibers (below 100nm) are predominant and probably represent collagen in early an stage of maturation.

b) Increased transcription of α_1 collagen I and TGFβ genes

Increased transcription of α_1 collagen gene in TSK mice (Jimenez et al, 1986) as well as alteration of cytokine production were previously reported in scleroderma patients (Loppnow et al 1989 and White et al 1992). Based on these informations, we have studied the level of transcription of α_1 collagen I TNFα, TGFβ, IL-4 and IL-6 genes in skin tissue. Replicate slot blots containing RNA from skin tissue were hybridized with probe specific for each of the above genes and for β actin gene and the intensity of the signal was quantitated by densitometry. An increased amount of α_1 collagen and TGFβ was observed in TSK compared to pa/pa mice (Figure 4).

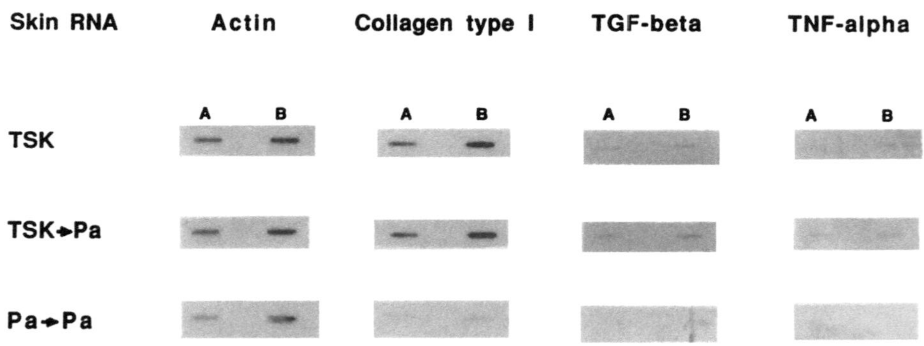

Figure 4. Slot blots of skin RNA hybridized with probes specific for β actin, α_1 collagen, TGFβ and TNFα genes. RNA from TSK from C57BL/6 pa infused with TSK bone marrow cells (TSK→pa) and from C57BL/6 pa/pa infused with C57BL/pa/pa bone marrow cells (pa→pa). A=1μg RNA, B=2μg RNA

An increase of the transcripts of these two genes was also observed in the skin of pa/pa mice harvested three months after infusion of TSK bone marrow cells. No significant changes in the level of TNFα and IL-6 transcripts were observed. IL-4 transcripts were not detectable in TSK or C57BL/6 pa/pa skin. These results indicate that the skin fibrosis observed in TSK mice and in pa/pa mice infused with TSK bone marrow cells is related to the activation of collagen and TGFβ genes. Actually, it has been shown that TGFβ causes a persistent increase in the steady state level of type I and III collagen transcripts (Abergel et al 1987).

c) Occurrence of autoantibodies

A significant increase in the titer of anti-topoisomerase I and RNA polymerase I autoantibodies and skin fibrosis was observed in 2 or 3 months after infusion of TSK bone marrow cells into pa/pa mice.

5. Conclusions

These results strongly suggest that immunocompetent cells play an important role in skin fibrosis of TSK mice. The induced alterations require a long prodromal period probably related to maturation of TSK bone marrow cells in pa/pa mice. Compared to the histopathological skin alterations observed in TSK mice, pa/pa mice infused with TSK bone marrow cells show a strong cellular infiltration in the mid dermis which is similar to early the inflammatory phases observed in morphea and SSc. These histopathological alterations are associated with increased transcription of α_1 collagen and TGFβ genes. This suggests that the inflammatory cells which are recruited in the skin probably secrete lymphokines which stimulate fibroblast proliferation and synthesis of collagen.

The occurrence of autoantibodies in pa/pa mice infused with TSK bone marrow cells preceded skin alterations. This probably reflects early maturation of B cells from bone marrow precursors genetically programmed to produce autoantibodies. However, the role of autoantibodies in the pathogenesis of skin sclerosis is not yet understood. It is also unclear whether they play a pathogenic role or represent an epiphenomenon that is a consequence of stimulation of clones by antigens released subsequent to tissue injury. Taken together, our data strongly suggest that autoimmune processes play a role in the occurrence of skin fibrosis, a major histopathological trait of scleroderma.

6. References

Abergel RP, Chu, ML, Bauer, EA, Uilto J, (1987) Regulation of collagen expression in cutaneous diseases with dermal fibrosis:evidence for pretranslational control. J Invest Dermatol 88:727-731.

Bocchieri MH, Heuriksen PD, Kasturi KN, Muryoi T, Bona CA, and Jimenez SA. (1991) Evidence for autoimmunity in tight skin (TSK/+) mouse model progressive systemic sclerosis. Arthritis Rheum 34:599-605.

Bona C. (1991) Postulates defining pathogenic autoantibodies and T cells. Autoimmunity 10:169-172.

Boros P, Chen J, Bona C, Unkeless JC (1990) Autoimmune mice make anti-Fcgamma receptor antibodies. J Exp Med 171:1581-1595.

Boros P, Muryoi T, Spiera H, Bona C, Unkeless JC (1993) Autoantibodies directed against different clones of FcγR are found in sera of autoimmune patients. J Immunol 750:2018-2029.

Boros P, Odin JA, Muryoi T, Masur SU, Bona C, and Unkeless JC (1991) IgM anti-FcγR autoantibodies trigger neutrophil degranulation. J Exp Med 175:1473-1482.

Flesichmajer R, Perlish, JS, Burgeson RE, Shaikh-Bahai F, and Timpl R (1990) Type I and Type III collagen interactions during fibrilliagenesis. Ann. NY-Acad Sci 580:161-176.

Fleischmajer R, Perlish JS, Reeves JRT (1977) Cellular infiltrates in scleroderma skin. Arthritis Rheum 20:975.

Green MC, Sweet HO, Bunker L.E (1976) Tight skin, a new mutation of the mouse causing excessive growth of connective tissue and skeleton. Am J Pathol 82:493-507.

Jaffee BD, and Claman HN (1983) Chronic graft-versus-host diseases (GVHD) a model for scleroderma. Cell Immunol. 77:1-12.

Jimenez SA, Williams Ch J, Meyers JC and Bashey R (1986) Increased expression of type I and type III procollagen genes in tight-skin (TSK) mouse fibroblasts. J Biol Chem 261-657-662.

LeRoy EC, (1989) The pathogenesis of systemic sclerosis. Clin Exp Rheumatol 7/S-3:135.

Loppnow H, Flad H-D, Durrbaum I, Musehold J, Fetting R, Ulmeer AJ, Herzbeck H, and Brandt E (1989) Detection of interleukin-1 with human dermal fibroblasts. Immunobiol 179:283-291.

Lublin FD, (1985) Adoptive transfer of murine relapsing experimental allergic encephalomyelitis. Ann Neurol 17:188-190

Muryoi T, Andre-Schwartz J, Saitoh Y, Daian Chr, Hall B, Dimitriu-Bona A, Schwartz R, Bona CA and Kasturi KN (1992) Self reactive repertoire of tight skin mice:immunochemical and molecular characterization of anti-cytoplasmic autoantibodies. Cell Immunol 144:43-54.

Muryoi T, Kasturi KN, Kafina MJ, Cram DS, Harrison LC, Sasaki T, and Bona CA (1992) Antitopoisomerase I monoclonal autoantibodies from scleroderma patients and tight skin mouse interact with similar epitopes. J Exp Med 175:1103-1109.

Muryoi T, Kasturi KN, Kafina MJ, Saitoh Y, Usuba O, Perlish JS, Fleischmayer R, and Bona C (1991) Self reactive repertoire of tight skin mouse:immunochemical and molecular characterization of anti-topoisomerase I autoantibodies. Autoimmunity 9:109-119.

Naparstek Y, Andre-Schwartz J, Manser T, Wysocki LJ, Breitman L, Stollar BD, Gefter M, and Schwartz RS, (1986) A single germline VH gene segment of normal A/J mice encodes autoantibodies characteristic of systemic lupus erythematosus. J Exp Med 164:614-626.

Osborn TG, Bauer NE, Ross SC (1983) The tight-skin mouse:physical and biochemical properties of the skin. J Rheumatol 10:793-796

Peltonen J, Kahari L, Uilto J, Jimenez SA, (1990) Increased expression of type VI collagen genes in systemic sclerosis. Arthritis Rheum 33:1829-1835.

Phelps RG, Daian C, Shibata S, Fleischmajer R, Bona CA (submitted) Induction of skin fibrosis and autoantibodies by infusion of tight skin mice immunocompetent cells into C57BL/6 pa/pa mice.

Serreze DV, Leiter EH, Worthen SM, and Shultz LD, (1988) NOD marrow stem cells adoptively transfer diabetes to resistant (NODxNON)F1 mice. Diabetes 37:252-55.

Shero, JH, Bordwell B, Rothfield NI, Earnshaw WC, (1986) High titers of autoantibodies to topoisomerase I (SCL-70) in sera from scleroderma patients. Science 231:737-740.

Shibata S, Muryoi T, Saito Y, Brumeanu T, Bona CA, Kasturi KN (1993) Immunochemical and molecular characterization of anti-RNA polymerase I autoantibodies produced by tight skin mice. J Clin Invest:in press.

Stetler DA, Rose KM, Wenger ME, Berlin CM, and Jacob S (1982)

Antibodies to distinct polypeptides of RNA polymerase I in sera from patients with rheumatic autoimmune diseases. Proc Natl Acad Sci USA 79:6499-7503.

Szegedi A, Boros P, Bona C, Kafina M, Unkeless JC (1993) An FcγR III specific autoantibody from a patient with progressive systemic sclerosis. Immunol Letters 35:69-76.

White-Needleman B, Wigley T-M, Stair, RW, (1992) Interleukin-1, interleukin-2, interleukin-4, interleukin-6, tumor necrosis factor alpha and interferon gamma levels in sera from patients with scleroderma. Arthritis Rheum 35:67-92

MRL/*lpr* mice as a model of autoimmune diseases: mutiple strategies for immune intervention

Ignacio Moreno de Alboran, Jose A. Gonzalo, Guido Kroemer and Carlos Martinez-A.
Centro Nacional de Biotecnologia, CSIC, Universidad Autónoma de Madrid, Campus de Cantoblanco, 28049 Madrid, Spain

INTRODUCTION

Autoaggression has a multifactorial and polygenetic etiology. The catalogue of putative autoimmunity-inducing factors is extremely heterogeneous, ranging from certain MHC haplotypes to TCR locus alleles, genetically determined defects in T cell maturation, macrophage defects, endogenous viruses, aberrations in the endocrine control of immune function and target organ defects to evironmental factors, including viral and bacterial infections [Möller 1990; Wick *et al.* 1993]. In support of this view, experimental autoimmunity can only be induced in certain, susceptible strains and virtually all animal models developing spontaneous autoimmune disease have been shown to rely on a polygenetic basis. Thus, it appears that a given autoimmune manifestation does not have exactly the same etiology in different individuals, but constitutes the final outcome of distinct, partially overlapping combinations of predisposing factors and/or external stimuli. Accordingly, for example, development of systemic lupus erythematosus (SLE) [Cohen & Eisenberg 1991; Theofilopoulos *et al.* 1989] or diabetes mellitus [Garchon *et al.* 1991; Prins *et al.* 1993; Todd *et al.* 1991] in various animal strains has rather different genetic bases.

T cells, B cells, as well as T and B cell-derived factors, play a pivotal role in generating various autoimmune diseases, both organ-specific and systemic. Systemic lupus erythematosus (SLE) exemplifies the systemic autoimmune disease *par excellence.* Several animal models have been used to study SLE-type

NATO ASI Series, Vol. H 80
Autoimmunity: Experimental Aspects
Edited by Moncef Zouali

syndromes, but probably the more extensively studied SLE animal model is the MRL/Mp-*lpr/lpr* strain of mice (abbreviated MRL/*lpr*). Lupus-prone inbred strains of the mouse provide an excellent model for human systemic lupus erythematosus (SLE) because most of the immunopathological abnormalities found in the human disease appear to be operative [Theofilopoulos & Dixon 1985; Theofilopoulos *et al.* 1989]. Both in SLE patients and in MRL/*lpr* mice abnormalities in T cell activation, and cytokine expression have been described, converting this strain into an interesting model to address the role of cytokines and to evaluate reagents targeted to the T cell receptor (TCR) in autoimmunity.

MRL/Mp-*lpr/lpr* mice exhibit an acute disseminated autoimmune disease characterized by hypergammaglobulinemia, anti-DNA and other antinuclear antibodies, circulating immune complexes, interstitial glomerulonephritis, necrotic polyarteritis, and a rheumatoid arthritis-like pathology accompanied by serum rheumatoid factors (Table 1).

TABLE 1: Clinical, phenotypic and functional alterations present in MRL/lpr mice

Clinical features	Phenotypic manifestations	Functional defects
Autoantibodies	Early thymic involution	Lack of IL.2 and IL.4 production
Arthritis	Lymphoid hyperplasia	Deficient T and B cell expansion
Arteritis	Depletion of B1 cell subset	Lack of Fas/Apo expression
Glomerulonephritis	Depletion of NK cells	Accelerated programmed cell death
SLE	Increased rate of CFU-S	Abnormal tyrosine phosphorylation
Limited life span	Increased M-CSF production	

The MRL/Mp-*lpr/lpr* strain carries the single-locus recessive *lpr* (lymphoproliferation) mutation that concerns the *Fas* gene (chr. 19), whose expression is deficient or abnormal [Watanabe *et al.* 1991; Watanabe-Fukunaga *et al.* 1992]. This mutation is genetically linked to a pleiotropic phenotype, including an angioimmunoblastic lymphadenopathy and a massive expansion of an unusual subset of T cells (Thy1$^+$ CD3^{dull} CD5$^+$ CD4$^-$CD8$^-$CD44$^+$) (reviewed in ref. [Cohen & Eisenberg 1991; Theofilopoulos *et al.* 1989]), associated with a systemic autoimmune disease, that is influenced by a set of additional, yet to be defined MRL/Mp "background" genes on chromosomes 7 and 12 [Theofilopoulos & Dixon 1985; Watson *et al.* 1992]. Phenotypically abnormal T cells that accumulate in the periphery also display the B cell marker B220 and express rearranged TCRα and TCRβ chains (α/βTCR/CD3^{dull}), among which Vβ8.2 and Vβ8.3 gene products account for 60% of the expressed TCR [Singer *et al.* 1986; Singer & Theofilopoulos 1990]. Individuals suffering from SLE also have increased numbers of α/βTCR/CD3^{+} CD4^{-} CD8^{-} lymphocytes that augment the production of anti-DNA autoantibodies *in vitro* [Marcos *et al.* 1988; Shivakumar *et al.* 1989].

Although the exact pathogenic contribution of the *lpr* cell remains unknown, it is clear that the *lpr* gene precipitates development of a generalized autoimmune disease when present on a lupus background. Adoptive bone marrow transfer experiments have demonstrated that hematopoietic stem cells from MRL/*lpr* mice carry the *lpr* defect, whereas the genotype of the thymus has no importance [Ikehara *et al.* 1991; Theofilopoulos *et al.* 1981]. Nonetheless, the presence of a thymus is necessary for the selective expansion of the abnormal *lpr* T cell subset whose exact cellular derivation remains obscure. Because *lpr* cells express a combination of immature (CD4^{-} CD8^{-}) and mature (α/βTCR/CD3^{+}) surface markers, it has been difficult to accommodate them in currently accepted schemes of intrathymic T cell maturation. Whereas some authors speculate that *lpr* cells represent developing T cells that fail to die in the thymus or "branch off" from the main pathway and are exported into the periphery (reviewed by [Strominger 1989]), others believe that *lpr* cells represent a relatively mature, postdeletional stage, given that both I-E- and Mls-related Vβ clonal deletions may be detected among these cells [Theofilopoulos *et al.* 1989]. RNA protection analysis of *lpr* T cells revealed an identical set of I-E tolerance-related clonal deletions (i.e., Vβ5.1, 11, 12 and 16) of the non-*lpr* congenic MRL/Mp-+/+ strain [Mountz *et al.* 1990; Singer *et al.* 1989]. It therefore has been proposed that *lpr* cells derive from a "loss pathway" where the accessory molecules CD4 and/or CD8 are down-regulated on cells that previously have expressed these molecules. In accord with this hypothesis, the crossing of MRL/*lpr* with a TCR transgenic mouse whose T-cell receptor is expressed in CD8^{+} cells, prevents the accumulation of "double negative" cells [Sidman *et al.* 1992]. It appears even more convincing that the backcross of MRL/*lpr* mice into a CD8-deficient background completely abrogates the development of the autoimmune disease as well as the lymphoid hyperplasia. Quite in contrast, the bakcross of MRL/*lpr* mice into a CD4$^{-/-}$ results in the exacerbation of accumulation of CD4^{-}CD8^{-} cells (T. Mak, personal communication).

The effect of exogenous IL-2 on the MRL/*lpr* model

The *lpr* cells are defective in interleukin-2 (IL-2) production, but transcribe cytokine genes involved in the inflammatory process (IFN-γ, TNF-α, TNF-β, and IL-6) [Murray *et al.* 1990]. It remains an ongoing conundrum whether abnormalities in the IL-2/IL-2R system reflect a level of immune dysfunction common to individuals suffering from similar or related autoimmune symptoms, and whether such imbalances in the IL-2/IL-2R system could contribute as primary coetiological factors to the initiation of autoimmunity [Kroemer *et al.* 1991]. We have recently reported the use of a virus vector lymphokine-delivery system for immunomodulating autoimmune disorders [Andreu-Sánchez *et al.* 1991; Gutierrez-Ramos *et al.* 1990b]. Human IL2 was delivered by infecting MRL/*lpr* mice with a vaccinia virus containing the human IL-2 gene under the control of the vaccinia p7.5 promotor (IL-2.VV) [Flexner *et al.* 1987]. Vaccinia virus can infect virtually all cells, since it does not require interaction with a specific viral receptor. In the host cell, the p7.5 promoter allows for transcription of the IL-2 gene (whose endogenous promoter is strictly tissue-specific and activation-dependent) both during the early and the late phase of infection. Prophylactic treatment of undiseased, 2- to 3-month-old MRL/*lpr* mice with IL-2 has a number of effects: **First,** it restores normal thymic differentiation, as illustrated by an expansion of the double positive population accompanied by increased numbers of mature thymocytes. **Second,** it impedes the accumulation of $CD3^+CD4^-CD8^-$ ("double negative") T cells in lymph nodes and spleen, and consequently restores the development of an equilibrated mature T cell population. **Third,** IL-2.VV leads to a normalization in the pattern of *TCR Vβ* gene expression displayed by mature T cells. **Fourth,** it decreases urine protein levels and immune complex deposition in the kidney, with a resultant absence of glomerulonephritis. **Fifth,** IL-2 causes a marked increase in longevity (from 195 to more than 400 days) [Gutierrez-Ramos *et al.* 1990a; Gutierrez-Ramos *et al.* 1990b].

We have speculated that the dramatic reduction in the abnormally expanded $CD3^+CD4^-CD8^-$ T-cell population following IL-2 therapy might be directly related to the amelioration and/or prevention of autoimmune disease in these mice. However, given the age-dependence of the IL-2/IL-2R dysregulation in the *lpr* syndrome [Murray *et al.* 1990], one important goal was to optimize the therapeutical windows for a succesfull IL-2-targeted therapy of the progression of autoimmune disease and lymphoproliferation developed by the MRL/*lpr* strain.

For this purpose, we administered IL-2 at different stages of disease progression via infection with IL-2.VV. The specificity of the IL-2 treatment was controlled studying littermate groups infected in parallel with vaccinia viruses expressing human IL-4 (that fails to interact with the murine IL-4 receptor). These studies revealed, that the beneficial IL-2 effect is only obtained when, IL-2.VV is inoculated well before the *lpr* phenomenon and associated autoimmune lesions begin to develop. When the treatment is inititated after 4 months of age, when the disease has already started to develop, no reduction of the serological or histomorphological signs of autoimmunity is observed and the life span of MRL-*lpr* mice is not prolonged [Gutierrez-Ramos *et al.* 1991]. Hence, IL-2 may serve as a prophylactic agent, but is probably not capable of reducing a developing or established autoimmune disease.

Studies regarding the signals and growth factors involved in the abnormal expansion of T-lymphocytes in this syndrome have uncovered a deficient induction of the IL-2/IL-2R pathway by antigenic stimulation [Simon *et al.* 1984; Via & Shearer 1988]. Although early investigations on T-cell function had correlated the deficient proliferative response present in *lpr* mice with reduced IL-2 production *in vitro* [Wofsy *et al.* 1981], it has been reported that, prior to the manifestation of the disease, MRL/*lpr* lymph node (LN) cells spontaneously produce IL-2 and proliferate *in vitro* [Murray *et al.* 1990; Weston *et al.* 1988; Weston *et al.* 1987]. It thus remains an open question whether IL-2 treatment starting at 2-3 months of age, actually reverses a beginning IL-2 defect or rather elevates IL-2 tissue levels over a (non-physiological?) threshold that impedes the manifestation of the *lpr* phenotype.

As to the mechanism by which IL-2 exerts its beneficial effect on MRL/*lpr* mice, three non-exclusive possibilities may be taken into account. **First**, IL-2 may normalize a deficient thymic differentiation, that could be due to an intrinsic IL-2 deficiency. Indeed, mice in which the IL-2 gene has been eliminated on both alleles via homologous recombination [Schorle *et al.* 1991], exhibit a clear-cut defect in thymic architecture and cellularity that becomes manifest more than 6 weeks after birth. Such mice also develop inflammatory symptoms reminescent of an autoimmune disease (T. Hünig, personal communication). **Second**, IL-2 causes the surge of T cells that transfer disease protection. Indeed, spleen cells recovered from MRL/*lpr* females successfully treated with IL-2 impede, upon intravenous injection into young MRL/*lpr* females, the development of lymphadenopathy and glomerulonephritis [M. de Alborán *et al.* 1992a; M. de

Alborán *et al.* 1992b]. $CD4^+$ T cells are more effective than $CD8^+$ cells. **Third,** as a final possibility, IL-2 might directly inhibit the generation of $CD3^+CD4^-CD8^-$ cells. This is suggested by *in vitro* experiments in which IL-2 interacts with the ß chain of the IL-2 receptor (IL-2Rß; p75) of an MRL/*lpr*-derived $CD3^+CD4^-CD8^-$ cell line [Gutierrez-Ramos *et al.* 1989], causing growth inhibition and apoptotic cell death (E. Lenardo et al., unpublished observation). However, it remains to be clarified whether the IL-2 levels produced by the IL-2.VV in the tissue attain levels high enough to allow for signalling via the intermediate affinity IL-2 receptor (IL-2Rß).

Effect of TCR Vß8-targeted drugs in the MRL/*lpr* mouse

Different approaches aimed at reducing the number of *lpr* T cells also lead to an improvement of autoimmunity (**Table 2**). Thus, neonatal thymectomy, anti-Thy-1.2 treatment, treatment with 5-azacytidine or the use of immunomodulating or immunosuppressive agents like cyclosporin-A (CsA), corticosteroids etc. prevent lymphoproliferation and delay disease onset in MRL/*lpr* mice by several months [Steinberg *et al.* 1980; Yoshida *et al.* 1990]. Similarly, antisera or mAbs directed against MHC molecules and T-cell markers i.e. anti-Thy1 and anti-CD4 inhibit the accumulation of *lpr* cells, and prevent or ameliorate the autoimmune disease [Santoro *et al.* 1983] . Although these treatments are able to restore normal levels of urine-protein values and to decrease the levels of autoantibodies, they are unable to prolong the mean survival of MRL/*lpr* mice, possibly due to their high intrinsic toxicity.

As mentioned in the introduction, phenotypically abnormal, "double negative" *lpr* cells express more than 50 % of Vß8.2 and Vß8.3 TCR gene products, whereas these Vß8 is found on less than 25% of phenotypically normal "single positive" T cells. This fact *per se* suggests the involvement of these particular TCR Vß chains in the autoimmune process. A further indication that this may be the case came from experiments in which irradiated, purified *lpr* cells were inoculated into young, undiseased syngeneic MRL/*lpr* recipients, leading to a significant reduction in disease parameters. Analysis of the TCR repertoire of protected animals revealed a specific elimination of Vß8.2 cells [M. de Alborán *et al.* 1992b]. Thus, the fact that *lpr* cells exhibit a marked bias of the utilization of some TCR Vβ8 gene products (Vß8.2 and Vß8.2), prompted us to evaluate the effect of two

Table 2: Multiple strategies for the prophylaxis of lupus in mice homozygous for the *lpr* mutation

Target	Treatment	Mechanism	Reference
T cells	injection of anti-Thy-1 or anti-CD4	non-selective elimination of pathogenic T cells	[Santoro *et al.* 1988; Wofsy *et al.* 1985]
	periodical injections of anti-Vß8 (F23.1)	elimination of TCR Vß8$^+$ T cells	[M. de Alborán *et al.* 1992a]
	periodical injections of staphylococcal enterotoxin B	clonal deletion of TCR Vß8$^+$ T cells	[Kim *et al.* 1991]
	vaccination with irradiated *lpr* cells	induction of T cells that transfer disease protection	[M. de Alborán *et al.* 1992b]
	injection of a recombinant	induction of CD8$^+$ T cells that	[Gutierrez-Ramos *et al.* 1990a]
	vaccinia virus producing IL-2	transfer disease protection	
	cylosporin A	inhibition of T cell help for IgG3 autoantibody formation	[Takahashi *et al.* 1991]
B cells	anti-IgM	B cell depletion	[Cerny *et al.* 1989]
	maternal administration of an anti-idiotype to the F23.1 antibody	induction of F23.1-like antibodies	[M. de Alborán *et al.* 1992a]
unknown	administration of azacytidine or linomide		[Spetz-Hagberg *et al.* 1990; Yoshida *et al.* 1990]

molecules targeted to all members of the Vß8 gene family (Vß8.1, 8.2, and 8.3), namely the monoclonal antibody F23.1 (a mouse IgG2a) and the murine superantigen *Staphylococcus aureus enterotoxin* B (SEB) (*vide infra*).

Prophylactic treatment with F23.1 attenuates both the *lpr* phenomenon and the autoimmune manifestations in the MRL/*lpr* mouse. However, to exert its beneficial effect, F23.1 has to be applied several times per week throughout the life span, starting relatively early in life (8 weeks). As an alternative strategy, the production of F23.1-like antibodies can be induced by maternal immunization with an F23.1 idiotype of F23.1 itself. This treatment leads to a stable amelioration of autoimmune disease markers throughout the lifetime of the offspring [M. de Alborán *et al.* 1992a].

Superantigens like SEB are known to produce *in vivo* a biphasic response involving an early deletion (12-20 hours after i.v injection; CD4<CD8) followed by the expansion (48-96 hours) of the superantigen reactive T cells (CD4=CD8), and another phase of clonal deletion, starting from day 4 after inoculation (CD4>CD8) [D'Adamio *et al.* 1993; Gonzalo *et al.* 1992; Kawabe & Ochi 1991; MacDonald *et al.* 1991]. The remaining superantigen-reactive cells are unable to respond *in vitro* to the same superantigen through a process not yet well characterized that is generically denoted as anergy [Kawabe & Ochi 1990; Lee & Vitetta 1992]. Peripheral (splenic, as well as lymph node-derived) $CD4^+$ and $CD8^+$ T cells from MRL/*lpr* are reactive to SEB and exhibit subsequent phases of clonal expansion and deletion, thus fully mimicking the response of "normal" T cells [Gonzalo *et al.* 1993] . Thus, although the *lpr* mutation should concern all T cells, the (auto)superantigen-induced clonal deletion of thymocytes [Singer *et al.* 1989; Singer & Theofilopoulos 1990], as well as the superantigen-driven clonal deletion and anergy of peripheral $CD4^+$ or $CD8^+$ T lymphocytes [Scott *et al.* 1993], are functioning normally in MRL/Mp-*lpr/lpr* mice [Gonzalo *et al.* 1993]. In strict contrast to this observation, abnormal "double negative" ($CD4^-CD8^-$) T cells are completely resistant to SEB-mediated deletion. In this context, it may be interesting that $CD4^-CD8^-$ T cells recovered from the normal bone marrow *ex vivo* (e.g. from BALB/c mice) have been shown to resist clonal deletion by superantigen and fail to exhibit TCR-downmodulation and anergy in response to anti-CD3 stimulation [Martínez-A. *et al.* 1993]. Thus, the non-responsiveness of *lpr* cells to the mitogenic and pro-deletional effect of SEB might be a particularity of this lineage not directly linked to the defective or altered expression of the Fas/Apo antigen.

In opposition to the resistance of $CD4^-CD8^-$ cells to SEB, their precursors appear to be fully susceptible to the induction of non-responsiveness. Treatment of mice that have not manifested the *lpr* phenotype (at 2 months of age) with SEB fully impedes the surge of $CD4^-CD8^-$ cells and the associated lymphoid hyperplasia (lymphadenopathy, splenomegaly) that is observed in PBS-injected littermates. The SEB-mediated inhibition of lymphoproliferation is accompanied by a significant amelioration of autoimmune parameters, thus pointing to the feasibility of employing superantigens for simple manipulation of the immune repertoire that result in the long term prophylaxis of autoimmune diseases [Gonzalo *et al.* 1993].

One unresolved question regards the fact that a single administration of SEB is sufficient to obtain a full protective effect, comparable to that caused by repeated SEB injections, whereas F23.1 has to be administered continuously. The mode of action by which SEB and F23.1 lead to the deletion of Vß8-expressing cells is completely different. Whereas, F23.1 causes a macrophage-dependent passive elimination of cells binding this IgG2a antibody at the surface, SEB drives an activation-dependent active cell death. In addition, SEB, not F23.1, leads to a transient, promiscuous cytokine release (including IL-1, IL-2, IL-4, IL-6, IL-10, interferon-γ, tumor necrosis factor ß and granulocyte macrophage colony stimulating factor). It remains to be determined whether the massive, though brief (< 6 hours after intravenous injection), secretion of cytokines into the circulation may contribute to a long-term change in immune function. In addition, SEB causes anergy among T cells not succumbing to clonal deletion, a fact that may also reduce the virulence of potentially autoaggressive $Vß8^+$ T cells.

CONCLUSION

Very different strategies may mitigate the systemic autoimmune disease and the associated lymphoid hyperplasia of MRL/*lpr* mice. Successfull immune interventions may be completely non-specific, aiming for instance at the elimination of most B and T cells, or involving, on a more subtle level, the administration of exogenous immunomodulators such as IL-2. Non-specific drugs interact with monomorphic structures (e.g. Thy1, CD4, CD3, cytokine receptors), whereas antigen-specific immunomodulators would be exlusively targeted to self-aggressive lymphocytes. At an intermediate level, agents targeted to products of variable (V) gene families of the antigen receptor may be considered as semi-specific drugs, since they lack true specificity (in the sense that they recognize domains of the antigen receptor not directly implicated in antigen recognition), at the time that they are restricted to minor lymphocyte populations. Two agents that are targeted to TCR Vß gene products that are overexpressed among putatively disease mediating T cells of the MRL/*lpr* strain (Vß8.2 and Vß8.3), a monoclonal antibody (F23.1) and a superantigen (SEB), have been successfully employed for the experimental prophylaxis of the autoimmune disorder.

Acknowledgments:

The present work was funded by CICyT and FISS.

References

Andreu-Sánchez, J. L., Moreno de Alborán, I., Marcos, M. A. R., Sánchez-Movilla, A., Martínez-A, C. & Kroemer, G. **(1991)** IL-2 abrogates the non-responsive state of T cells expressing a forbidden TCR repertoire and induces autoimmune disease in neonatally thymectomized mice. *J. Exp. Med.* **173**, 1323-1329.

Cerny, A., Kimota, M., Hugin, A. W., Merino, R. & Izui, S. **(1989)** Anti-IgM-treatment of C57BL/6-lpr/lpr mice. Depletion of B cells reduces lpr gene-induced lymphoproliferation and mononuclear cell vasculitis. *Clin. Exp. Immunol.* **77**, 124-129.

Cohen, P. L. & Eisenberg, R. A. **(1991)** Lpr and gld: Single gene models of systemic autoimmunity and lymphoproliferative disease. *Annu. Rev. Immunol.* **9**, 243-269.

D'Adamio, L., Awad, K. M. & Reinherz, E. L. **(1993)** Thymic and peripheral apoptosis of antigen-specific T cells might cooperate in establishing self tolerance. *Eur. J. Immunol.* **23**, 747-753.

Flexner, C., Hügin, A. & Moss, B. **(1987)** Prevention of vaccinia virus infection in immunodeficient mice by vector-directed IL-2 expression. *Nature* **330**, 259-262.

Garchon, H.-J., P., B., Eloy, L. & Bach, J.-F. **(1991)** Identification and mapping to chromosome 1 of a susceptibility locus for periinsulitis in non-obes diabetic mice. *Nature* **353**, 260-262.

Gonzalo, J. A., Moreno de Alborán, I., Alés-Martínez, J. E., Martínez-A., C. & Kroemer, G. **(1992)** Expansion and clonal deletion of peripheral T cells induced by bacterial superantigen is independent of the interleukin 2 pathway. *Eur. J. Immunol.* **22**, 1007-1011.

Gutierrez-Ramos, J. C., Andreu, J. L., Marcos, M. A. R., Vegazo, I. R. & Martínez-A., C. **(1991)** IL-2 treatment with IL-2/vaccinia recombinant virus leads to serologic, histologic and phenotypic normalization of autoimmune MRL/lpr-lpr mice. *Autoimmunity* **10**, 15-25.

Gutierrez-Ramos, J. C., Andreu, J. L., Moreno de Alboran, I., Rodriguez, J., Leonard, E., Kroemer, G., Marcos, M. A. R. & Martínez-A, C. **(1990a)** Insights into autoimmunity: from classical models to current perspectives. *Immunol. Rev.* **118**, 73-101.

Gutierrez-Ramos, J. C., Andreu, J. L., Revilla, Y., Viñuela, E. & Martínez-A., C. **(1990b)** Recovery from autoimmunity of MRL/lpr mice after infection with an interleukin-2/vaccinia recombinant virus. *Nature* **346**, 271-274.

Gutierrez-Ramos, J. C., Pezzi, L., Palacios, R. & Martínez-A., C. **(1989)** Expression of the p75 interleukin 2-binding protein on CD3+4-8-Tac- cells from autoimmune MRL/Mp-lpr/lpr mice. *Eur. J. Immunol.* **19**, 201-204.

Ikehara, S., Kawamura, M., Takao, F., Inaba, M., Yasumizu, R., Than, S., Hisha, H., Sugiura, K., Koide, Y., Yoshida, T. O., Ida, T., Imura, H. & Good, R. A. **(1991)** Organ-specific and systemic autoimmune diseases originate from defects in hematopoietic stem cells. *Proc. Natl. Acad. Sci. USA* **87**, 8341-8344.

Kawabe, Y. & Ochi, A. **(1990)** Selective anergy of Vß8+, CD4+ T cells in Staphylocccus enterotoxin B-primed mice. *J. Exp. Med.* **172**, 1065-1070.

Kawabe, Y. & Ochi, A. **(1991)** Programmed cell death and extrathymic reduction of $Vß8^+$ $CD4^+$ T cells in mice tolerant to Staphylococcus aureus enterotoxin B. *Nature* **349**, 245-248.

Kim, C., Siminovitch, K. A. & Ochi, A. **(1991)** Reduction of lupus nephritis in MRL/*lpr* mice by a bacterial superantigen treatment. *J. Exp. Med.* **174**, 1431-1437.

Kroemer, G., Andreu-Sánchex, J. L., Gonzalo, J. A., Gutierrez-Ramos, J. C. & Martínez-A., C. **(1991)** Interleukin-2, autotolerance, and autoimmunity. *Adv. Immunol.* **150**, 147-235.

Lee, W. T. & Vitetta, E. S. **(1992)** Memory T cells are anergic to the superantigen staphylococcal enterotoxin B. *J. Exp. Med.* **176**, 575-579.

M. de Alborán, I., Gonzalo, J. A., Kroemer, G., Leonardo, E., Marcos, A. R. & Martínez-A., C. **(1992a)** Attenuation of autoimmune disease and lymphocyte accumulation in MRL/lpr mice by treatment with anti-Vß8 antibodies. *Eur. J. Immunol.* **22**, 2153-2158.

M. de Alborán, I., Gutiérrez, J. C., Gonzalo, J. A., Andreu, J. L., Marcos, A. R., Kroemer, G. & Martínez-A., C. **(1992b)** *lpr* T cells vaccinate against lupus in MRL/*lpr* mice. *Eur. J. Immunol.* **22**, 1089-1093.

MacDonald, H. R., Baschieri, S. & Lees, R. K. **(1991)** Clonal expansion precedes anergy and death of $Vß8^+$ peripheral T cells responding to staphylococcal enterotoxin B in vivo. *Eur. J. Immunol.* **21** 1963-1966.

Marcos, M. A. R., Toribio, M. L., de la Hera, A., Márquez, C., Gaspar, M. L. & Martínez-A., C. **(1988)** Mutual cell interactions and the selection of immune repertories: Implication in autoimmunity. *Immunol. Today* **9**, 204-209.

Martínez-A., C., Marcos, M. A. R., M. de Alborán, I., Alsonso, J. M., de Cid, R., Kroemer, G. & Coutinho, A. (1993) Functional double-negative T cells in the periphery express TCR Vß gene products that are deleted in single-positive T cells. *Eur. J. Immunol.* **23**, 250-254.

Möller, G. E. (Ed.) (1990) Aspects in Autoimmunity. *Immunol Rev. Vol. 122*

Mountz, J. D., Smith, T. M. & Toth, K. S. (1990) Altered expression of self-reactive T cell receptor Vß regions in autoimmune mice. *J. Immunol.* **144**, 2159-2166.

Murray, L. J., Lee, R. & Martens, C. (1990) In vivo cytokine gene expression in T cell subsets of the autoimmune MRL/Mp-lpr/lpr mouse. *Eur. J. Immunol.* **20**, 163-170.

Prins, J. B., Todd, J. A., Rodrigues, N. R., Ghosh, S., Hogarth, P. M., Wicker, L. S., Gaffney, E., Podolin, P. L., Fischer, P. A., Sirotina, A. & Peterson, L. B. (1993) Linkage on chromosome 3 of autoimmune diabetes and defective Fc receptor for IgG in NOD mice. *Science* **260**, 695-698.

Santoro, T. J., Luger, T. A., Raveche, E. S., Smolen, J. S., J.J., O. & Steinberg, A. D. (1983) In vitro correction of the interleukin 2 defect of autoimmune mice. *Eur. J. Immunol.* **13**, 601-604.

Santoro, T. J., Portanova, J. P. & Kotzin, B. L. (1988) The contribution of L3T4+ T cells to lymphoproliferation and autoantibody production in MRL-lpr/lpr mice. *J. Exp. Med.* **167**, 1713-1718.

Schorle, H., Holtschke, T., Hünig, T., Schimpl, A. & Horak, I. (1991) Development and function of T cell in mice rendered interleukin-2 deficient by gene targeting. *Nature* **239**, 845-847

Scott, D. E., Kisch, W. J. & Steinberg, A. D. (1993) Studies of T cell deletion and T cell anergy following in vivo administration of SEB to normal and lupus-prone mice. *J. Immunol.* **150**, 664-672.

Shivakumar, S., Tsokos, G. C. & Datta, S. K. (1989) T cell receptor α/ß expressing double negative ($CD4^-$/$CD8^-$) and $CD4^+$ T helper cells in humans augment the production of pathogenic anti-DNA autoantibodies associated with lupus nephritis. *J. Immunol.* **143**, 103-112.

Sidman, C. L., Marshall, J. D. & von Boehmer, H. (1992) Transgenic T cell receptor interactions in the lymphoproliferative and autoimmune syndromes of lpr and gld mutant mice. *Eur. J. Immunol.* **22**, 499-504.

Simon, M. M., Moller, D. R., Saltini, C., Kirby, M. & Crystal, R. G. (1984) Quantitative studies on T-cell functions in MRL/Mp-*lpr/lpr* mice. *Clin. Immunol. Immunopathol.* **33**, 39-53.

Singer, P. A., Balderas, R. S., McEvilly, R. J., Bobardt, M. & Theofilopoulos, A. N. **(1989)** Tolerance-related Vß clonal deletions in normal CD4-8-, TCR-a/ß+ and abnormal lpr and gld cell populations. *J. Exp. Med.* **170**, 1869-1877.

Singer, P. A., McEvilly, R. J., Dixon, F. J. & Theophilopoulos, A. N. **(1986)** Clonal diversity and T-cell receptor ß-chain variable gene expression in enlarged lymph nodes of MRL/*lpr* lupus mice. *Proc. Natl. Acad. Sci. USA* **83**, 7018.

Singer, P. A. & Theofilopoulos, A. N. **(1990)** T-cell receptor v beta repertoire expression in murine models of SLE. *Immunol. Reviews* **118**, 103-127.

Spetz-Hagberg, A.-L., Goldschmidt, T. J., Stålhandske, T. & Larsson-Sciard, E. L. **(1990)** Amelioration of intrathymic T cell development and peripheral T cell reactivities in autoimmune mice undergoing therapy with a novel immunomodulator. *Int. Immunol.* **2**, 645-650.

Steinberg, A. D., Roths, J. B., Murphy, E. D., Steinberg, R. T. & Raveche, E. S. **(1980)** Effects of thymectomy and androgen administration upon the autoimmune disease of MRL/Mp-*lpr/lpr* mice. *J. Immunol.* **125**, 871-873.

Strominger, J. L. **(1989)** Developmental biology of T cell receptors. Science:**244**, 943-950.

Takahashi, S., Nose, M., Sasaki, J., Yamamoto, T. & Kyogoku, M. **(1991)** IgG3 production in MRL/lpr mice is responsible for development of lupus nephritis. *J. Immunol.* **147**, 515-519.

Theofilopoulos, A. & Dixon, F. J. **(1985)** Murine models of SLE. *Adv. Immunol.* **37**, 269-390.

Theofilopoulos, A. N., Balderas, R. S., Shawler, D. L., Lee, S. & Dixon, F. J. **(1981)** The influence of thymic genotype on the SLE-like disease and T cell proliferation of MRL/MP-*lpr/lpr* mice. *J. Exp. Med.* **153**, 1405-1420.

Theofilopoulos, A. N., Kofler, R., Singer, P. A. & Dixon, F. J. **(1989)** Molecular genetics of murine lupus models. *Adv. Immunol.* **46**, 61-109.

Todd, J. A., Aitman, T. J., Cornall, R. J., Ghosh, S., Hall, J. R. S., Hearne, C. M., Knight, A. M., Love, J. M., McAleer, M. A., Prins, J.-B., Rodrigues, N., Lathrop, M., Pressey, A., DeLarato, N. H., Peterson, L. B. & Wicker, L. S. **(1991)** Genetic analysis of autoimmune type 1 diabetes mellitus in mice. *Nature* **351**, 542-547.

Via, C. S. & Shearer, G. M. **(1988)** Functional heterogeneity of L3T4+ T cells in MRL-lpr/lpr mice. L3T4+ T cells suppress major histocompatibility complex-self-restricted L3T4+ T helper cell function in association with autoimmunity. *J. Exp. Med.* **168**, 2165-2181.

Watanabe, T., Sakai, Y., Miyawaki, S., Atsuko, S., Koiwai, O. & Chno, K. **(1991)** A molecular genetic linkage map of mouse chromosome 19, including the lpr, Ly-44 and Tdt genes. *Biochem. Genet.* **29**, 325-331.

Watanabe-Fukunaga, R., Brannan, C. I., Copeland, N. G., Jenkins, N. A. & Nagata, S. **(1992)** Lymphoproliferation disorder in mice explained by defects in Fas antigen that mediates apoptosis. *Nature* **356**, 314-317.

Watson, M. L., Rao, Y. K., Gilkeson, G. S., Ruiz, P., Eicher, E. M., Pisetsky, D. S., Matsuzawa, A., Rochelle, J. M. & Seldin, M. F. **(1992)** Genetic analysis of MRL-*lpr* mice: relationship of the Fas apoptosis gene to disease manifestations and renal disease-modifying loci. *J. Exp. Med.* **176**, 1645-1656.

Weston, K. M., Ju, S.-T., Lu, C. Y. & Sy, M.-S. **(1988)** Autorreactive T cells in MRL/*lpr* mice. Characterization of the lymphokines produced and analysis of antigene presenting cells required. *J.Immunol.* **141** 1941-1948.

Weston, K. M., Yeh, E. T. H. & Man-Sun, S. **(1987)** Autoreactivity accelerates the development of autoimmunity and lymphoproliferation in MRL/MP-*lpr/lpr* mice. *J. Immunol.* **139**, 734-742.

Wick, G., Hus, Y., Schwartz, S. & Kroemer, G. **(1993)** The immune-hypothalamo-pituitary-adrenal axis in autoimmune diseases. *Endocrine Rev.* in press,

Wofsy, D., Ledbetter, J. A., Hendler, P. L. & Seaman, W. E. **(1985)** Treatment of murine lupus with monoclonal anti-T cell antibody. *J. Immunol.* **134**, 852-857.

Wofsy, D., Murphy, E. D., Roths, J. B., Dauphine, M. J., Kipper, S. B. & Talal, N. **(1981)** Deficient interleukin-2 activity in MRL/Mp and C57BL/6J mice bearing the *lpr* gene. *J. Exp. Med.* **154**, 1671-1680.

Yoshida, H., Yoshida, M., Merino, R., Shibata, T. & Izui, S. **(1990)** 5-azacytidine inhibits the *lpr* gene-induced lymphadenopathy and acceleration of lupus-like syndrome in MRL/MpJ-*lpr/lpr* mice. *Eur. J. Immunol.* **20** 1989-1993.

Major histocompatibility binding peptides: an experimental approach towards specific immunotherapy of autoimmune diseases

Marca H.M. Wauben and Willem van Eden
Institute of Infectious Diseases and Immunology
Department of Immunology
Faculty of Veterinary Science, University of Utrecht
Yalelaan 1, P.O. Box 80165
3508TD Utrecht, The Netherlands

New approaches for immune intervention in autoimmune diseases

The close association between certain autoimmune diseases and particular MHC alleles, and the restrictions in the diversity of the early autoreactive T cell responses, as shown in several animal models, has targeted the research of immune intervention in autoimmune diseases to the molecular level of T cell recognition. Antigen-specific T cell triggering occurs via engagement of the T cell receptor by the MHC-antigen complex (Janeway 1992). There are several possible ways to interfere with the formation of a T cell activating triad consisting of T cell receptor-antigen-MHC. In experimental animal models successful disease intervention with antibodies has been shown, *e.g.* by anti-T cell receptor antibodies, and, at the MHC level, by blocking with anti-MHC class II antibodies (Reviewed in Martin et al. 1992; Wraith et al., 1989). However, in general the problems of anaphylactic and host anti-idiotypic responses to administered antibodies remains and largely precludes repeated treatments with such reagents.

An alternative approach to prevent the formation of a T cell stimulatory triad is by using synthetic peptides. The association between certain MHC alleles and autoimmune diseases can be used to develop new therapies based on synthetic peptides tailored to specifically interact with the MHC molecules involved in the

NATO ASI Series, Vol. H 80
Autoimmunity: Experimental Aspects
Edited by Moncef Zouali

disease process. It has been shown that synthetic peptides designed to passively block antigen presentation in the context of a particular MHC molecule can inhibit T cell activation restricted by that MHC molecule both *in vitro* and *in vivo* (Reviewed in Adorini 1992; Adorini and Nagy 1990). With this approach it should be possible to specifically target the therapy to the restriction element of the disease, thereby leaving other MHC molecules free for presentation of antigens. Recently, it has been shown that altering the ligand of a T cell receptor by using a peptide analogue of the original T cell epitope, can induce unresponsiveness to subsequent stimulation with the original epitope (Sloan-Lancaster et al. 1993). Although in this situation the formation of a triad between T cell receptor-antigen-MHC of the autoreactive T cell was not prevented, the outcome of the triad formation, being anergy induction of the autoreactive T cells without or with altered production of cytokines, offers good possibilities for immune intervention at the T cell level. Furthermore, this approach will be highly specific because only T cells specific for a certain epitope and sensitive for the analogue peptide will be anergized. However, the major drawback at this moment is the lack of information concerning the antigen specificity of autoreactive T cells in most spontaneous autoimmune diseases.

Competition for MHC binding

$CD4^+$ T lymphocytes recognize their processed antigen in the context of particular MHC class II molecules (Unanue 1984; Schwarz 1985; Buus et al. 1987; Möller 1987). Because a given individual expresses only a limited number of different MHC class II molecules, it should be clear that to ensure T cell immunity, each MHC class II molecule should be able to bind a large variety of different peptides. Therefore, a low degree of selectivity was predicted for the interaction between peptides and MHC class II molecules. According to this, it appeared that interactions between peptides and MHC class II molecules usually tolerate about 70-90% of single amino acid substitutions in the peptide molecule without appreciable effects (Sette et al. 1987; Jardetzky et al. 1990; Gautam et al. 1992). Nevertheless, although the definition of MHC class II binding peptide motifs is quite broad, it has been demonstrated that peptides capable of binding to a particular MHC molecule share structural motifs detectable at the level of primary amino acid sequence (Hunt et al. 1992; Rudensky et al. 1992; Sette et al. 1988; Sette et al. 1989; O'Sullivan et al. 1991; Leighton et al. 1991).

An important consequence of the relatively broad-range specificity of peptide-MHC interactions is that peptides derived from different antigens may compete for binding to the same MHC molecule. That peptide competion for MHC binding actually occurs has first been demonstrated *in vitro*, by showing that unrelated antigens binding to the same MHC molecule can compete with each other for presentation to T cells (Werdelin 1982; Rock and Benacerraf 1984; Guillet et al. 1986). It has become clear that, peptide competition for MHC class II binding does not only occur *in vitro* but also *in vivo* (Lamont et al. 1990; Adorini et al. 1988; Muller et al. 1990; Guery et al. 1992). The inhibition of T cell responses to a peptide or even an entire protein by administration of a competitor peptide strictly correlates with the capacity of the competitor peptide to bind to the same MHC restriction element as the antigen. Furthermore, it appeared that the inhibition of T cell responses was associated with *in vivo* inhibition of antigen presentation and with the occupancy of MHC class II binding sites by the competitor peptide (Guery et al. 1992). Moreover, by blocking antigen presentation to class II restricted T helper cells, MHC class II binding competitor peptides also prevented efficiently and selectively the induction of T cell dependent primary and secondary *in vivo* antibody responses (Guery et al. 1993). The extent of inhibition of T cell proliferation depended on the molar ratio between antigen and competitor, and the stability of the competitor peptide *in vivo* (Lamont et al. 1990; Adorini et al. 1988; Muller et al. 1990; Guery et al. 1992).

Inhibition of T cell responses by antigenically related MHC binding peptides

Inhibition of T cell activation by injecting MHC class II binding competitor peptides could have several possible explanations, apart from competition for antigen presentation. If the competitor peptide is structurally related to the antigenic determinant, antigen-specific rather than MHC-specific mechanisms may be involved in inhibition of T cell activation. In this respect novel concepts have been put forward (Sloan-Lancaster et al. 1993; DeMagistris et al. 1992; Racioppi et al. 1993). It has been observed that analogue peptides of a T cell epitope can inhibit *in vitro* proliferation of T cells specific for the particular epitope up to 1000-fold more effectively when compared with unrelated peptides with similar MHC binding affinity, by engaging the T cell receptor in a non-stimulatory fashion (DeMagistris et al. 1992). Since no intracellular signalling events were detected, the superior inhibitory activity of structurally related peptides was ascribed to an antagonistic principle (DeMagistris et al. 1992). However, recently it has been shown that there was more than mere passive antagonism, because it appeared that the antagonistic peptide inhibited early biochemical events required for T cell proliferation (Ruppert et al. 1993). Furthermore, evidence has been obtained for differential T cell signalling to be dependent on the nature of the peptide used to engage the T cell receptor (Sloan-Lancaster et al. 1993; Racioppi et al. 1993; Evavold & Allen 1991). It has been shown that analogue peptides of an antigenic determinant are capable of altering the quality of T cell signalling which resulted in different effector responses, *e.g.* induction of production of certain lymphokines while no T cell proliferation is induced, or alteration of the lymphokine profile (Racioppi et al. 1993; Evavold and Allen 1991). These findings have resulted in the definition of a new class of T cell receptor ligands with the properties of mixed agonists/antagonists (Racioppi et al. 1993). In favor of this new theory, that the affinity of interaction between the T cell receptor and the MHC-peptide ligand dictates the outcome of T cell effector activities, are recent findings (1993), showing that peptide analogues of a T cell epitope that are unable to stimulate clonal proliferation or cytokine production can render the T cells unresponsive to subsequent stimulation with the original peptide. It was concluded that alteration of the ligand of the T cell receptor by using a peptide analogue resulted in an altered signalling leading to anergy induction in T cells (Sloan-Lancaster et al. 1993). From all these different studies it can be concluded that the process of T cell recognition and activation is very complex.

Modulation of experimental autoimmunity by MHC binding peptides

Blocking of the antigen presentation function of MHC class II molecules by competitor peptides with high MHC binding affinity has been proposed as a potential immunotherapy for the treatment of MHC-associated diseases. In our studies and in several others the induction of experimentally induced autoimmunity can be inhibited very efficiently by coadministration of MHC class II binding peptides. The mechanistic principles involved in the inhibition of the disease process are gradually becoming apparent. It is already clear that only in a minority of the studies it can be concluded that disease inhibition was due to mere competition for MHC binding. Because of the different mechanistic principles that could be involved in the *in vivo* inhibition of the disease process, the studies can be divided into two groups based on the presence or absence of a structural relationship between the competitor peptide and the disease inducing antigen.

Disease-nonrelated competitor peptides

Although in studies using disease-nonrelated competitor peptides, the involvement of antigen-specific mechanisms in the inhibition of the disease process was excluded, mechanisms other than MHC blockade may have contributed to the successful intervention. As shown in Figure 1, theoretically, immunogenic disease-nonrelated competitor peptides may inhibit responses of autoreactive T cells not only by MHC blockade but also by distraction of the autoimmune T cell response caused by clonal dominance of the T cell response to the competitor peptide. Furthermore, it may well be possible that cytokines produced by T cells in response to recognition of the competitor peptides may cause a local bystander suppression thereby interfering in the triggering of auto-aggressive T cell responses.

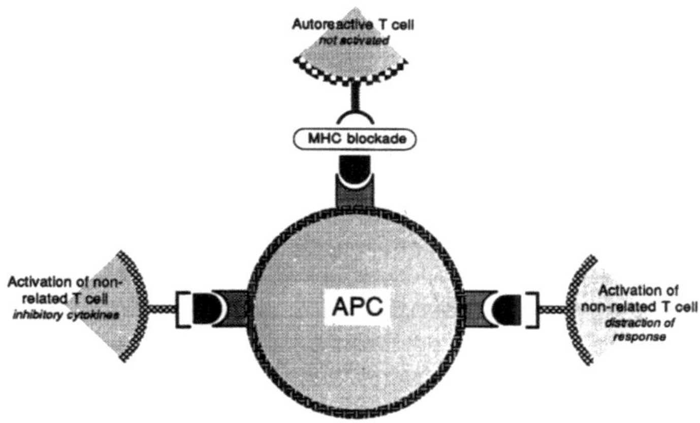

Figure 1: Disease-nonrelated competitor peptide

We have shown that both peptide-induced experimental autoimmune encephalomyelitis (EAE) and entire myelin basic protein (MBP)-induced EAE can be inhibited by coimmunization with either immunogenic or nonimmunogenic disease-nonrelated competitor peptides (Wauben et al. 1992, Wauben et al. submitted). It appeared that the relative efficiency of inhibition of proliferative responses of disease inducing T cells *in vitro*, and of disease development *in vivo*, was fully in agreement with the relative MHC binding affinity of the competitor peptide. Also in peptide induced EAE in mice, and in myosin-induced experimental myocarditis, coimmunization with nonimmunogenic competitor peptides could inhibit the disease process (Lamont et al. 1990; Gautam et al. 1992; Smith and Allen 1991). Recently, it has been shown that besides inhibition of these experimentally induced autoimmune models, nonimmunogenic competitor peptides can also prevent the onset of spontaneous MHC-linked autoimmune disease, namely in non-obese diabetic (NOD) mice (Hurtenbach et al. 1993). Although, in the studies using immunogenic disease-nonrelated competitors it can not fully be excluded that activation of autoreactive T cells was inhibited because of the immunogenicity of the competitor peptides, the successful disease inhibition by nonimmunogenic disease-nonrelated peptides (Lamont et al. 1990; Gautam et al. 1992; Smith and Allen 1991; Hurtenbach et al. 1993; Wauben et al. submitted), and the reversal of successful disease inhibition by competitor peptides by increasing the MHC binding affinity of the disease inducing peptide (Wauben et al. submitted), provided good evidence for MHC blockade as a valid principle for inhibition of autoimmunity.

Disease-related competitor peptides

Coimmunization with competitor peptides designed on the basis of known autoimmune disease-associated T cell epitopes can inhibit very efficiently induction of adjuvant arthritis (AA), EAE, and collagen type II-induced arthritis (Wauben et al. 1992; Sakai et al. 1989; Wraith et al. 1989; Smilek et al. 1991; Myers et al. 1993; Wauben et al. submitted). Especially in the case of AA, we now have demonstrated the extremely powerful potential of disease-related peptide analogues to inhibit disease, even when induced with a very immunogenic and complex antigen such as whole Mycobacteria. On the basis of the data obtained in AA it was clear that MHC blockade was not the sole mechanism of the successful intervention in the disease process. Because of the close structural relationship between the competitor and the disease inducing counterpart it was likely that antigen-specific mechanisms contributed to successful inhibition of AA. As in our study, several other studies have shown that peptide analogues antigenically related to a particular T cell epitope were relatively superior in competitively inhibiting T cell proliferation to that epitope as compared to non-related competitor peptides with comparable or even stronger MHC binding affinities (DeMagistris et al. 1992; Ruppert et al. 1993; Alexander et al. 1993). Similarly, such superior effect of disease-related competitor peptides was also seen *in vivo* in the inhibition of autoimmune diseases, and this effect was sometimes even more pronounced than anticipated from the *in vitro* experiments (Wauben et al. 1992; Sakai et al. 1989). Although these findings suggest that superior disease inhibitory activity is a correlate of superior *in vitro* functional competitive activity unrelated to the relative degree of MHC binding affinity, in reality it appeared to be more complicated. In EAE it has been shown that a peptide analogue of an encephalitogenic epitope was stimulatory for encephalitogenic T cells *in vitro*, but upon *in vivo* immunization it appeared to be nonimmunogenic and nonencephalitogenic. Interestingly, this peptide analogue could prevent EAE induction when administered before or at the time of immunization with the original encephalitogenic peptide (Wraith et al. 1989; Smilek et al. 1991). Furthermore, it was demonstrated that, although the development of the disease was prevented very efficiently, T cells reactive against the autoimmune T cell epitopes were not inhibited and sometimes even stimulated by coimmunization with the disease-related competitor peptide (Wauben et al. submitted; Sakai et al. 1989; Wraith et al. 1989; Smilek et al. 1991). This contrasts with the results obtained by MHC blockade with disease-nonrelated competitor peptides. The prominent disease inhibitory activity of disease-related competitor peptides may well offer unique possibilities for disease-specific immune intervention. It is possible that disease-related competitors exert their superior disease inhibitory qualities by integrating both MHC competitor qualities and antigen specific modulatory qualities in a synergistic manner. Such peptides, in which MHC competition and antigen specific immunomodulation are integrated, have been termed **competitor-modulator** peptides (Wauben et al. 1992 a&b). However, the exact mechanism of the immunomodulatory qualities of such peptides is still unclear.

Immunomodulatory mechanisms of competitor-modulator peptides

As shown in Figure 2, besides the three mechanistic principles described for inhibition of autoreactive T cell activation by immunogenic disease-nonrelated competitor peptides at least three other mechanistic principles are possible in the case of disease-related competitors. One possibility is that the structurally related competitor peptide in the context of an MHC class II molecule can still engage the TCR of the autoreactive T cell but that TCR occupation does not lead to any signal transduction (DeMagistris et al. 1992). Based on crystal structure studies, a tentative model has recently been put forward suggesting that MHC class II molecules, each consisting of an α and β chain, are present at the cell surface as dimers (Brown et al. 1993). One could speculate that the formation of mixed MHC dimers, containing antigenic and antagonistic peptides, is a prerequisite for inhibition of T cell activation. Alternatively, it has been shown recently that T cell receptor engagement by structurally related peptides can selectively modulate certain T cell effector activities leading to *e.g.* different cytokine production (Racioppi et al. 1993; Evavold and Allen 1991). Based on these results the "agonist/antagonist" concept has been introduced in which a ligand-related variation in the quality of TCR signaling events and their integration with other signaling processes has been proposed (Racioppi et al. 1993). One can imagine that alterations in the cytokine profile induced by disease-related agonist/antagonist peptides may change the nature of the autoreactive T cells involved, thereby interfere in the development of the disease process. Furthermore it has been shown that single residue substitutions within a TCR ligand can induce clonal anergy (Sloan-Lancaster et al. 1993). This indicates that *in vivo* administration of such agonist/antagonist peptides might lead to a lasting decrease in autoimmune disease due to anergy induction amongst the autoreactive T cells. Another, rather speculative, possibility could be that upon T cell receptor engagement by the disease-related competitor peptide the physiology of the T cells changes in such a way that subsequent engagement with the disease-inducing epitope would trigger apoptosis rather than activation.

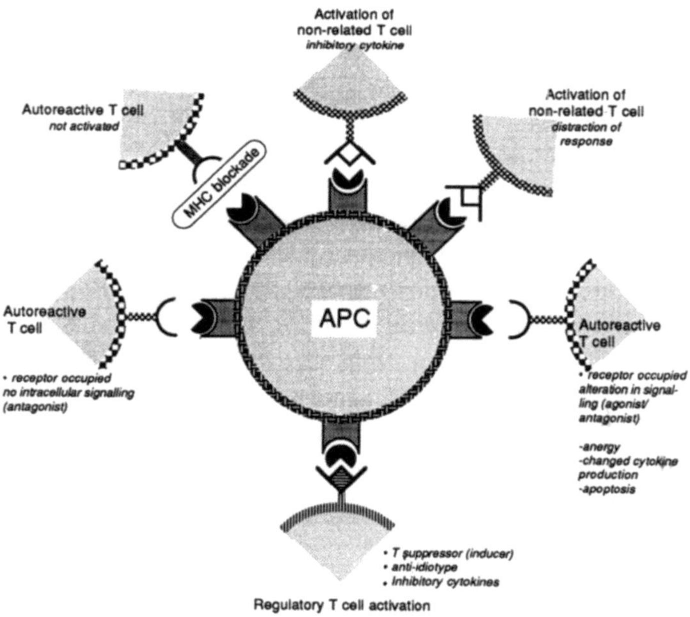

Figure 2: Disease-related competitor peptide

Analysis of structure-function relationships revealed that for both the antagonist and agonist/antagonist peptides, their functional activity was dependent on which TCR contact residue in the original epitope was altered (Sloan-Lancaster et al. 1993; Alexander et al. 1993; Ostrov et al. 1993). Moreover, their respective mechanistic principles were regarded to be very similar and based on low-affinity TCR-peptide interactions (Ruppert et al. 1993; Travers 1993). Therefore, one might doubt whether the antagonist concept and the agonist/antagonist concept are fundamentally different, rather than being different interpretations of the same phenomenon.

One of the possible cellular events that may contribute to mechanisms of peripheral tolerance is direct T-T cell interaction (Mitchison 1992). In other words, the selective expansion of certain T cells based on their specificity for critical antigens, may well be counteracted by the consequential induction of regulatory T cells. These regulatory T cells may include various cells, -the still elusive T suppressor cells, suppressor inducer cells, cytotoxic T cells or T cells with regulatory cytokine profiles. Therefore, instead of the described direct effects of structurally related peptides on autoreactive T cells, it may well be possible that the inhibition of activation of such T cells was due to regulatory T cells triggered by the disease-related competitor peptide. It is conceivable that the disease-related competitor peptide induces antigen-specific suppressor or suppressor inducer T cells. However, the mechanisms by which such T cells suppress the triggering of autoreactive T cells are still under debate (Bloom et al. 1992; Dorf et al. 1992). Alternatively, it is possible that the disease-related competitor peptide induces anti-idiotypic T cells with specificity for the autoreactive T cells. In several experimental autoimmune models, it has been demonstrated that T cell vaccination, a procedure involving immunization with whole T cells, can induce anti-idiotypic T cells leading to resistance to autoimmunity (Neilson and Philips 1982; Holoshitz et al. 1983; Lider et al. 1987; Lider 1988; Cohen 1989). The underlying principle of T cell vaccination has been suggested to rely on the induction of anti-TCR peptide-specific T cells. Accordingly, successes have been reported of EAE inhibition by vaccination with TCR peptides (Vandenbark et al. 1989; Howell et al. 1989; Offner et al. 1990). Recently, it has been shown that indeed TCR-derived peptides can induce MHC class II-restricted T-T cell interactions (Broeren et al. submitted). Furthermore, it has been demonstrated that T cells recognizing different antigens can induce a crossreactive immunity against shared T cell receptor idiotopes (Offner et al. 1991). Because of the close structural relationship between the competitor peptide and the disease-inducing peptide it is possible that T cells recognizing the competitor peptide evoke anti-idiotypic T cell responses against idiotopes shared with the autoreactive T cells. Such anti-idiotypic T cell responses could then suppress the activity of the autoreactive T cells. From rats protected for AA by immunization with a disease-related competitor peptide, we isolated several T cell clones, which recognize both the analogue and the original disease-associated peptide (Wauben et al. 1992; Wauben et al. 1993). Whether these T cells are relevant for disease suppression by one of these above mentioned mechanisms is unclear at present. However, the dramatic effect of the disease-related analogue peptide in AA, and the fact that the isolated T cell clones represent the dominant and recurrent specificity (Wauben et al. 1993), seem to argue in favor of such a regulatory role.

Conclusion

MHC class II binding peptides administered *in vivo* can be effective and selective inhibitors of T cell activation. Passive therapies aimed at inhibition of autoimmune T cell activation by blocking the disease-associated MHC molecule have been shown to be successful in the prevention of EAE, experimental myocarditis, and spontaneous non-obese diabetes in mice. Since in some human autoimmune disorders predisposition occurs in association with particular MHC molecules, added to the fact that autoantigens involved in such diseases are largely unknown, the approach of competitor peptides blocking a specific MHC gene product could be a novel immunosuppressive strategy. However, it is evident that actively induced specific effector mechanisms may have intrinsic advantages over passive forms of MHC competition. Activation of specific effector mechanisms may obviate the need and the risk of continuous infusion or frequent injection schedules of competitor peptides. Furthermore, actively induced antigen-specific effector mechanisms may be less dependent on peptide stability, may induce persistent effects, and may not have the unwanted side-effect of some degree of general immunosuppression. As argued before, the integration of both MHC competition and specific immunomodulation in a single "competitor-modulator" peptide may offer attractive possibilities for active intervention strategies in autoimmune diseases. For the clinical application of such strategies it will be of prime interest to further explore the nature of the autoantigens involved in human disorders, and to unravel the *in vivo* mode of action of the inhibition of experimental autoimmunity by disease-related MHC binding competitor peptides.

ACKNOWLEDGMENTS
We thank Prof. L. Adorini (Milan) for helpful discussions. This work is supported by the European Concerted Action on Immunotherapy of Chronic Arthritis experimental basis.

REFERENCES

Adorini L (1992) Selective suppression of T cell activation by administration of MHC class II binding peptides. Transpl. Proc. Suppl.2 24:14-16.

Adorini L and Nagy Z (1990) Inhibition of T cell activation by MHC blockade. Int. Rev. Immunol.6:23-35.

Adorini L, Muller S, Cardinaux F, Lehmann PV, Falcioni F, and Nagy ZA (1988) *In vivo* competition between self peptides and foreign antigens in T cell activation.Nature 334:623-625.

Alexander J, Snoke K, Sidney J, Wall M, Southwood S, Oseroff C, Arrhenius T,Gaeta FCA, Colon SM, Grey HM, and Sette A (1993) Functional consequences of engagement of the T cell receptor by low affinity ligands. J. Immunol. 150:1-7.

Bloom BR, Salgame P, and Diamond B (1992) Revisiting and revising suppressor T cells. Imm. Today. 13:131-136.

Brown JH, Jardetsky TS, Gorga JC, Stern LJ, Urban RG, Strominger JL, Wiley DC (1993). Three-dimensional structure of the human class II histocompatibility antigen HLA-DR1. Nature 364: 33-39.

Buus S, Sette A, and Grey HM (1987) The interaction between protein derived immunogenic peptides and Ia. Immunol. Rev. 98:115-141.

Cohen IR (1989) T cell vaccination and suppression of autoimmune disease. Prog. Immunol. VII. (Melchers et al. Eds) Springer Verlag, Berlin. pp.867-873.

DeMagistris MT, Alexander J, Coggeshall M, Altman A, Gaeta FCA, Grey HM, and Sette A (1992) Antigen analog-Major histocompatibility complexes act as antagonists at the T cell receptor. Cell. 68:625-634.

Dorf ME, Kuchroo VK, and Collins M (1992) Suppressor T cells: some answers but more questions. Imm. Today. 13:241-243.

Evavold BD and Allen PM (1991) Separation of Il4 production from Th cell proliferation by an altered T cell receptor ligand. Science. 251:1308-1310.

Gautam AM, Pearson CI, Smilek DE, Steinman L, and McDevitt HO (1992) A polyalanine peptide with only five native myelin basic protein residues induces autoimmune encephalomyelitis. J. Exp. Med. 176:605-609.

Gautam AM, Pearson CI, Sinha AA, Smilek DE, Steinman L, and McDevitt HO (1992) Inhibition of experimental autoimmune encephalomyelitis by a nonimmunogenic non-self peptide that binds to I-Au1. J. Immunol. 148:3049-3054.

Guery JC, Sette A, Leighton J, Dragomir A, and Adorini L (1992) Selective immunosuppression by administration of major histocompatibility complex (MHC) class II-binding peptides. I. Evidence for *in vivo* MHC blockade preventing T cell activation. J. Exp. Med. 175: 1345-1352.

Guery JC, Neagu M, Rodriguez-Tarduchy G, and Adorini L (1993) Selective immunosuppression by administration of major histocompatibility complex class II binding peptides. II. Preventive inhibition of primary and secondary *in vivo* antibody responses. J. Exp. Med. 177:1461-1468.

Guillet JG, Lai MZ, Briner TJ, Smith JA, and Gefter ML (1986) Interaction of peptide antigens and class II major histocompatibility complex antigens. Nature. 324:260-262.

Holoshitz J, Frenkel A, Ben-Nun A, and IR Cohen (1983) Autoimmune encephalomyelitis (EAE) mediated and prevented by T lymphocyte lines directed against diverse antigenic determinants of myelin basic protein. Vaccination is determinant specific. J. Immunol. 131:2810-2815.

Howell MD, Winters ST, Olee T, Powell HC, Carlo DJ, and Brostoff SW (1989) Vaccination against experimental allergic encephalomyelitis with T cell receptor peptides. Science 246:668-670.

Hunt DF, Michel H, Dickinson TA, Shabanowitz J, Cox AL, Sakaguchi K, Appella E, Grey HM and Sette A (1992) Peptides presented to the immune system by the murine class II major histocompatibility complex molecule I-A^{d}. Science. 256:1817-1820.

Hurtenbach U, Lier E, Adorini L, and Nagy ZA (1993) Prevention of autoimmune diabetes in Non-Obese Diabetic mice by treatment with a class II major histocompatibility complex-blocking peptide. J. Exp. Med. 177:1499-1504.

Janeway CA (1992) The T cell receptor multicomponent signalling machine: CD4/CD8 coreceptors and CD45 in T cell activation. Annu. Rev. Immunol. 10:645-674.

Jardetzky TS, Gorga JC, Busch R, Rothbard J, Strominger JL, and Wiley DC (1990) Peptide binding to HLA-DR1: a peptide with most residues substituted to alanine retains MHC binding. EMBO J. 9:1797-1803.

Lamont AG, Powell MF, Colon SM, Miles C, Grey HM, and Sette A (1990) The use of peptide analogs with improved stability and MHC binding capacity to inhibit antigen presentation *in vitro* and *in vivo*. J. Immunol. 144:2493-2498.

Lamont AG, Sette A, Fujinami R, Colon SM, Miles C, and Grey HM (1990) Inhibition of experimental encephalomyelitis induction in SJL/J mice by using a peptide with high affinity for IA^s molecules. J. Immunol. 145:1687-1693.

Leighton J, Sette A, Sidney J, Appella E, Ehrhardt C, Fuchs S, and Adorini L (1991) Comparison of structural requirements for interaction of the same peptide with $I\text{-}E^k$ and $I\text{-}E^d$ molecules in the activation of MHC class II restricted T cells. J. Immunol. 147:198-204.

Lider O, Karin N, Shinitzki M, and Cohen IR (1987) Therapeutic vaccination against adjuvant arthritis using autoimmune T cells treated with hydrostatic pressure. Proc. Natl. Acad. Sci.(USA). 84:4577-4580.

Lider O (1988) Anti-idiotypic network induced by T cell vaccination against experimental autoimmune encephalomyelitis. Science. 239:181-183.

Martin R, McFarland HF, and McFarlin DE (1992) Immunological aspects of demyelinating diseases. Annu. Rev. Immunol. 10:153-187.

Mitchison NA (1992) Specialisation, tolerance, memory, competition, latency, and strife among T cells. Annu. Rev. Immunol. 10:1-12.

Myers LK, Rosloniec EF, Seyer JM, Stuart JM, and Kang AH (1993) A synthetic peptide analogue of a determinant of type II collagen prevents the onset of collagen-induced arthritis. J. Immunol. 150:4652-4658.

Möller G (Ed.) (1987) Antigenic requirements for activation of MHC restricted responses. Immunol. Rev. 98:1-187.

Muller S, Adorini L, Juretic A, and Nagy ZA (1990) Selective *in vivo* inhibition of T cell activation by class II MHC binding peptides administered in soluble form. J. Immunol. 145:4006-4011.

Neilson EG, and Philips SM (1982) Suppression of interstitial nephritis by auto-anti-idiotypic immunity. J. Exp. Med. 155:179-185.

Offner H, Hashim GA, and Vandenbark AA (1991) T cell receptor peptide therapy triggers autoregulation of experimental encephalomyelitis. Science. 251:430-432.

Offner H, Vainiene M, Gold DP, Morrison WJ, Wang RY, Hashim GA, and Vandenbark AA (1991) Protection against experimental encephalomyelitis, idiotypic autoregulation induced by a nonencephalitogenic T cell clone expressing a cross-reactive T cell receptor V gene. J. Immunol. 146:4165-4172.

O'Sullivan D, Sidney J, Del Guercio M, Colon SM, and Sette A (1991) Truncation analysis of several DR binding epitopes. J. Immunol. 146:1240-1246.

Ostrov D, Krieger J, Sidney J, Sette A, and Concannon P (1993) T cell receptor antagonism mediated by interaction between TCR junctional residues and peptide antigen analogues. J. Immunol. 150:4277-4283.

Racioppi L, Ronchese F, Matis LA, and Germain RN (1993) Peptide-major histocompatibility complex class II complexes with mixed agonist/antagonist properties provide evidence for ligand-related differences in T cell receptor-dependent intracellular signaling. J. Exp. Med. 177:1047-1060.

Rock KL, and Benacerraf B (1984) Selective modification of a private I-A allostimulating determinant(s) upon association of antigen with an antigen-presenting cell. J. Exp. Med. 159:1238-1252.

Rudensky AY, Preston-Hurlburt P, Al-Ramadi BK, Rothbard J, and Janeway CA (1992) Truncation variants of peptides isolated from MHC class II molecules suggest sequence motifs. Nature. 359:429-431.

Ruppert J, Alexander J, Snoke K, Coggeshall M, Herbert E, McKenzie D, Grey HM, and Sette A (1993) Effect of T cell receptor antagonism on interaction between T cells and antigen presenting cells and on T cell signalling events. Proc. Natl. Acad. Sci. USA. 90:2671-2675.

Sakai K, Zamvil SS, Mitchell DJ, Hodgkinson S, Rothbard JB, and Steinman L (1989) Prevention of experimental encephalomyelitis with peptides that block interaction of T cells with MHC proteins. Proc. Natl. Acad. Sci. USA. 86:9470-9474.

Schwartz RH (1985) T lymphocyte recognition of antigen in association with gene products of the major histocompatibility complex. Annu. Rev. Immunol. 3:239-263.

Sette A, Buus S, Colon S, Smith JA, Miles C, and Grey HM (1987) Structural characteristics of an antigen required for its interaction with Ia and recognition by T cells. Nature 328:395-399.

Sette A, Buus S, Colon S, Miles C, and Grey H (1988) I-A^d binding peptides derived from unrelated protein antigens share a common structural motif. J. Immunol. 141:45-48.

Sette A, Adorini L, Apella E, Colon SM, Miles G, Tanaka S, Ehrhardt C, Doria G, Nagy ZA, Buus S, and Grey H (1989) Structural requirements for the interaction between peptide antigens and I-E^d molecules. J. Immunol. 143:3289-3294.

Sloan-Lancaster J, Evavold BD, and Allen PM (1993) Induction of T cell anergy by altered T cell receptor ligand on live antigen presenting cells. Nature. 363:156-159.

Smilek DE, Wraith DC, Hodgkinson S, Dwivedy S, Steinman L, and McDevitt HO (1991) A single amino acid change in a myelin basic protein peptide confers the capacity to prevent rather than induce experimental autoimmune encephalomyelitis. Proc. Natl. Acad. Sci. USA. 88:9633-9637.

Smith SC, and PM Allen (1991) Myosin-induced acute myocarditis is a T cell mediated disease. J. Immunol.147:2141-2147.

Travers P (1993) Immunological agnosia. Nature 363:117-118.

Unanue ER (1984) Antigen presenting function of macrophages. Annu. Rev. Immunol. 2:395 - 410

Vandenbark AA, Hashim G, and Offner H (1989) Immunization with a synthetic T cell receptor V-region peptide protects against experimental autoimmune encephalomyelitis. Nature 341:541-544.

Wauben MHM, Boog CJP, van der Zee R, and van Eden W (1992a) Towards peptide immunotherapy in rheumatoid arthritis: competitor-modulator concept. J. Autoimmunity. 5A:205-208.

Wauben MHM, Boog CJP, van der Zee R, Joosten I, Schlief A, van Eden W (1992b) Disease inhibition by major histocompatibility complex binding peptide analogues of disease associated epitopes: More than blocking alone. J. Exp. Med. 176: 667-677.

Wauben MHM, van der Zee R, Joosten I, Boog CJP, van Dijk AMC, Holewijn MC, Meloen RH, van Eden W (1993) A peptide variant of an arthritis related T cell epitope induces T cells that recognize this epitope as a synthetic peptide but not in its naturally processed form. J. Immunol. 150: 5722-5730

Werdelin O (1982) Chemically related antigens compete for presentation by accessory cells to T cells. J. Immunol. 129:1883-1891.

Wraith DC, McDevitt HO, Steinman L, and Acha-Orbea H (1989) T cell recognition as the target for immune intervention in autoimmune disease. Cell. 57:709-715.

Wraith DC, Smilek DE, Mitchell DJ, Steinman L, and McDevitt HO (1989) Antigen recognition in autoimmune encephalomyelitis and the potential for peptide mediated immunotherapy. Cell. 59:247-255.

Type-I (insulin-dependent) and type-II (non-insulin-dependent) diabetes mellitus in BC1[(NODx*Mus spretus*)F1xNOD] mice

Masakazu Hattori[1], Eiji Yamato[1], Masahiro Fukuda[1], Maria Petruzzelli[1], Alix Weaver[2], Armand MacMurray[2], Eric S. Lander[2], and Verne M. Chapman[3]
[1]Section on Immunology and Immunogenetics, Joslin Diabetes Center, One Joslin Place, Boston, MA 02215; [2]Whitehead Institute for Biomedical Research, Nine Cambridge Center, Cambridge, MA 02142; and [3]Department of Molecular and Cellular Biology, Roswell Park Cancer Institute, Elm and Carlton Streets, Buffalo, N.Y. 14263, U. S. A.

Genealogy of the NOD Mouse and its Related Strains

The nonobese diabetic (NOD), nonobese nondiabetic (NON), cataract Shionogi (CTS), ICR-L-Ishibe (ILI) and ICR-I-Ishibe (III) mouse strains have been used in our experiments. The NOD, NON, CTS and ILI mice were originally derived from outbred Institute of Cancer Research (ICR) mice (Figure 1). One mouse with cataracts and small eyes was found in outbred ICR mice and an inbred cataract Shionogi (CTS) strain characterized by cataracts and microphthalmia was established (Ohtori et al, 1968). Since cataracts are often observed in diabetic patients, two sublines were separated at the sixth generation during the process of establishing the CTS strain according to fasting blood glucose levels; one with euglycemia (fasting blood glucose: approximately 100 mg/dl) and the other with slight hyperglycemia (fasting blood glucose: approximately 150 mg/dl). One female mouse in the former line spontaneously developed diabetes at the 20th generation and the inbred NOD strain was established by selective breeding for diabetes. From the latter line the inbred non-obese nondiabetic (NON) strain was established. More details on the genealogy is described in Ikegami et al, 1988. Inbred ILI and III mice were established directly from outbred ICR mice.

NATO ASI Series, Vol. H 80
Autoimmunity: Experimental Aspects
Edited by Moncef Zouali

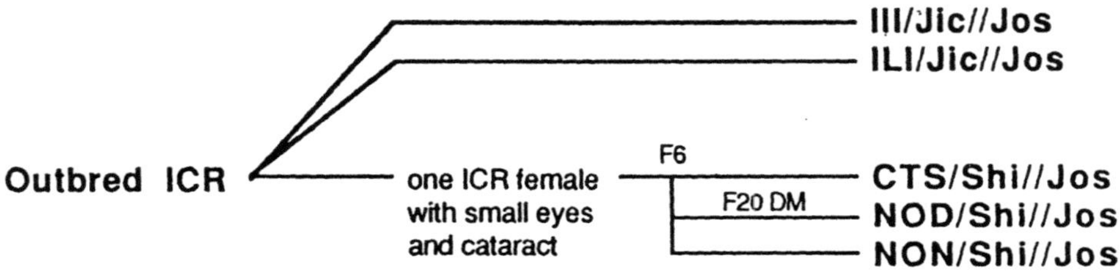

Figure 1. Genealogy of the NOD mouse and its related strains.

Mus spretus Spain

The mice which provided the bulk of the genome of most laboratory mouse strains are descendants of the western European house mouse, *Mus domesticus*. They were separated from other wild mice of subspecies approximately 1 to 2.5 million years ago. Since then, the ancestors of laboratory mice have been bred within the same species. From the evolutionary point of view, the genetic background of the laboratory mouse strains including NOD, NON, ILI, BALB/c, C3H and C57BL/6 is close to each other. This close genetic background causes difficulty in finding polymorphisms between inbred laboratory mouse strains. Wild mice, *Mus spretus Spain*, have been found to represent an interesting source of biochemical polymorphisms at many loci, compared with the inbred laboratory mouse strains. In our study *Mus spretus* did not develop obesity, diabetes, hypertrophy of islets and insulitis at 12 months of age when the animals were fed on Agway 3000 mouse chow (protein 22%, fat 5%).

Criteria used for diabetes and insulitis

The criteria used for diabetes mellitus were: blood glucose >250 mg/dl at onset,persisting hyperglycemia (more than 1 week), and HbA1 (hemoglobin A1) >6% (normal range in mice: 3%<HbA1<4%)

Sections of pancreas from parental, F1 and backcross animals were stained with Hematoxylin & Eosin and subjected to microscopic observation. Insulitis was defined as destruction of islet β cells by infiltrating immune cells. Each islet was graded according to the criteria (0-4) listed below.

Grade 0: No infiltrating immune cells
Grade 1: Peri-insulitis; Immune cells adjacent to islets but do not enter the islets (pre-stage leading to insulitis).
Grade 2: Infiltrating immune cells occupy 25% of the islet area.
Grade 3: Infiltrating cells occupy 25-50% of the islet area.
Grade 4: Infiltrating cells occupy >50% of the islet

After grading 15-30 islets, the most severe grade was used for the degree of insulitis.

Insulitis and Diabetes in the NOD Mouse

Type-I (insulin-dependent) diabetes of man, the BB rat and the NOD mouse appears to result from autoimmune β cell destruction in a setting of genetic predisposition. The NOD mouse strain was established at Shionogi Aburahi Laboratories in Shiga-ken, Japan (Makino et al, 1980). The NOD mouse develops type-I diabetes secondary to β cell destruction by infiltrating immune cells (insulitis). Insulitis appears as early as 5 weeks of age. The immune cells invade islets and are often seen adjacent to the ducts and blood vessels. The incidence of insulitis is 90% in NOD females and 70% in NOD males at 9 weeks of age. The incidence and degree of insulitis increase with age, resulting in lack of intrinsic insulin secretion, hyperglycemia, loss of body weight and ketosis as seen in human insulin-dependent diabetes. The NOD mouse usually develops type-I diabetes by 6 months of age. The cumulative incidence of diabetes in our NOD/Shi//Jos colony is 70-90% in NOD females and 20-50% in NOD males at 7 months of age.

Characterization of Islet Infiltrating Cells Isolated from the Insulitis Lesion of the Diabetic NOD Female Mice

The infiltrating cell populations and their proportions in the insulitis lesions of the NOD mouse were examined serologically. When islets isolated from diabetic NOD female mice were examined shortly after the onset of diabetes, many had infiltrates of lymphoid and monocytic cells. Islet infiltrating cells were collected after placing isolated islets in a tissue culture plate overnight (T cells: 54%, B cells: 4%, macrophages: 12.6%, anti-asialo GM1-positive cells: 24.8% and endocrine cells: 4.6%). By dual staining flow cytometric analysis of the islet T-cells from the insulitis lesions, $CD4^+$ T-cells represented 65% and $CD8^+$ T-cells 35%.

Insulitis and T-cells of the NOD Mouse

Infiltration of the islets of Langerhans by immune cells (insulitis) is a well-recognized feature of type-I diabetes both in human subjects (Gepts W, 1965) and in rodents with spontaneous diabetes (Makino et al, 1980; Like AA and Rossini, 1984). This insulitis lesion suggests that cell-mediated immune processes are responsible for the destruction of islet β cells. T-cells play an important role in the development of insulitis and diabetes. Athymic nude NOD mice did not develop insulitis and diabetes (Makino et al, 1986). Transfer of splenocytes from adult NOD mice induced insulitis but not diabetes in the athymic nude NOD mice. Following the transfer of adult NOD splenocytes, administration of human IL-2 induced diabetes in the nude NOD mice. Injection of splenocytes from diabetic NOD mice induced diabetes in irradiated young NOD mice (Wicker et al, 1986). More recent studies suggest that induction of insulitis and diabetes in NOD mice is mediated by $CD4^+$ (L3T4) T-cells (Haskins and McDuffie, 1991; Nakano et al, 1991; Bradley et al, 1992) or both $CD4^+$ and $CD8^+$ T-cells (Miller et al, 1988, O'Reilly et al, 1991).

Role of Macrophages and Cytokines in Islet β Cell Destruction

Cells infiltrating the islets of the NOD mouse consist of T-cells (approximately 50%), macrophages (18%), plasma cells (3%) and unidentified cells

(25%, stained with rabbit anti-asialo GM1 antibody). Recent studies indicate that macrophages may play an important role in the initiation of islet β cell destruction. Silica particles inhibit selectively macrophage function *in vivo* (Brosnan CF et al, 1981) and has been shown to prevent the development of diabetes in BB rats (Oschilewski U et al, 1985). Administration of silica particles prevents the development of insulitis and diabetes in NOD mice (Charlton et al, 1988, Lee et al, 1988). Administration of a monoclonal antibody against the adhesion-promoting receptor on macrophages also prevents the development of diabetes (Hutchings et al, 1990).

Macrophages appear to exert their functions in at least two ways as antigen-presenting cells and effector cells. An antigen(s) on islet β cells may be processed by macrophages and presented to $CD4^+$ helper T cells in a MHC restricted manner (Ziegler and Unanue, 1981). Then, $CD4^+$ helper T cells may be activated by antigen-presenting cells and secrete cytokines, such as interleukin-1 (IL-1) (Adams, 1982; Durum SK et al, 1985) and tumor necrosis factor (TNF) (Beutler et al, 1985), and simultaneously lymphokines such as interferon-γ (INF-γ) resulting in activation of macrophages (Murray et al, 1985). Activated macrophages may secrete IL-1 and simultaneously produce oxygen free radicals (Horio et al, 1989 and 1993). IL-1 can impair the islet β cell activity *in vitro* (Bendtzen et al, 1986; Palmer et al, 1989). IL-1, TNF and INF-γ have synergistically cytotoxic effects on islet cells *in vitro* (Pukel et al, 1988).

Influential Factors on the Development of Type-I Diabetes

Viruses. Type-I diabetes is a multifactorial disease, and its development appears to be influenced by genetic and environmental factors. Viruses have been implicated in the pathogenesis of β cell destruction in type-I diabetes. In the early studies on association of viruses with diabetes, viruses were regarded as agents causing diabetes. However, the appearance of animal models for spontaneous type-I diabetes, the BB rat and the NOD mouse raised under viral-free conditions, has changed the concept on the role of viruses in the development of type-I diabetes. Thus, viruses may be divided into two groups, those that induce or accelerate the development of type-I diabetes and those that suppress its development (Karasik and Hattori, 1993).

Diets and chronic stress. Semisynthetic diets in which natural proteins were replaced by l-amino acids prevented the development of diabetes in the BB rat (Elliot and Martin, 1984; Scott et al, 1985). More animals that received the semisynthetic diet plus 1% skin milk or 1% gliadin developed diabetes than did animals fed with the semisynthetic diet alone, suggesting the precipitation of diabetes by dietary protein (Elliot and Martin, 1984). A semipurified diet (AIN-76) also could prevent the development of diabetes in the BB rat (Scott et al, 1985) and the NOD mouse (Coleman et al, 1990). The AIN-76 diet increased the weight of the thymus and the total number of white blood cells, doubled the ratio of thymic helper T cells to suppressor T cells, and decreased the proportion of suppressor T cells in the spleen and thymus in the BB rat. Pregestimil is a hypoallergic infant formula based on enzymatically hydrolyzed casein from cow's milk further treated with charcoal to remove antigenic peptides. Pregestimil ingestion by NOD mice reduced islet damage (Elliot et al, 1988) and completely prevented the development of diabetes in animals up to 40 weeks of age (Coleman et al, 1990). When a chloroform-methanol extract of the natural ingredient Old Guilford 96 mouse diet was added to Pregestimil, the incidence of diabetes in the NOD mice reached 75% (Coleman et al, 1990).

Environmental stress, especially chronic moderate stress can accelerate the development of diabetes in the BB rat (Lehman et al, 1991). BB rats exposed to daily chronic moderate stress (rotation, vibration and restraint) over a 14-week period developed diabetes more often than did control animals (70-80% in the stressed group vs. 50% in the control group). BB rats exposed to the stress showed larger adrenal glands and smaller thymuses and spleens than the control animals.

Non-MHC Diabetogenic Genes in the NOD Mouse

Breeding studies of NOD with C3H/He, NON, BALB/c, and B10 mice showed 8-15% of diabetes incidence in BC1 animals and indicate that the development of diabetes is controlled by at least two or three recessive major genes (Hattori et al, 1986; Prochazka, et al, 1987; Wicker et al, 1987). Ten linkage groups have been reported and defined as Idd-1 - Idd-10 (Ghosh et al, 1993) (Table 1). Idd-1 is linked to the major histocompatibility complex (MHC) on chromosome 17 (Hattori et al, 1986; Prochazka, et al, 1987; Wicker et al, 1987). Idd-2 is linked to the Thy-1 locus on chromosome 9 (Prochazka, et al, 1987). Idd-3 is linked to an interval between IL-2 and Tshb on chromosome 3 (Todd et al, 1991), Idd-4 linked to an interval between Acrb and Mpo on chromosome 11 (Todd et al, 1991) and Idd-5 linked to the region of Il-1r1 and Lsh/Ity/Bcg locus in BC1 of NOD with B10.NOD-H-2 congenic strain (Cornall et al, 1991), and to the region of Bcl-2 and D1Nds2 locus in F2 of NOD with B6 on chromosome 1 (Garchon et al, 1991).

Table 1. Linkage groups to Type-I diabetes in the NOD mouse

MHC-linked gene(s)

Idd-1: MHC-linked gene (idd-1) on chromosome 17

MHC class II: Unique I-Aβ, lack of I-E molecules

The exact localization of the MHC-linked diabetogenic gene(s) is still unknown.

Non-MHC-linked genes

Idd-2 on chromosome 9 implicated in the development of diabetes

Idd-3 on chromosome 3 implicated in the development of insulitis

Idd-4 on chromosome 11 implicated in the aging factor for the development of diabetes.

Idd-5 on chromosome 1 implicated in the development of peri-insulitis, insulitis and diabetes

Idd-6 on chromosome 6 implicated in the development of diabetes

Idd-7 on chromosome 7 implicated in the development of diabetes

Idd-8 on chromosome 14 implicated in the development of diabetes

Idd-9 on chromosome 4 implicated in the development of diabetes

Idd-10 on chromosome 3 implicated in the development of diabetes and insulitis

Several questions have been raised in the current genetic analysis of non-MHC-linked diabetogenic genes in the NOD mouse. First, the above-mentioned linkages are not necessarily found in all the breeding studies, depending on the combinations of the breeding mouse strains with the NOD mouse. Secondly, there are more linkages found than the number of potential diabetogenic genes. Thirdly, the linkage groups (or intervals) are too great to identify the real loci. For instance, the distance of the linkage group on chromosome 3 is 35cM. To identify the diabetogenic genes, it is essential to reach within 1cM. Fourth, in the previous breeding studies inbred laboratory mouse strains have been used as a partner of the NOD mouse (Dietrich et al, 1992). Polymorphisms between the laboratory inbred strains are much less than those between laboratory inbred and wild mouse strains.

MHC-linked Type-I diabetogenic gene(s) in the NOD Mouse

The inheritance of insulitis in NOD mice is controlled by a single recessive gene in the presence of an MHC-linked diabetogenic gene(s) (Hattori et al, 1986; Wicker et al, 1987). The development of diabetes, however, is controlled by several genes. One of the diabetogenic genes is tightly linked to the major histocompatibility complex (MHC) on chromosome 17. With the use of a panel of monoclonal antibodies and restriction fragment length polymorphism (RFLP) analysis, it has been shown that the NOD mouse possesses a recombinant class I (K^d, D^b: H-2^{g7}) and unique MHC class II molecules with no expression of surface I-E molecules and different I-A molecules from any known I-A. Lack of surface I-E molecules is due to absence of messenger RNA for I-Eα (Hattori et al, 1986). Sequence studies indicate that the I-A_β chain is unique in NOD mice (Acha-Orbea et al, 1987). The amino acid residue at position 57 of the I-A_β chain is serine in NOD mice and aspartic acid in nondiabetic inbred laboratory mouse strains. It is suggested that the unique sequence of I-A_β allele in NOD mice makes the animals susceptible to diabetes (the MHC-linked diabetogenic gene). The unique MHC class II of the NOD mouse is derived from its original strain, the ICR mouse but not the result of recent mutation (Ikegami et al, 1990).

Breeding studies in crosses of NOD with C57BL/6 [B6(I-Eα of d) transgenic mice expressing I-E suggest that I-Eα gene expression prevents the development of insulitis (Nishimoto, 1987). Recent studies of transgenic mice indicate that introduction of normal I-A_α and I-A_β of k into NOD mice can prevent the development of insulitis (Miyazaki et al, 1990; Slattery et al, 1990, Lund et al,

1990). In the breeding studies of NOD (lack of I-E molecules) with NOD MHC congenic strains expressing I-E molecules, F1 hybrids heterozygous for MHC developed diabetes and insulitis (Podolin et al, 1993). F1 hybrids (MHC: nod/b) of NOD with the NOD. B10-MHC congenic strain (NOD background with B10's MHC: b haplotype) developed insulitis and diabetes (Wicker et al, 1989). These findings in the transgenic and congenic studies need to be confirmed in a **homologous recombination state.** The cataract Shionogi (CTS) mouse is identical to the NOD mouse at the MHC class II I-A and I-E regions, but differs at the MHC class I-K and -D and class III regions (homologous recombination) (Ikegami et al, 1990; Lund et al, 1993). Makino et al (1991) reported that NOD. CTS-MHC congenic mice homozygous for MHC cts/cts developed insulitis and diabetes, suggesting another diabetogenic gene outside the MHC class-II region. The exact location of the MHC-linked type-I diabetogenic gene(s) is still unknown.

Recessive nature of diabetes and insulitis in F1 generations of NOD with C3H/He, NON, BALB/c, C57BL/6, and *Mus spretus*

To examine the genetic mode of inheritance of diabetes and insulitis, breeding studies of NOD with nondiabetic mouse strains were performed. NOD mice were mated with C3H/He, NON, BALB/c, C57BL/6 (B6) and *Mus spretus Spain* to produce F1 animals (Table 2).

Table 2. Incidence of diabetes and insulitis in F1 animals of NOD with *Mus spretus Spain*, C57BL/6 (B6), C3H/He, NON, and BALB/c at 1 year of age.

	NODx*Spretus*	NODxB6	NODxC3H/He	NODxNON	NODxBALB/c
Diabetes					
Female	0/17 (0%)	0/75 (0%)	0/37 (0%)	0/24 (0%)	0/23 (0%)
Male	0/7 (0%)	0/49 (0%)	0/42 (0%)	0/38 (0%)	0/13 (0%)
Insulitis					
Female	0/5 (0%)	0/75 (0%)	0/37 (0%)	0/24 (0%)	0/7 (0%)
Male	0/5 (0%)	0/49 (0%)	-	-	-

Then, F1 animals were backcrossed to NOD mice reciprocally. F1 males of NOD with *Mus spretus* were infertile. Therefore, only female F1 of NOD with *Mus spretus* were backcrossed to NOD males. The F1 and backcross (BC) animals were screened for the development of diabetes weekly for one year, and then subjected to histological examination for insulitis (islet β cell destruction by infiltrating lymphocytes and macrophages). The penetrance of diabetes in our NOD/Jos colony is 70% in female NOD mice and less than 20% in NOD males. As shown in Table 2, the incidence of diabetes and insulitis in F1 animals of the five breeding studies was 0%. This result suggests that the development of diabetes and insulitis is "recessive".

Regardless of NOD MHC Homozygosity and Heterozygosity, a Single Recessive "Non-MHC" Gene Determines the Development of Insulitis in NOD Mice

To study the genetic control of insulitis in NOD mice, we performed breeding studies in crosses of NOD with ILI, NON and C3H/He mice. The ILI mouse serologically shared the same MHC class I and class II molecules with the NOD mouse (Hattori et al, 1990). The ILI mice, however, do not develop insulitis and diabetes. The incidence of insulitis was 0% in F1 generations and 40% (8/20) in female BC1 [(NODxILI)F1xNOD] at 9 weeks of age, 50% (12/24) in female BC1 [(NONxNOD)F1xNOD] and 48% (14/25) in female BC1 [(C3H/He)F1x NOD] at 1 year of age (Table 3). The backcross animals were typed for the MHC to investigate correlation between the development of insulitis and MHC haplotypes. BC1[(NODxILI)F1xNOD] were always homozygous for NOD's MHC(nod/nod). Of the female animals with insulitis in BC1[(NONxNOD)F1xNOD] and BC1 [(C3H/He)F1x NOD], approximately half the animals were MHC heterozygous. The results indicate that a single recessive "non-MHC" gene determines the development of insulitis regardless of NOD MHC homozygosity or heterozygosity. F2 (NODxNON) animals homozygous for NON's MHC (non/non) did not develop insulitis.

Breeding studies of NOD with C3H/He, NON and BALB/c mice

The incidence of diabetes and insulitis in BC1 animals of NOD with C3H/He, NON, BALB/c, C57BL/6 and *Mus spretus Spain* is listed in Table 4. Both female

Table 3. Development of insulitis in female crosses of NOD with ILI, NON and C3H/He mice

	MHC	F1			Backcross to NOD		
		ILIxNOD	NONxNOD	C3H/HexNOD	ILIxNOD@	NONxNOD	C3HxNOD
Insulitis (+)	Hm	-	-	-	8*	7	8
	Ht	0	0	0	-	5	4
Insulitis (-)	Hm	-	-	-	12	7	5
	Ht	7	10	37	-	5	8

*: Number of animals examined.
@: Crosses of ILI with NOD were 9-15 weeks of age. The other crosses were 1 year of age.
Hm=Homozygosity for NOD MHC; Ht=Heterozygosity.

and male BC1 animals of NOD with C3H/He, NON and BALB/c developed diabetes. The incidence of diabetes in the three breeding studies of NOD with C3H/He, NON and BALB/c was 12.1-15.6% in female BC1 mice and 1.7-3.4% in male BC1 mice, suggesting a contribution of two or three recessive genes to the development of diabetes. Histological examination revealed that all the diabetic BC1 animals had insulitis. Among the diabetic and nondiabetic BC1 animals in the three breeding studies, 42-50% of the female BC1 animals developed insulitis. This incidence of insulitis suggests a single recessive gene contributing to the development of insulitis.

Breeding study with NOD with C57BL/6

The incidence of diabetes in BC1 animals of NOD with C57BL/6 was 3.3% in females and 0% in males (Table 4). The diabetic females had insulitis. Among the diabetic and nondiabetic BC1 animals, 24% of the female and 10% male BC1 mice developed insulitis. As compared to the breeding studies of NOD with C3H/He, NON and BALB/c, the lower incidence of diabetes in the BC1 of NOD with C57BL/6 suggests the role of one resistant or suppressive gene present in C57BL/6 mice.

Table 4. Incidence of diabetes and insulitis in the 1st backcross animals of NOD with *Mus spretus Spain,* C57BL/6 (B6), C3H/He, NON and BALB/c at 1 year of age (backcrossed to NOD).

	NODx*Spretus*	NODxB6	NODxC3H/He	NODxNON	NODxBALB/c
Diabetes					
Female	5/170 (3%)	3/92 (3.3%)	12/77 (15.6%)	12/99 (12.1%)	11/91 (12.1%)
Male	16/130 (12%)	0/63 (0%)	1/58 (1.7%)	2/58 (3.4%)	1/50 (2%)
Diabetes with insulitis					
Female	0/3 (0%)	3/3 (100%)	12/12 (100%)	12/12 (100%)	11/11 (100%)
Male	0/14 (0%)	0/63 (0%)	1/1 (100%)	2/2 (100%)	1/1 (100%)
Insulitis with grade >2#					
Female	7/43 (16.3%)**	22/93 (24%)	4/28 (50%)	12/24 (48%)	8/19 (42%)
Male	2/42 (4.8%)	5/49 (10%)	-	-	-

#: Grade 2 of insulitis indicates that 25% of the islets area is damaged by infiltrating immune cells.
**: In addition to the 7 mice with grade > 2, 6 mice showed peri-insulitis (lymphocytic infiltration adjacent to the islets). Peri-insulitis is a pre-stage leading to insulitis. If the 6 mice were included in the animals with insulitis, the incidence would be 30% (13/43).

In fact, such a resistant gene was found in the hybrids of C57BL/6 with New Zealand Black (NZB) mice (Ozaki et al, 1983). The NZB mouse developed Systemic Lupus Erythematosus and autoimmune hemolytic anemia with anti-erythrocyte autoantibodies. By introducing C57BL/6 background into NZB, a single dominant modifying gene from the C57BL mouse could suppress the autoantibody production in the hybrids. Such a suppressive gene in the C57BL/6 mouse may influence the development of diabetes and insulitis in the hybrids of NOD with C57BL/6.

Breeding study of NOD with *Mus spretus*: Occurrence of type-II diabetes mellitus (NIDDM) and pretype-I diabetes mellitus (IDDM)

Type-II diabetes mellitus (NIDDM). The incidence of diabetes and insulitis in BC1[(NODx*Mus spretus*)F1xNOD] was 3% (5/170) in females and 12% (16/130) in males (Table 4). The incidence of diabetes in female and male BC1[(NODx*Mus spretus*)F1xNOD] mice is opposite to that in BC1 animals of NOD with C3H/He, NON and BALB/c (The incidence of diabetes: 12.1-15.6% in females and 1.7-3.4% in males). To date we have histologically examined 3 female and 14 male diabetic BC1 animals of NOD with *Mus spretus* for the development of insulitis. Surprisingly, none of the 17 diabetic BC1 females and males showed insulitis. The islets from the diabetic BC1 animals were not destroyed by infiltrating immune cells and rather hypertrophied (Figure 2). The diabetic BC1 animals showed hyperinsulinemia (average: 155 μU/ml in BC1, 27 μU/ml in NOD and 17 μU/ml in *Mus spretus* at 1 year of age) and insulin-resistance. The pathological and metabolic features resemble those of Type-II (non-insulin-dependent) diabetes mellitus (NIDDM). We still need to characterize the nature of the diabetes without insulitis in BC1[(NODx*Mus spretus*)F1xNOD]. The diabetic BC1 animals should be examined for abnormality in insulin, insulin-receptor, glucose-transporter, anti-insulin and anti-insulin-receptor antibodies.

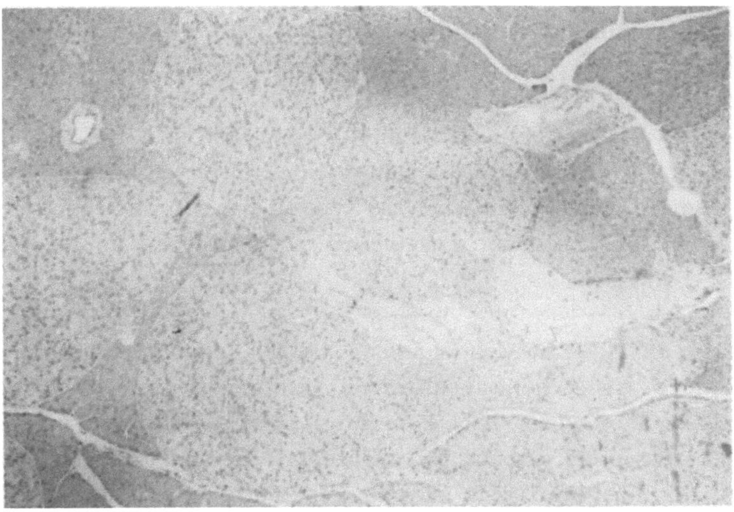

Figure 2. Hypertrophied islets in a diabetic BC1 male [(NODx*Mus spretus*)F1xNOD] at 1 year of age. Note the large islets and the absence of lymphocytic infiltration. Hematoxylin and eosin (x80). Immunohistochemical studies showed that the major component of the hypertrophied islets was β cells.

Pretype-I diabetes mellitus (IDDM). In addition to the studies on the diabetic BC1[(NODx*Mus spretus*)F1xNOD], the nondiabetic BC1 animals were histologically examined for the development of insulitis (Table 4). Forty-three female BC1 (40 nondiabetic and 3 diabetic) and 42 males (28 nondiabetic and 14 diabetic) were examined. The incidence of insulitis was 16.3% (7/43) in BC1 females and 4.8% (2/42) in BC1 males. Among the female BC1 animals, 6 mice showed peri-insulitis (grade 1: immune cells adjacent to islets but not invading the islets, pre-stage leading to insulitis). If the 6 mice with peri-insulitis (grade 1) were included in the animals with insulitis (> grade 2), the incidence of peri-insulitis (pre-stage leading to insulitis) and insulitis would be 30% (13/43) in BC1 females. The incidence of insulitis is similar to that in BC1 of NOD with C57BL/6.

In the present breeding study of BC1[(NODx*Mus spretus*)F1xNOD], female and male BC1 animals developed pre-insulitis and insulitis (pretype-I diabetes) but did not develop overt type-I diabetes. This observation was consistent with the recent report by Avner's group (De Gouyon et al, 1993) When the BC1 animals are repeatedly backcrossed to NOD mice, the backcross animals should develop more severe insulitis and type-I diabetes (IDDM). To develop overt diabetes, at least 90% of the total islets need to be damaged by infiltrating immune cells. Thus, we have observed two different types of diabetes mellitus (pre-type-I diabetes: pre-IDDM, type-II diabetes: NIDDM) in BC1[(NODx*Mus spretus*)F1xNOD] (Figure 3). In human diabetes, increased odds ratio has been found in type-I diabetic children when type-II diabetes was reported in their relatives, suggesting a common diabetogenic gene(s) in type-I and type-II diabetes (Dahlquist et al, 1989).

Male BC1[(NODx*Mus spretus*)F1xNOD] developed diabetes with hypertrophied islets, hyperinsulinemia, insulin-resistance and absence of insulitis, resembling type-II diabetes mellitus. DNA samples were prepared from the liver tissues of 15 diabetic BC1 males and 28 nondiabetic BC1 males and subjected to linkage analysis of diabetes. Forty microsatellite loci on 20 chromosomes were tested for linkage analysis in our preliminary study. Our data suggest a possibility of linkage of type-II diabetes in BC1 [(NODx*Mus spretus*)F1xNOD] to D1Mit 17 (approximately 82 cM) on chromosome 1 (χ^2=5.6, p=0.018). This P value is still insignificant with corrections. The diabetes in male BC1

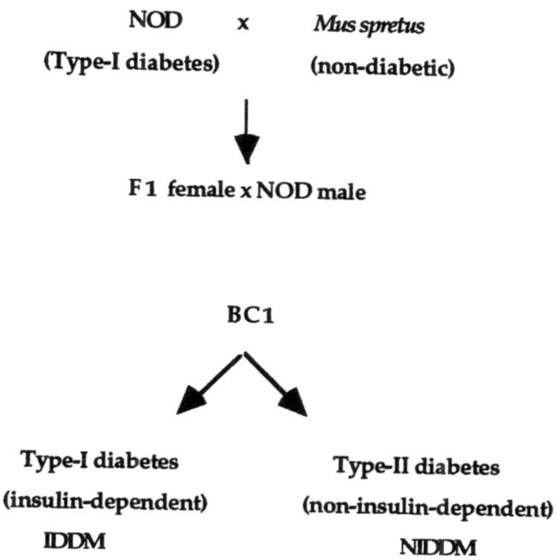

Figure 3. Development of type-I and -II diabetes in BC1[(NODx*Mus spretus*)F1xNOD]

Possible linkage of type-II diabetes in male BC1 males BC1[(NODx*Mus spretus*)F1xNOD] to D1Mit17 on chromosome 1

[(NODx*Mus spretus*)F1xNOD] was unlinked to the MHC on chromosome 17.

Type-II diabetes found in male BC1 [(NODx*Mus spretus*)F1xNOD] has two different aspects in linkage analysis, as compared to type-I diabetes found in the previous breeding studies of NOD with C3H/He (Hattori et al, 1986), B10 (Wicker et al, 1987) and NON (Porchazka et al, 1987). First, the type-II diabetes in the BC1 males was unlinked to the MHC on chromosome 17 (Hattori et al, 1992), while type-I diabetes in other BC1 animals is linked to the MHC. Secondly, type-II diabetes in male BC1 [(NODx*Mus spretus*)F1xNOD] was possibly linked to D1Mit 17 locus at a distal location on chromosome 1, while Idd-5 in type-I diabetes was linked to the region of Il-1r1 and Lsh/Ity/Bcg locus in BC1 of NOD with B10.NOD-H-2 congenic strain (Cornall et al, 1991), and to the region of Bcl-2 and D1Nds2 locus in F2 of NOD with B6 (Garchon et al, 1991) on chromosome 1. Thus, our present data and the previous reports indicate that there are two linkage groups related to type-I diabetes and one linkage group related to type-II diabetes on chromosome 1.

The number of type-II diabetic animals in our present breeding study of NOD with *Mus spretus* is still small to confirm the linkage of type-II diabetes to D1Mit 17 locus. Therefore, we should produce more BC1 [(NODx*Mus spretus*)F1xNOD] and collect a large number of diabetic animals (100 diabetic mice). We should also continue to examine other microsatellite markers around D1Mit 17 locus and candidate genes related to insulin action and glucose metabolism. Insulin receptor substrate-1 (IRS-1) is a major substrate of insulin receptor tyrosine kinase (Sun XJ et al, 1991; White MF et al, 1988, Araki et al, 1993) and located on mouse chromosome 1 (White and Kahn, personal communication). Breeding studies of NOD with Mus *spretus* should provide a new clue to better understanding the pathogenesis of type I and type II diabetes mellitus.

Conclusions

The nonobese diabetic (NOD) mouse develops autoimmune type-I (insulin-dependent) diabetes mellitus secondary to islet β cell destruction by infiltrating immune cells. The clinical and pathological features of diabetes in the NOD mouse are similar to those in human type I diabetes. Type I diabetes is a multifactorial disease in a setting of genetic predisposition and its development influenced by diet, viruses and other environmental factors. Genetic analyses have demonstrated ten linkage groups (Idd-1 ~ Idd-10) to type-I diabetes in the NOD mouse (Hattori et al, 1986; Prochazka et al, 1987; Wicker et al, 1987; Todd et al, 1991; Prins et al, 1993; Ghosh et al, 1993).

In our genetic analysis NOD mice were mated with *Mus spretus Spain* (wild mouse strain). *Mus spretus* represent polymorphisms at many loci and are suitable for linkage analysis. BC1[(NODx*Mus spretus*)F1xNOD] developed pre-type-I diabetes mellitus (IDDM) and type-II (non-insulin-dependent) diabetes mellitus (NIDDM). The diabetic BC1 mice showed large islets, hyperinsulinemia and insulin-resistance. The wild mouse strain *Mus spretus* did not develop diabetes, obesity, hypertrophy of islets and insulitis at 12 months of age. Our present study indicates a possibility of linkage of type-II diabetes in male BC1[(NODx*Mus spretus*)F1xNOD] to D1Mit 17 locus at approximately 82 cM from the centromere on chromosome 1, while Idd-5 in type-I diabetes was linked to the region of Il-1r1 and Lsh/Ity/Bcg locus in BC1 of NOD with B10.NOD-H-2 congenic strain (Cornall et al, 1991), and to the region of Bcl-2 and D1Nds2 locus in F2 of NOD with B6 on chromosome 1 (Garchon et al, 1991). Thus, the data in the present study and the previous reports indicate that there are two linkage groups related to type-I diabetes and one linkage group related to type-II diabetes on chromosome 1. In human diabetes, increased odds ratio has been found in type-I diabetic children when type-II diabetes was reported in their relatives, suggesting a common diabetogenic gene(s) in type-I and type-II diabetes (Dahlquist et al, 1989).

Acknowledgements

We thank Dr. C. Ronald Kahn at the Joslin Diabetes Center for his continued support for our research. This study was supported by grants from Juvenile Diabetes Foundation International (190341), Procept, Inc., Shionogi & Co., LTD., NIH RO1 DK43613 and NIH 2 P30 DK36836 (Diabetes and Endocrinology Research Center grant). E.Y. and M.F. were the recipients of Postdoctoral Fellowship from the Juvenile Diabetes Foundation International.

References

Acha-Orbea, H., McDevitt, H.O. (1987) The first external domain of the non obese diabetic mouse class II I-Aβ chain is unique. Proc. Natl.Acad.Sci.USA 84,2435-2439

Adams, D. (1982) Molecules, membranes and macrophage activation. Immunol Today 3,285-287

Araki, E., Sun, X.J., Haag, B., Chuang, L.M., Zhang, Y., Yang-Feng, T.L., White, M.F., Kahn, C.R. (1993) Human skeletal muscle insulin receptor substrate-1. Diabetes 42,1041-1054

Bendtzen, K., Mandrup-Poulsen, T., Nerup, J., Nielsen, J.H., Dinarello, C.A., Svenson, M. (1986) Cytotoxicity of human pI 7 interleukin-1 for pancreatic islets of Langerhans. Science 232,1545-1547

Beutler, B., Greenwald, D., Hulmes, J.D., Chang, M., Pan, Y.C.E., Mathison, J., Ulevitch, R., Cerami, A. (1985) Identity of tumor necrosis factor and macrophages-secreted factor cachectin. Nature 316,552-554

Boerwinkle: E., Xiong, W., Fourest, E., Chan, L. (1989) Rapid typing of tandemly repeated hypervariable loci by the polymerase chain reaction: Application to the apolipoprotein B 3' hypervariable region. Proc. Natl. Acad. Sci. USA 86,212-216

Bradley, B.J., Haskins, K., LaRosa, F.G., Lafferty, K.J. (1992) CD8 T cells are not required for islet destruction induced by a CD4+ islet-specific T cell clone. Diabetes 41,1603-1608

Brosnan, C.F., Bornstein, M.B., Bloom, B.R. (1981) The effects of macrophage depletion on the clinical and pathological expression of experimental allergic encephalomyelitis. J Immunol. 126,614-620

Charlton, B., Bacelj, A., Mandel, T.E. (1988) Administration of silica particles or anti-Lyt2 antibody prevents β-cell destruction in NOD mice given cyclophosphamide. Diabetes 37,930-935

Coleman, D.L., Kuzava, J.E., Leiter, E.H. (1990) Effect of diet on incidence of diabetes in nonobese diabetic mice. Diabetes 39,432-436

Cornall, R.J., Prins, J-B., Todd, J.A., Pressey, A., DeLarato, N.H., Wicker, L.S., Peterson, L.B. (1991) Type 1 diabetes in mice is linked to the interleukin-1 receptor and Lsh/Ity/Bcg genes on chromosome 1. Nature 353,262-265

Dahlquist, G., Blom, L., Tuvemo, T., Nystrom, Sandstrom, A., Wall, S. (1989) The Swedish childhood diabetes study - Results from a nine year case register and a one year case-refernt study indicating that Type 1 (insulin-dependent) diabetes mellitus is associated with both Type 2 (non-insulin-dependent) diabetes mellitus and autoimmune disorders. Diabetologia 32,2-6

De Gouyon, B., Melanitou, F., Richard, M.F., Requarth, M., Hahn, I.H., Guenet, J.L., Demenais, F., Julier, C., Lathrop, G.M., Boitard, C., Avner, P. (1993) Genetic analysis of diabetes and insulitis in an interspecific cross of the nonobese diabetic mouse with *Mus spretus*. Proc. Natl. Acad. Sci. USA 90,1877-1881

Dietrich, W., Katz, H., Lincoln, S.E., Shin, H.S., Friedman, J., Dracopoli, N., Lander, E.S. (1992) A genetic map of the mouse suitable for typing intraspecific crosses. Genetics 131,423-447

Durum, S.K., Schmidt, J.A., Oppenheim, J.J. (1985) Interleukin-1: an immunological perspective. Ann. Rev. Immunol. 3,263-287

Elliot, R.B., Martin, J.M. (1984) Dietary protein: a trigger of insulin-dependent diabetes in the BB rat? Diabetologia 26,297-299

Elliott, R.B., Reddy, S.N., Bibby, N.J., Kida, K. (1988) Dietary prevention of diabetes in the non-obese diabetic mouse. Diabetologia 31,62-64

Garchon, H-J., Bedossa, P., Eloy, L., Bach, J-F. Identification and mapping to chromosome 1 of a susceptibility locus for periinsulitis in non-obese diabetic mice (1991) Nature 353,260-262

Gepts, W (1965) Pathologic anatomy of the pancreas in juvenile diabetes mellitus. Diabetes 14,619-633

Ghosh, S., Palmer, S.M., Rodrigues, N.R., Cordell, H.J., Hearne, C.M., Cornall, R.J., Prins, J-B., McShane, P., Lathrop, G.M., Peterson, L.B., Wicker, L.S., Todd, J.A. (1993) Polygenic control of autoimmune diabetes in nonobese diabetic mice. Nature Genetics 4,404-409

Haskins, K. and McDuffie, M. (1991) Acceleration of diabetes in young NOD mice with a $CD4^+$ islet-specific T cell clone. Science 249,1433-1436

Hattori, M., Buse, J.B., Jakson, R.A., Glimcher, L., Dorf, M.E., Minami, M., Makino, S., Moriwaki, K., Kuzuya, H., Imura, H., Strauss, W.M., Seidman, J.G., Eisenbarth, G.S. (1986) The NOD mouse: recessive diabetogenic gene in the major histocompatibility complex. Science 231,733-735

Hattori, M., Fukuda, M., Ichikawa, T., Katoh, H., Makino, S. (1990) A single recessive "non-MHC" diabetogenic gene in nonobese diabetic (NOD) mice. J Autoimmunity 3,1-10

Hattori,M, Yamato, E., Hirokawa, K.J., Petruzzelli, M., Makino, S., Chapman, V.M. (1992) Male backcross mice of NOD with *Mus spretus Spain* predominantly develop diabetes regardless of MHC homozygosity and heterozygosity. Diabetes 41(Suppl. 1),93A.

Horio, F., Fukuda, M., Bonner-Weir, S., Hattori, M. (1989) Free radical oxygen scavengers (superoxide dismutase and catalase) prevent the development of insulitis in NOD mice. Clinical Research 37,451A.

Horio, F., Fukuda, M., Katoh, H., Petruzzelli, M., Yano, N., Rittershaus, Bonner-Weir, S., Hattori, M. (1993) Reactive oxygen intermediates in autoimmune islet cell destruction of the NOD mouse by peritoneal exudate cells (rich in macrophages) but not T cells. Diabetologia (in press)

Hutchings, P., Rosen, H., O'Reilly, L., Simpson, E., Gordon. S., Cooke, A. (1990) Transfer of diabetes in mice prevented by blockade of adhesion-promoting receptor on macrophages. Nature 348,639-642

Ikegami, H., Makino, S., Harada, M., Eisenbarth, G.S., Hattori, M. (1988) The cataract shionogi mouse, a sister strain of the non-obesediabetic mouse: similar class II but different class I gene products. Diabetologia 31,254-258

Ikegami, H., Eisenbarth, G.S., Hattori, M. (1990) MHC-linked diabetogenic gene of the NOD mouse: analysis of genomic DNA amplified by polymerase reaction chain. J. Clin. Invest. 85,18-24

Karasik, A., Hattori, M. (1993) Use of animal models in the study of diabetes. In Joslin's Textbook of Diabetes. G. Weir and C.R. Kahn (eds), Lea and Febiger (in press)

Lee, K., Amano, K., Yoon, J. (1988) Evidence for initial involvement of macrophage in the development of insulitis in NOD mice. Diabetes 37,989-991

Lehman, C.D., Rodin, J., McEwen, B., Brinton, R. (1991) Impact of environmental stress on the expression of insulin-dependent diabetes mellitus. Behavioral Neuroscience 105,241-245

Like, A.A., Rossini, A.A. (1984) Spontaneous autoimmune diabetes mellitus in the BioBreeding/Worcester rat. Surv Synth Pathol Res. 3,131-138

Lund, T., O'Reilly, L., Hutchings, P., Kanagawa, O., Simpson, E., Gravely, R., Chandler, P., Dyson, J., Picard, J.K., Edwards, A., Kioussis, D., Cooke, A. (1990) Prevention of insulin-dependent diabetes mellitus in non-obese diabetic mice by transgenes encoding modified I-A β-chain or normal I-E α-chain. Nature 345,727-729

Lund, T., Shaikh, S., Kendall, E., Campbell, R.D., Hattori, M., Makino, S., Cooke, A. (1993) RFLP analysis of the MHC class III region defines unique haplotypes for the non-obese diabetic, cataract Shionogi and the non-obese non-diabetic mouse strains. Diabetologia 36,727-733

Makino, S., Kunimoto, K., Muraoka, Y., Mizushima, Y., Katagiri, K., Tochino, Y. (1980) Breeding of a non-obese diabetic strain of mice. Exp Anim 29,1-13

Makino, S., Harada, M., Kishimoto, Y., Hayashi, Y. (1986) Absence of insulitis and overt diabetes in athymic nude mice with NOD genetic background. Exp. Anim. 35,495-498

Makino, S., Kishimoto, Y., Kunimoto, K., Kawaguchi, J., Uchida, K. (1991) Localization of the MHC-linked diabetogenic genes of the NOD mouse by using the congenic strains. Diabetes Res. Clin. Pract. 14,S40 (Abstract)

Miller, B.J., Appel, M.C., O'Neil, J.J., Wicker, L.S. (1988) Both the $Lyt2^+$ and L3T4+ T cell subset are required for the transfer of diabetes in nonobese diabetic mice. J. Immunol. 140,52-58

Miyazaki, T., Uno, M., Uehara, M., Kikutani, H., Kishimoto, T., Kimoto, M., Nishimoto, H., Miyazaki, J., Yamamura, K. (1990) Direct evidence for the contribution of the unique $I\text{-}A^{nod}$ to the development of insulitis in non-obese diabetic mice. 345,722-724

Murray, H.W., Spitalny, G.L., Nathan, C.F. (1985) Activation of mouse peritoneal macrophages *in vitro* and *in vivo* by interferon. J. Immunol. 134,1619-1622

Nakano, N., Kikutani, H., Nishimoto, H., Kishimoto, T. (1991) T cell receptor V gene usage of islet β cell reactive T cells is not restricted in nonobese diabetic mice. J. Exp. Med. 173,1091-1097

Nishimoto, H., Kikutani, K., Yamamura, K., Kishimoto, T. (1987) Prevention of autoimmune insulitis by expression of I-E molecules in NOD mice. Nature 328,432-434

Ohtori, H., Yoshida, T., Inuta, T. (1968) "Small eyes and cataract", a new dominant mutation in the mouse. Exp. Animal. 17,91-96

O'Reilly, L.A., Huchings, P.R., Crocker, P.R., Simpson, E., Lund, T., Kioussis, D., Takei, F., Baird, J., Cooke, A. (1991) Characterization of pancreatic islet cell infiltrates in NOD mice: effect of cell transfer and transgenic expression. Eur. J. Immunol. 21,1171-1180

Oschilewski, U., Kiesel, U., Kolb, H. (1985) Administration of silica prevents diabetes in BB rats, Diabetes 34,197-199

Ozaki, S., Honda, H., Maruyama, N., Hirose, S., Hamaoki, M., Sato, M., Shirai, T. (1983) Genetic regulation of erythrocyte autoantibody production in New Zealand black mice. Immunogenetics 18,241-254

Palmer, J.P., Helqvist, S., Spinas, G.A., Molvig, J., Mandrup-Poulsen, T., Anderson, H.V., Nerup, J. (1989) Interaction of β-cell activity and IL-1 concentration and exposure time in isolated rat islets of Langerhans. Diabetes 28,1211-1216

Podolin, P., Pressey, A., DeLarato, N.H., Fischer, P.A., Peterson, L.B., Wicker, L.S. (1993) $I\text{-}E^+$ nonobese diabetic mice develop insulitis and diabetes. J. Exp. Med. 178,793-803

Prins, J.B., Todd, J.A., Rodrigues, N.R., Ghosh, S., Hogarth, P.M., Wicker, L.S., Gaffney, E., Podolin, P.L., Fischer, P.A., Sirotina, A., Peterson, L.B. (1993) Linkage on chromosome 3 of autoimmune diabetes and defective Fc receptor for IgG in NOD mice. Science 260,695-698

Prochazka, M., Leiter, E.H., Serrez, D.V., Colman, D.G. (1987) Three recessive loci required for insulin-dependent diabetes mellitus in nonobese diabetic mice. Science 237,286-289

Pukel, C., Baquerizo, H., Rabinovitch, A. (1988) Destruction of rat islet cell monolayers by cytokines: Synergistic interactions of interferon-γ, tumor necrosis factor, lymphotoxin, and interleukin-1. Diabetes 37,133-136

Scott, F.W., Mongeau, R., Kardish, M., et al (1985) Diet can prevent diabetes in the BB rat. Diabetes 34,1059-1062

Slattery, R.N., Kjer-Nielsen, L., Allison, J., Charlton, B., Mandel, T.E., Miller, J.F.A.P. (1990) Prevention of diabetes in non-obese diabetic I-A^k transgenic mice. Nature 345,724-726

Sun, X.J., Rothenberg, P.L., Kahn, C.R., Backer, J.M., Araki, E., Wilden, P., Cahill, D.A., Goldstein, B.J., White, M.F. (1991) Structure of the insulin receptor substrate IRS-1 defines a unique signal transduction protein. Nature (Lond) 352,73-77

Todd, J.A., Aitman, T.J., Cornall, R.J., Ghosh, S., Hall, J.R.S., Hearne, C.M., Knight, A.M., Love, J.M., McAleer, M.A., Prins, J-B., Rodrigues, N., Lathrop, M., Pressey, A., DeLarato, N.H., Peterson, L.B., Wicker, L.S. (1991) Genetic analysis of autoimmune type 1 diabetes mellitus in mice. Nature 351,542-547

Wicker, L.S., Miller, B.J., Mullen, Y. (1986) Transfer of autoimmune diabetes mellitus with splenocytes from non-obese diabetic (NOD) mice. Diabetes 35,855-860

Wicker, L.S., Miller, B.J., Cocker, L.Z., Mcnally, S.E., Scott, S., Mullen, Y., Appel, M.C. (1987) Genetic control of diabetes and insulitis in the nonobese diabetic (NOD) mouse. J. Exp. Med. 165,1639-1654

Wicker, L.S., Miller, B.J., Fischer, P.A., Pressey, A., Peterson, L.B. (1989) Genetic control of diabetes and insulitis in the nonobese diabetic mouse. Pedigree analysis of a diabetic H-$2^{nod/b}$ heterozygote. J. Immunol. 142,81-784

Wicker, L.S., Appel, M.C., Dotta, F., Pressey, A., Miller, B.J., DeLarato, N.H., Fischer, P.A., Boltz, R.C., Peterson, L.B. (1992) Autoimmune syndrome in major histocompatibility complex (MHC) congenic strains of nonobese diabetic (NOD) mice. The NOD MHC is dominant for insulitis and cyclophosphamide-induced diabetes. J. Exp. Med. 176,67-77

White, M.F., Livingstone, J.N., Backer, J.M., Lauris, V., Dull, T.J., Ullrich, A., Kahn, C.R. (1988) Mutation of the insulin receptor at tyrosine 960 inhibits sjgnal transmission but does not affect its tyrosine kinase activity. Cell 54,641-649

Ziegler, K., Unanue, E.R. (1981) Identification of a macrophage antigen-processing event required for I-region-restricted antigen presentation to lymphocytes. J. Immunol. 127,1869-1875

Subject Index

M

N

O

P

R

S

NATO ASI Series H

Vol. 1: **Biology and Molecular Biology of Plant-Pathogen Interactions.** Edited by J.A. Bailey. 415 pages. 1986.

Vol. 2: **Glial-Neuronal Communication in Development and Regeneration.** Edited by H.H. Althaus and W. Seifert. 865 pages. 1987.

Vol. 3: **Nicotinic Acetylcholine Receptor: Structure and Function.** Edited by A. Maelicke. 489 pages. 1986.

Vol. 4: **Recognition in Microbe-Plant Symbiotic and Pathogenic Interactions.** Edited by B. Lugtenberg. 449 pages. 1986.

Vol. 5: **Mesenchymal-Epithelial Interactions in Neural Development.** Edited by J. R. Wolff, J. Sievers, and M. Berry. 428 pages. 1987.

Vol. 6: **Molecular Mechanisms of Desensitization to Signal Molecules.** Edited by T M. Konijn, P J. M. Van Haastert, H. Van der Starre, H. Van der Wel, and M.D. Houslay. 336 pages. 1987.

Vol. 7: **Gangliosides and Modulation of Neuronal Functions.** Edited by H. Rahmann. 647 pages. 1987.

Vol. 8: **Molecular and Cellular Aspects of Erythropoietin and Erythropoiesis.** Edited by I.N. Rich. 460 pages. 1987.

Vol. 9: **Modification of Cell to Cell Signals During Normal and Pathological Aging.** Edited by S. Govoni and F. Battaini. 297 pages. 1987.

Vol. 10: **Plant Hormone Receptors.** Edited by D. Klämbt. 319 pages. 1987.

Vol. 11: **Host-Parasite Cellular and Molecular Interactions in Protozoal Infections.** Edited by K.-P. Chang and D. Snary. 425 pages. 1987.

Vol. 12: **The Cell Surface in Signal Transduction.** Edited by E. Wagner, H. Greppin, and B. Millet. 243 pages. 1987.

Vol. 13: **Toxicology of Pesticides: Experimental, Clinical and Regulatory Perspectives.** Edited by L.G. Costa, C.L. Galli, and S.D. Murphy. 320 pages. 1987.

Vol. 14: **Genetics of Translation. New Approaches.** Edited by M.F. Tuite, M. Picard, and M. Bolotin-Fukuhara. 524 pages. 1988.

Vol. 15: **Photosensitisation. Molecular, Cellular and Medical Aspects.** Edited by G. Moreno, R. H. Pottier, and T. G. Truscott. 521 pages. 1988.

Vol. 16: **Membrane Biogenesis.** Edited by J.A.F Op den Kamp. 477 pages. 1988.

Vol. 17: **Cell to Cell Signals in Plant, Animal and Microbial Symbiosis.** Edited by S. Scannerini, D. Smith, P. Bonfante-Fasolo, and V. Gianinazzi-Pearson. 414 pages. 1988.

Vol. 18: **Plant Cell Biotechnology.** Edited by M.S.S. Pais, F. Mavituna, and J. M. Novais. 500 pages. 1988.

Vol. 19: **Modulation of Synaptic Transmission and Plasticity in Nervous Systems.** Edited by G. Hertting and H.-C. Spatz. 457 pages. 1988.

Vol. 20: **Amino Acid Availability and Brain Function in Health and Disease.** Edited by G. Huether. 487 pages. 1988.

NATO ASI Series H

Vol. 21: **Cellular and Molecular Basis of Synaptic Transmission.** Edited by H. Zimmermann. 547 pages. 1988.

Vol. 22: **Neural Development and Regeneration. Cellular and Molecular Aspects.** Edited by A. Gorio, J. R. Perez-Polo, J. de Vellis, and B. Haber. 711 pages. 1988.

Vol. 23: **The Semiotics of Cellular Communication in the Immune System.** Edited by E.E. Sercarz, F. Celada, N.A. Mitchison, and T. Tada. 326 pages. 1988.

Vol. 24: **Bacteria, Complement and the Phagocytic Cell.** Edited by F. C. Cabello und C. Pruzzo. 372 pages. 1988.

Vol. 25: **Nicotinic Acetylcholine Receptors in the Nervous System.** Edited by F. Clementi, C. Gotti, and E. Sher. 424 pages. 1988.

Vol. 26: **Cell to Cell Signals in Mammalian Development.** Edited by S.W. de Laat, J.G. Bluemink, and C.L. Mummery. 322 pages. 1989.

Vol. 27: **Phytotoxins and Plant Pathogenesis.** Edited by A. Graniti, R. D. Durbin, and A. Ballio. 508 pages. 1989.

Vol. 28: **Vascular Wilt Diseases of Plants. Basic Studies and Control.** Edited by E. C. Tjamos and C. H. Beckman. 590 pages. 1989.

Vol. 29: **Receptors, Membrane Transport and Signal Transduction.** Edited by A. E. Evangelopoulos, J. P. Changeux, L. Packer, T. G. Sotiroudis, and K.W.A. Wirtz. 387 pages. 1989.

Vol. 30: **Effects of Mineral Dusts on Cells.** Edited by B.T. Mossman and R.O. Begin. 470 pages. 1989.

Vol. 31: **Neurobiology of the Inner Retina.** Edited by R. Weiler and N.N. Osborne. 529 pages. 1989.

Vol. 32: **Molecular Biology of Neuroreceptors and Ion Channels.** Edited by A. Maelicke. 675 pages. 1989.

Vol. 33: **Regulatory Mechanisms of Neuron to Vessel Communication in Brain.** Edited by F. Battaini, S. Govoni, M.S. Magnoni, and M. Trabucchi. 416 pages. 1989.

Vol. 34: **Vectors asTools for the Study of Normal and Abnormal Growth and Differentiation.** Edited by H. Lother, R. Dernick, and W. Ostertag. 477 pages. 1989.

Vol. 35: **Cell Separation in Plants: Physiology, Biochemistry and Molecular Biology.** Edited by D. J. Osborne and M. B. Jackson. 449 pages. 1989.

Vol. 36: **Signal Molecules in Plants and Plant-Microbe Interactions.** Edited by B.J.J. Lugtenberg. 425 pages. 1989.

Vol. 37: **Tin-Based Antitumour Drugs.** Edited by M. Gielen. 226 pages. 1990.

Vol. 38: **The Molecular Biology of Autoimmune Disease.** Edited by A.G. Demaine, J-P. Banga, and A.M. McGregor. 404 pages. 1990.

NATO ASI Series H

Vol. 39: **Chemosensory Information Processing.**
Edited by D. Schild. 403 pages. 1990.

Vol. 40: **Dynamics and Biogenesis of Membranes.**
Edited by J. A. F. Op den Kamp. 367 pages. 1990.

Vol. 41: **Recognition and Response in Plant-Virus Interactions.**
Edited by R. S. S. Fraser. 467 pages. 1990.

Vol. 42: **Biomechanics of Active Movement and Deformation of Cells.**
Edited by N. Akkas. 524 pages. 1990.

Vol. 43: **Cellular and Molecular Biology of Myelination.**
Edited by G. Jeserich, H. H. Althaus, and T. V. Waehneldt. 565 pages. 1990.

Vol. 44: **Activation and Desensitization of Transducing Pathways.**
Edited by T. M. Konijn, M. D. Houslay, and P. J. M. Van Haastert. 336 pages. 1990.

Vol. 45: **Mechanism of Fertilization: Plants to Humans.**
Edited by B. Dale. 710 pages. 1990.

Vol .46: **Parallels in Cell to Cell Junctions in Plants and Animals.**
Edited by A. W Robards, W. J . Lucas, J . D. Pitts, H . J . Jongsma, and D. C. Spray. 296 pages. 1990.

Vol. 47: **Signal Perception and Transduction in Higher Plants.**
Edited by R. Ranjeva and A. M. Boudet. 357 pages. 1990.

Vol. 48: **Calcium Transport and Intracellular Calcium Homeostasis.**
Edited by D. Pansu and F. Bronner. 456 pages. 1990.

Vol. 49: **Post-Transcriptional Control of Gene Expression.**
Edited by J. E. G. McCarthy and M. F. Tuite. 671 pages. 1990.

Vol. 50: **Phytochrome Properties and Biological Action.**
Edited by B. Thomas and C. B. Johnson. 337 pages. 1991.

Vol. 51: **Cell to Cell Signals in Plants and Animals.**
Edited by V. Neuhoff and J. Friend. 404 pages. 1991.

Vol. 52: **Biological Signal Transduction.**
Edited by E. M . Ross and K . W. A. Wirtz. 560 pages. 1991.

Vol. 53: **Fungal Cell Wall and Immune Response.**
Edited by J. P. Latge and D. Boucias. 472 pages. 1991.

Vol. 54: **The Early Effects of Radiation on DNA.**
Edited by E. M. Fielden and P. O'Neill. 448 pages. 1991.

Vol. 55: **The Translational Apparatus of Photosynthetic Organelles.**
Edited by R. Mache, E. Stutz, and A. R. Subramanian. 260 pages. 1991.

Vol. 56: **Cellular Regulation by Protein Phosphorylation.**
Edited by L. M. G. Heilmeyer, Jr. 520 pages. 1991.

NATO ASI Series H

Vol. 57: **Molecular Techniques in Taxonomy.**
Edited by G. M. Hewitt, A. W. B. Johnston, and J. P. W. Young. 420 pages. 1991.

Vol. 58: **Neurocytochemical Methods.**
Edited by A. Calas and D. Eugene. 352 pages. 1991.

Vol. 59: **Molecular Evolution of the Major Histocompatibility Complex.**
Edited by J. Klein and D. Klein. 522 pages. 1991.

Vol. 60: **Intracellular Regulation of Ion Channels.**
Edited by M. Morad and Z. Agus. 261 pages. 1992.

Vol. 61: **Prader-Willi Syndrome and Other Chromosome 15q Deletion Disorders.**
Edited by S. B. Cassidy. 277 pages. 1992.

Vol. 62: **Endocytosis. From Cell Biology to Health, Disease and Therapie.**
Edited by P. J. Courtoy. 547 pages. 1992.

Vol. 63: **Dynamics of Membrane Assembly.**
Edited by J. A. F. Op den Kamp. 402 pages. 1992.

Vol. 64: **Mechanics of Swelling. From Clays to Living Cells and Tissues.**
Edited by T. K. Karalis. 802 pages. 1992.

Vol. 65: **Bacteriocins, Microcins and Lantibiotics.**
Edited by R. James, C. Lazdunski, and F. Pattus. 530 pages. 1992.

Vol. 66: **Theoretical and Experimental Insights into Immunology.**
Edited by A. S. Perelson and G. Weisbuch. 497 pages. 1992.

Vol. 67: **Flow Cytometry. New Developments.**
Edited by A. Jacquemin-Sablon. 1993.

Vol. 68: **Biomarkers. Research and Application in the Assessment of Environmental Health.** Edited by D. B. Peakall and L. R. Shugart. 138 pages. 1993.

Vol. 69: **Molecular Biology and its Application to Medical Mycology.**
Edited by B. Maresca, G. S. Kobayashi, and H. Yamaguchi. 271 pages. 1993.

Vol. 70: **Phospholipids and Signal Transmission.**
Edited by R. Massarelli, L. A. Horrocks, J. N. Kanfer, and K. Löffelholz. 448 pages. 1993.

Vol. 71: **Protein Synthesis and Targeting in Yeast.**
Edited by A. J. P. Brown, M. F. Tuite, and J. E. G. McCarthy. 425 pages. 1993.

Vol. 72: **Chromosome Segregation and Aneuploidy.**
Edited by B. K. Vig. 425 pages. 1993.

Vol. 73: **Human Apolipoprotein Mutants III. In Diagnosis and Treatment.**
Edited by C. R. Sirtori, G. Franceschini, B. H. Brewer Jr. 302 pages. 1993.

NATO ASI Series H

Vol. 74: **Molecular Mechanisms of Membrane Traffic.**
Edited by D. J. Morré, K. E. Howell, and J. J. M. Bergeron. 429 pages. 1993.

Vol. 75: **Cancer Therapy. Differentiation, Immunomodulation and Angiogenesis.**
Edited by N. D'Alessandro, E. Mihich, L. Rausa, H. Tapiero, and T. R.Tritton. 299 pages. 1993.

Vol. 76: **Tyrosine Phosphorylation/Dephosphorylation and Downstream Signalling.**
Edited by L. M. G. Heilmeyer Jr. 388 pages. 1993.

Vol. 77: **Ataxia-Telangiectasia.** Edited by R. A. Gatti, R. B. Painter. 306 pages. 1993.

Vol. 78: **Toxoplasmosis.** Edited by J. E. Smith. 272 pages. 1993.

Vol. 79: **Cellular Mechanisms of Sensory Processing. The Somatosensory System.**
Edited by L. Urban. 514 pages. 1994.

Vol. 80: **Autoimmunity: Experimental Aspects.**
Edited by M. Zouali. 318 pages. 1994.

Springer-Verlag and the Environment

We at Springer-Verlag firmly believe that an international science publisher has a special obligation to the environment, and our corporate policies consistently reflect this conviction.

We also expect our business partners – paper mills, printers, packaging manufacturers, etc. – to commit themselves to using environmentally friendly materials and production processes.

The paper in this book is made from low- or no-chlorine pulp and is acid free, in conformance with international standards for paper permanency.

MIX
Papier aus verantwortungsvollen Quellen
Paper from responsible sources
FSC® C105338

If you have any concerns about our products,
you can contact us on
ProductSafety@springernature.com

In case Publisher is established outside the EU,
the EU authorized representative is:
Springer Nature Customer Service Center GmbH
Europaplatz 3, 69115 Heidelberg, Germany

Printed by Libri Plureos GmbH
in Hamburg, Germany